Green Energy and Technology

More information about this series at http://www.springer.com/series/8059

Rocco Papa · Romano Fistola
Carmela Gargiulo
Editors

Smart Planning: Sustainability and Mobility in the Age of Change

Springer

Editors
Rocco Papa
DICEA
University of Naples Federico II
Naples
Italy

Carmela Gargiulo
DICEA
University of Naples Federico II
Naples
Italy

Romano Fistola
Department of Engineering
University of Sannio
Benevento
Italy

ISSN 1865-3529 ISSN 1865-3537 (electronic)
Green Energy and Technology
ISBN 978-3-030-08512-4 ISBN 978-3-319-77682-8 (eBook)
https://doi.org/10.1007/978-3-319-77682-8

Printed on acid-free paper

This Springer imprint is published by the registered company Springer International Publishing AG part of Springer Nature
The registered company address is: Gewerbestrasse 11, 6330 Cham, Switzerland

Preface

This volume presents a collection of contributions on a subject of considerable interest in the ambit of studies on managing urban change. The main objective is to show that *smartness* in managing territorial changes may be implemented with urban and area interventions which ultimately aim at sustainability, pursued by suitable measures on the mobility of people, goods and information. This underlying assumption provides the structure for contributions within two large thematic areas: sustainability and mobility, which are used to map out a new approach to urban planning in Italy. We were particularly pleased to be able to involve in the editorial project the main research groups active in the field of urban sciences from the various schools of engineering operative in Italy. This book offers an overview of sustainability by urban planning scholars and provides an up-to-date review of urban mobility in the context of urban planning—topics that are of considerable interest in the development of smart cities. Environmental sustainability is universally recognized as a fundamental condition for any urban policy or urban management activity, while mobility is essential for the survival of complex urban systems. The new opportunities offered by innovations in the mobility of people, goods and information, as well as radically changing interactions and activities, are transforming cities. Including contributions by urban planning scholars, this book provides an up-to-date picture of the latest studies and innovative policies and practices in Italy, of particular interest due to its spatial, functional and social peculiarities. Sustainability and mobility must form the basis of "smart planning"—a new dimension of urban planning linked to two main innovations: procedural innovation in managing territorial change, and technological innovation in the generation, processing and distribution of data (big data) for the creation of new "digital environments" such as GIS, BIM, models of augmented and mixed reality, useful for describing changes in human settlements in real time. The contributions are structured as follows: the innovative methodology is first described, and procedures and tools are then proposed for urban interventions with specific reference to real cases within the Italian context. As already highlighted in the volume entitled: "Smart Energy in the Smart City", published in the same series in which this publication represents a natural evolution, the Italian context represents, also in this case, a test bench of major interest due to such specific aspects as geography, socio-economic variability between the North and South of the country, differentiated local development potential, climate and exposure to various conditions of risk for urban systems.

The editors wish to express their gratitude to Springer for its professional assistance and in particular to Mr. Pierpaolo Riva who supported this publication.

Naples, Italy — Rocco Papa
Benevento, Italy — Romano Fistola
Naples, Italy — Carmela Gargiulo

Contents

Part I
Sustainability

Knowledge of Places: An Ontological Analysis of the Social Level in the City

Rossella Stufano, Dino Borri, Domenico Camarda and Stefano Borgo

Abstract The present paper proposes to enrich standard methodologies to interpret places with information coming from other forms of place interpretation and description. We develop this proposal investigating geographical places since these are complex spatial environments well suited for the exploitation of different paradigms. The new approach we explore is based on ontological analysis. This approach, we believe, is very useful to integrate a cognitive stand within the traditional analytical and organizational views of complex spatial environments, in particular aiming to facilitate decision-making processes. The overall rationale of this paper is twofold. From the one hand, the introduction of ontological levels is rather useful for organizing the modeling of complex systems. On the other hand, while these levels are informative, our understanding of space cannot be reduced to the ontological elements per se since they lack the contextual perspective. Therefore, deeper studies and research are needed to develop formal frameworks that can completely integrate standard and ontological methodologies for general modeling purposes.

1 Introduction

Places are landscapes seen from far away, are cities lived from the inside, are cities imagined from the outside, are ecological ecosystems and much more. This richness of the notion has always been a challenge for the modeler which has available only limited methodologies and modeling techniques and, yet, is asked to identify and manage a large variety of information and viewpoints (Borri and Scandale 2005; Casey 1997; Cresswell 2004).

R. Stufano (✉) · D. Borri · D. Camarda
Politecnico di Bari, Bari, Italy
e-mail: mariarosaria.stufanomelone@gmail.com

S. Borgo
Laboratory for Applied Ontology, ISTC-CNR, Trento, Italy

R. Papa et al. (eds.), *Smart Planning: Sustainability and Mobility in the Age of Change*, Green Energy and Technology,
https://doi.org/10.1007/978-3-319-77682-8_1

We pick up this challenge by focusing our attention on *lived places*. To understand lived places one has first to develop a notion of physical place which in turn relies on a concept of space and a concept of place. Each of these concepts has different declinations and for each declination there is a possible definition. Still, none of them is simple and space here is not just a 3-dimensional geometrical entity (Ballatore 2012; Freksa et al. 2014). From a cognitive or design perspective, for example, space itself is something that develops and that changes with the agent.

Lived places are even more complex: at the same time an interpreted space, a reasoned space, a space that raises feelings, the result of an aesthetic fruition and so on. In short, a lived place is a space with multiple qualifications some of which depend on mental images, thus on agents with their architecture of cognitive processes. The essence of place lies in the quality of being somewhere specific, the knowing that one (the agent) is "here" rather than "there" (Rapoport 1977). For example, enclosure (or better the status of *being enclosed*) becomes an important aspect of the making of a place.

As said, we understand places mostly through cognitive contexts. We can interpret our being in a space as an objective proposition according to geometrical rules and topological boundaries. Nonetheless, our *being in it* is defined only via a richer, experiential description. Every single person in a place has a subjective point of view and it is that point of view that characterizes that place as such. Points of view and contexts are results coming out from a historical—cognitive—cultural process. The 'subjective knowledge' of space is a core element for the representation of places, and a representation vary from subject to subject and even across one's life (Orr 1992). "Knowledge of a place—where you are and where you come from—is intertwined with knowledge of who you are. Landscape, in other words, shapes mindscape." (Orr 1992). Even more challenging, in geographical large-scale activities of spatial organization places are formed by the integration of material objects (as construction of the physical elements belonging to the organized space), intangible objects (as projection of an organized space), technical knowledge (expertise related to different scientific areas) and non-technical knowledge (diffused, often tacit, knowledge).

This generality and broadness of the notion of place brings up several problems which are difficult to solve and even to represent. These problems cover linguistic, perceptive, educational as well as descriptive (completeness—incompleteness of representations) and communicative (in transmitting knowledge) issues.

In terms of participatory knowledge of spatial organizations, the dialectic action among the different languages involved in place representation is fundamental. Every expert and non-expert perspective which participates in the representational effort has its own metrics and refers to a particular knowledge structure and meaning.

In literature there are many attempts to get a definition of representation of space/place, e.g., see Cresswell (2004). Research in applied ontology (Borgo and Masolo 2010) has developed methodologies and tools to move forward in this direction. One advantage of the ontological approach is that ontologies are typically specified in languages that abstract away from data structures and implementation strategies. In computer and information science, an ontology is a technical term denoting a conceptual artifact that is designed for a purpose, which is to enable the modeling of knowledge about some domain, real or imagined (Gruber 1993).

Furthermore, the typical ontology languages are based on formal semantics to improve understanding and interoperability across systems and users.

2 Urban Planning

Today's awareness of the complexity of social and natural environments implies that in using state-of-the-art techniques to model such systems we must accept a dramatic, and perhaps discouraging, level of uncertainty. The traditional deterministic and quantitative approaches to urban planning and design in risky contexts increasingly fall short of expectations in environmental domains; and this is now widely recognized (McConnell 2010).

Planning tries to manage complexity as the result of a recurring interaction between collective knowledge and the desired results: a position that requires sharing as the foundation of a necessary political dimension of contemporary design (Formato and Russo 2014). An urban project as a plan or as a strategy has to evolve over time, it can't be frozen (Gregotti 2004). The planner, like the urban designer, has always to look at changes in the territory and to adapt the plan to the different relationships between built space and complex urban organization. Architecture, social sciences or anthropology have an active integral role in the thinking and the development of urban projects: in this anticipation game, a city is a relational system that must be thought of as a whole, not a mere composition of districts (Ingallina 2007). For these reasons a rich and reliable modeling methodology for places is an essential starting point for the planner.

When a planning process begins, one should remember that existing urban spaces, beyond their architectonic flavor, often are the result of anonymous creators, i.e., collective agents that happen to live and act in them without particular constraints. More and more frequently constraints on design, implementation, and management responsibilities are enforced to meet expectations of populations and this holds for architectural as well as non-architectural urban spaces. In recent times, during the conception of integrated design and planning, there is an evident trend toward a growing participation of non-expert agents a matching growing retraction of the role of expert agents. This trend coexists with an increasing awareness of the complexity and the dynamical structure of spatial organizations.

This evolution towards the integration of expertise and inhabitant knowledge, generally seen as positive, is important to recognise that sophistication and ingenuity have to coexist, and that the real risk is lack of communication and language mismatch. These are the critical source of inadequate planning and decisions.

3 Carving up Geographical Places

Humans live, move in and sense complex spatial environments using different paradigms. Their interaction with space is sophisticated. It continuously changes over time and relies on a variety of information types that can be classified in as many types as topology, geometry, dynamics, affordance, society, culture and so on. Perhaps due to the richness of this interaction, humans are not aware of how their understanding of and interacting with space is realized. Ontological analysis, the study of what is at the core of our view on reality, can help to recognize, clarify, discriminate, and organize the essential elements and features of places that is crucial to humans in terms of objects, properties and processes. Searching for a general framework where to discover and organize this kind of information, we can list a few levels that, without aiming to be exhaustive, are extremely informative for our analysis of lived places: the spatial, artifactual, cognitive, social, cultural and processual levels. These levels, in turn, can be subdivided in finer levels as we can see for some of these cases.

3.1 *The Spatial Level*

This is perhaps the most studied since it is in large part independent of the subjective perspective and thus can be more easily elaborated in terms of ontological analysis and formal representation (Schön 1983; Bateman et al. 2007). Here we can recognize the *mereological* level within which one understands space in terms of parts, e.g., recognizing the distinction between an area and its neighborhood. A second level is the *topological* one within which one understands space in terms of contact and unity, e.g., recognizing the contiguity between neighborhoods as well as the (spatial) unity of a city. Another level is the *geometrical* one where space is understood in terms of shapes, e.g., recognizing that the shape of a city is constrained by that of the valley where it is located. Finally, the *geographical* level in which one understands space in terms of spatial locations and their descriptions, e.g., distinguishing being along a river or having a radial disposition in space.

3.2 *The Artifact Level*

This is the level where one recognizes the physical realm and how human activities can change it. Here we have the *material* level where one understands space in terms of materiality, e.g., seeing the presence of wood, concrete, water. Then the *structural* level that allows to understand space in terms of qualified components, e.g., distinguishing natural vs manmade and a residential area vs a production area. The *artifactual* level adds an intentional aspect to the environment (Borgo 2007), e.g., looking at a garden as an intentionally modified environment. The next level is the *functional* one where one understands space in terms of functionality, e.g., recognizing a building as a shelter. Finally, the *production* level looks at entities as manipulators, e.g., seeing a farm as production site.

3.3 *The Cognitive Level*

At this level the specific capabilities of humans take the lead. The basic *cognitive* level allows to understand space in terms of experience, e.g., perceiving how to move across objects in space. Instead, the *representation* level leads to understand space in abstract terms, e.g., perceiving the relationships among areas in an airport. The *observation* level is where one understands space in terms of how things in it do or may change, e.g., perceiving the change in the transportation system. Next, we have the *phenomenological* level where one understands space as a moving entity, e.g. perceiving a city in its evolving process. The *perspectival* level allows to understand space from a perspectival viewpoint, e.g., differentiating a square depending from where one is looking at it. At the *conceptual* level space is seen as a collection of realized concepts, e.g., perceiving space as the manifestation of entities that manifest a natural and/or an artificial nature. Finally, the *action* level where one understands space as an entity in which to act, e.g. perceiving the changes that one can enforce on things.

4 The Case Study

Participatory mechanisms of knowledge about and for spatial organizations rely on facilitation agents (as knowledge engineers) who organize dedicated knowledge basis about the issue to be dealt with. Their presence aims to develop true 'participatory conversation' (for 'reflexive conversation' on a problematic situation see Schon (1983), among the participating agents, often translated into verbal texts transcripts ('verbal protocols') by the same agents formulating the interventions in the conversation or by verbalizing agents, with analysis of the results (qualitative or quantitative methods: semantic, statistical, ontological methods, etc.). This is clearly a complex structure, entirely subject to possible biases (like inconsistencies, inadequacies, misunderstandings, etc.) towards the paradoxical Babel effect for the development of plural knowledge.

It is fundamental to translate and disambiguate the different languages belonging to the different expert disciplines as well as those used to convey the non-expert knowledge (the non-expert knowledge has to be at least critic and critically formed). This is to guarantee that the participatory conversations between the various technical and not technical actors is productive and individuates the more convenient spatial frame avoiding the babel risk.

Our ontological analysis of places starts from the data collected for the making of Taranto strategic plan toward 2065. Data were collected via a series of nine community-based, interactive processes of knowledge exchange, aimed at building future scenarios for the new plan. The interactive processes of knowledge exchange were carried out in Taranto, a city in southern Italy, during spring and summer 2014. They were carried out to support policies and decisions on urban socioeconomic as well as environmental domains and organized as a sequence of face-to-face brainstorming forums aimed at cooperatively singling out strategic lines to build alternative development scenarios. From a methodological point of view, it was a 2-step *scenario-building* activity (Borgo et al. 2011; Khakee et al. 2002). First, participants were invited to report problems they faced in their town districts. Then, each participant was invited to generate a reflection about the future of the district, particularly concerning expectations of future changes. Such sessions were organized in all town districts, indoor or outdoor, with participants divided in groups each of them sitting around a dedicated desk. A municipal representative coordinated each desk without taking part in the generative session, she/he had only the task of transcribing in linear charts concerns, problems, expectations and desires presented by the participants at the desk.

In order to manage the results in real time (synthesis and refinement), the interactive process was supported by the use of conceptual maps drawn using dedicated software tools (Decision explorer, Inspiration) (Khakee et al. 2002; Heft 2013) (Fig. 1). This resulted in a real/virtual hybridization of the process, following well-established research trends, as reported in a number of case studies (Borgo et al. 2011; Heft 2013). Results achieved during the nine organized meetings were very different from one another. In particular, the first meeting was organized in the Città Vecchia (inner city district) with its great historical, environmental and cultural resources as well as significant environmental, physical and social degradation problems. In the Città Vecchia session the citizen participation was very high. About 80 stakeholders joined form different societal domains: residents, merchants, students, tourists, visitors. These participants, gathered around 6 desks and about 150 instances were collected.

The database collected during this session resulted interestingly rich and articulated. For this reason, it is a significant and valuable source for the present research effort.

The process naturally leads to two general types of instances: (a) contextual problems; (b) future visions. These two set of instances are about each quarter of the city as well as about the city in general. From an analysis of the data emerges that the environment is the most recurring issue in the groups. It is present in community problems and/or expectations, but also in the perceptions of the physical reality of the city. The industrial problem, on the contrary, is often absent from the discussed issues.

A first common character across the groups is the natural environment that persists in many city representations, so apparently resisting the consequences of an industrial culture. A second one is related to a structural relationship that the city has with the sea, intended as an element of both union and communication. A third is the

Fig. 1 Cognitive map example (excerpt)

potential of the city as tourist attraction which is linked to different characters in relation to the different peculiarities of the area. The industrial problem often seems idiosyncratically absent from protocols, but it is difficult that findings can be used for strategies disregarding industrial relations.

There was an almost total absence of participants in the forum session held in the industrial district, perhaps given the disillusionment with past policymaking. Other issues are related to the inadequacy of urban and metropolitan linkages to the city center, as well as related to the recovery of many illegal settlements. Other instances are about the inadequacy of the urban services and about the inadequacy of metropolitan connections. Further instances are about the recovery of unauthorized coastal settlements.

The session held in the inner city was quite complex. It was held with a hybrid computer-based and traditional, rather conflictual interaction among the participants. Outcomes showed a clear prevalence of visions on mere problems: important issues were the unstopped relationship with the sea (for touristic aims and/or city infrastracturing) and the enhancement of Taranto as archaeological and historical center (Magna Graecia colonization).

5 Analyzing the Social Level

Once the theoretical framework has been outlined and the case study introduced, the next step is to analyze the levels one at a time. Generally speaking, this analysis is useful to identify if every level is well structured or if it is necessary to model additional information. A third step in this research line will be to characterize the specific sub-levels.

Admittedly, the data for the Taranto case is not meant for our research objectives since it was collected during the participation activities belonging to the strategy planning process well before the preparation of our research study. Nevertheless it is useful to start from this data for a first delineation of 'objects', 'attributes' and 'relations' populating the different levels. We decide to start from the social ontological level since the material collected in the Taranto case study is very rich from this perspective. We also believe that the material could help to analyze the cultural level but this has not been evaluated yet. It seems however clear that the material collected in Taranto is not suitable for an analysis of the remaining levels, e.g., cognitive and spatial.

First of all, since 'social' has a broad connotation, we have to limit the boundary of this first analysis. The data we have focus primarily on social reference points, i.e., the elements that people use to identify an area or to navigate across the city. Note that at the social level it can be difficult to elicit the distinction between formal and informal knowledge because the social knowledge is principally informal, tacit and implicit. For this reason, the work in this paper is preliminary under several aspects.

The first analysis singles out references to places and landmarks for the relevance they have in social practices, and lists the relevant entities and some relationships that emerged during the discussions. Another modeling aspect is given by the use that inhabitants make of those places and the social habits they (implicitly or explicitly) express. Finally, the collected entities should be classified in ontological terms following a foundational ontology like DOLCE (Borgo and Masolo 2010).

From the data at the social level it clearly emerges that the objects of the city are not just mere building, locations and landmarks. These terms are used to indicate complex cognitive objects enriched with a set of different meanings/signifiers that can be contextual. A building can acquire different meanings depending on the time and even on the person that mentions it. This analysis shows that at some point it will be necessary to elaborate complex definitions for such entities.

Surprisingly, several objects that a technician would image essential key points for the sociality were not mentioned in the discussions, for instance: the San Cataldo cathedral, the S. Domenico church, the doric columns, the Aragonese castle, Fontana square, and even the town hall. Instead we find mentioned more general *places* and *landmarks* like the beachfront, the bathing establishments, the area of Porta Napoli, the new Acropolis, the island and the piers. These entities show that there is a distinction between *objects* which are taken at their face value (like buildings of low interest, the cruise ships and, from some aspects, the sea itself) and objects that somehow *mean* something else to the inhabitants. Indeed, the same term can occur during the discussions with different meanings, e.g., as a landmark and as a functional building. Many entities have special social *roles* like the city itself as the capital of Magna Grecia, the convergence point (a place where important roads and the railway lines converge), the service center or the old city seen as an eco-museum. Another layer of meanings is introduced by reference to the *service* level that includes public services like restaurants, cafes and shops, bathing establishments (this time the term denotes the service, not the physical entity nor the landmark), the university and the pedestrian network across the city (mainly identified in special areas like the waterfront). Finally, the data provide desired features which can be understood in different ways: functional objects, reassessment of existing objects, services, norms or generic topics. Here we find: work to do, areas to be reclassified (e.g., to be closed to traffic), primary and secondary infrastructures, regulation of public spaces.

After analysing these results with the help of the adopted ontology perspective, we obtain a network of meanings that holds for distinct object types. For instance, let us apply our network to the San Cataldo cathedral, a building which was not part of the entities discussed in the meetings. This building, being part of an architectonic type, has a relationship to space in terms of occupancy of an area, separation of inside vs outside, shape and location. The material building (the church) itself is a material object that affects the surrounding elements (e.g. in terms of size and occlusion, and even its color affects the perception of the area), has a particular internal subdivision and its functional classification determines the intended as well as the unintended uses. These are all distinct meanings that the church-building cognitively stimulates when one refers to it. This network of meanings is enriched by the roles a building like San Cataldo plays, e.g., in the hierarchical ordering (in this case, relatively to the catholic organization this is not a simple church but a cathedral) and in the social habits (a safe place, a place where to do some activities like pray, relax, hide) with all the interconnection across these roles. All these meanings have a position in the ontology and by collecting and classifying these meanings, we build an explicit representation of the tacit network about San Cataldo and similar buildings. Via the ontological system, we can then explain which meanings are triggered in a discussion and even identify those that form bundles, i.e., are intrinsically interconnected, from the inhabitants' perspective. In this way we elicit the local "culture", the way inhabitants tend to understand their own place, and even what they (perhaps unconsciously) consider to be the primary relationships between the elements in it.

6 About the Analytical Frame

In the participatory knowledge of (urban) architectural spaces there are plenty of descriptions (as shown in the Taranto protocols) generally oriented to three identifying dimensions: by name, by function and by localization. Apparently there are no hierarchies among these except (perhaps) the dominance of the name which is both a function and a localization. However, there are relationships and redundancies, in the sense that the name turn out to indicate also a function. In fact, if an agent in a participatory knowledge forum cites the name 'lungomare' (promenade) in Taranto, he immediately expresses the function of (urban) architectural space along the sea. The fact that often the promenade is accompanied by a personal (e.g., Nazario Sauro waterfront) rather than toponymic (e.g., seafront promenade) identifier expresses only indirectly the localization dimension (when there are more versions of the space), that needs a qualifier of person or place to be specified.

The outcomes of participatory conversations in both architectural and non-architectural space organizations can quickly provide site and agent positioning knowledge (for varied situations: transformative, emotional, evaluative, orientational, etc.) whose level of structuring in relational terms is usually low. It is so low that it cannot be a basis for behavioural performance in material (operational cognition: design-realization) and immaterial (cultural cognition: case comparison, structural critique) manipulation of spaces characterized by a sufficient degree of integrated social acceptance (which is obviously the only type of valuation possible within knowledge systems integrating expertise and common sense knowledge).

The problem here is important from both a substantial and a methodological point of view and calls for a solution and/or representation in order to give robustness and operational utility rather than fragility and uselessness to the participatory knowledge of space organizations.

The complex cognitive dialectic between expert knowledge and non-expert knowledge is characterized by 'cognitive balances' between the different cognitive subsystem involved. This is framed in the anthropology of an even more complex cognitive milieu whose structures are formed by distinguished relations between the different agents (human and not-human, biotic and abiotic in the socio-technical systems). Those relations are developed dynamically and in contingent situations that could be analyzed via a theory and a method based on ontological analysis.

There is a strong perspectival aspect in the way we live in places, a kind of description (mostly implicit) of the place that includes what are for us the relevant elements in it and their relationships. Thus, a perspective provided by a place is an information entity that contains: a (typically partial) description of the place, what there is in it and how the place is evolving (e.g. things moving, leaving or arriving, agents acting and transforming them etc.) and possibly the potential interactions between the inhabitants and what is in the place. A place is grounded, as opposed to a generic location, is a context that refers to one or more actual/existing entities.

A place is however also something that goes beyond the single agent. It is the result of many shared links across entities that, as we argued with the example of the San Cataldo cathedral, can be elicited via ontological analysis. The classification of what we use for understanding places in general and the actual place that we are experiencing is then a powerful tool for a more comprehensive representation of places. For this reason, we insist that the analysis has to include the physical elements (e.g. locations and objects), the material components and layout (e.g. enclosed spaces, object distribution); the agentive figures (e.g. habitants, organizations, social roles), the relationships across them and the objects (e.g. generic dependences and actual goal or habits).

In this paper we have only started this complex ontological analysis with which we aim to unravel the complex knowledge that forms the city as lived place.

References

Ballatore A (2012) Prolegomena for an ontology of place. Advancing Geogr Inf Sci Past Next Twenty Years 91

Bateman J, Borgo S, Lüttich K, Masolo C, Mossakowski T (2007) Ontological modularity and spatial diversity. Spat Cogn Comput 7(1):97–128

Borgo S (2007) How formal ontology can help civil engineers. Stud Comput Intell 61:37–45

Borgo S, Masolo C (2010) Ontological foundations of DOLCE. In: Theory and applications of ontology: computer applications. Springer, Berlin, pp 279–295

Borgo S, Franssen M, Garbacz P, Kitamura Y, Mizoguchi R, Vermaas PE (2011) Technical artifact: an integrated perspective. In: Frontiers in artificial intelligence and applications, vol 229. IOS Press, pp 3–15

Borri D, Scandale L (2005) Aspetti cognitivi e organizzativi di campagne politiche come strumenti di visioning e pianificazione strategica. Evidenze dal caso dei 'Forum per Bari'. In: Bruzzo A, Occelli S (eds) Le Relazioni tra Conoscenza e Innovazione nello Sviluppo dei Territori. Milano, Franco Angeli, pp 100–121

Casey ES (1997) The fate of place: a philosophical history. University of California Press, Berkeley

Cresswell T (2004) Place: a short introduction. Short introductions to geography. Wiley, Oxford

Formato E, Russo M (2014) Re-use/re-cycle territories, TeMA INPUT 2014 Print ISSN 1970-9889, e-ISSN 1970-9870

Freksa C, Nebel B, Hegarty M, Barkowsky T (eds) (2014) Spatial cognition IX. Springer, Heidelberg

Gregotti V (2004) L'architettura del Realismo Critico. Laterza, Roma

Gruber TR (1993) A translation approach to portable ontology specifications. Know Acquisition 5(2):199–220

Heft H (2013) Environment, cognition, and culture: reconsidering the cognitive map. J Environ Psychol 33:14–25. https://doi.org/10.1016/j.jenvp.2012.09.002

Ingallina P (2007) L'attractivité des territoires: regards croisés, Ministère de l'Ecologie, de l'Energie, du Développement Durable et de l'Aménagement du Territoire Ministère du Logement Direction Générale de l'Aménagement, du Logement et de la Nature PUCA—Plan Urbanisme Construction Architecture

Khakee A, Barbanente A, Camarda D, Puglisi M (2002) With or without? Comparative study of preparing participatory scenarios using computer-aided and traditional brainstorming. J Future Res 6:45–64

McConnell (2010) Understanding policy success: Rethinking public policy. Pacgrave Macmilliam

Orr DW (1992) Ecological literacy: education and the transition to a postmodern world. Suny Press, New York
Rapoport A (1977) Human aspects of urban form: towards a man-environment approach to urban form and design. Pergamon, Oxford
Schön DA (1983) The reflexive practitioner. Basic Books, New York

Smart Land: Regeneration and Sustainability in Lost Scenarios and New Performances

Donatella Cialdea

Abstract Urban Regeneration should have a role of integration between Planning and Design activities; moreover, it very often looks like a single project with detailed proposals and performances, without a general vision of the area. It is therefore necessary to build a reference for the design choices, which primarily concern the city but increasingly involve the surrounding territory. This paper explores the relationship between the plurality of factors that exist on the settlements and productive activities assets. Our work aims to deal with every aspect of the Regeneration potential, which is an opportunity for enrichment of urban planning, especially in the cases of the Regions—as Molise is—in which really cities do not exist but there is a "continuum" between adjoining Municipalities in a mainly rural territory.

1 Introduction

This study concerns the relationship between sustainability and landscape in connection with the urban and territorial regeneration. It focuses on natural infrastructures, which together with the artificial infrastructures, define the land asset, and moreover they define landscape features of the territory on which urban planning is implemented. It is a large-scale project for the New Landscape Plan for the Molise Region, in the South of Italy, as an experimental model of application of the recent Italian Code for Cultural and Landscape Heritage (Decreto Legislativo 42/2004), belonging the European Landscape Convention (CoE 2000, 2008). In fact, it exhibits a part of the work carried out for the Molise Region, under the Agreement by the Region with the "L.a.co.s.t.a. Laboratory" of the University of Molise, directed by the writer (Regione Molise 2011). Specific analyses, which are presented in this paper, aim of promoting a better environmental, social and economic sustainability in the planning of the hydrographical district, especially for the unique regional river that is the

D. Cialdea (✉)
L.a.Co.S.T.a. Laboratory, D.I.B.T. Department, University of Molise, Campobasso, Italy
e-mail: cialdea@unimol.it

R. Papa et al. (eds.), *Smart Planning: Sustainability and Mobility in the Age of Change*, Green Energy and Technology,
https://doi.org/10.1007/978-3-319-77682-8_2

Biferno River. A vast debate is actually in progress about the contemporary design of territorial transformation scenarios around the river's stream, involving the enhancement of the landscape and local development. The case analyzed, in fact, concerns the main stream of the Molise region.

This course was analyzed in relationship between natural resources and infrastructures: the water course is the connector element and generator of landscape performances and it is linked to the productive activities of the surrounding territory, or better, of the "territories" it passes through.

Rivers are generally at the heart of internal areas in the relationship with the tradition of productive activities such as elements resilience of the landscape (Regione Molise 2015; SNAI 2014); in their surroundings it is possible to describe new approaches for the regeneration of the territories involved in these waterways (Ingaramo and Voghera 2016).

Rivers establish relationship with the city and landscape and they are able to create a more or less extended area of influence. So, in our study, we identify some relevant topics useful to design these areas, which are important in the research on the landscape in order to identify the transformation's quality.

Moreover in these areas we could approach the analyses of lost scenario and new performance. How do you can re-think the relationship between settlements and their surroundings, remembering that everything is inserted in its environment that has its own life? We aimed to underline the Landscape Heritage transformation: how much you can keep the ancient form, already itself layered, allowing the new buildings realization, that inevitably will create a new plot? Territories sometimes seems not be able to receive new entries and sometimes instead soft turns, creating new contexts. In the variety of different perspectives, through which you can analyze the relationship between man and environment, the landscape occupies a specific space, because of the importance that it can assumes in determining the environmental quality.

Landscape means all signs that characterize a territory, also the memory of its past and present functions. So a lot of factors will be kept into account. No word has ever been wider than the word landscape, so much so that it can be affirmed that "in the landscape you can enter from different doors" (Socco 1999): it is a good that can be considered and analyzed according to different points of view and with different purposes. But in reality the term includes everything that surrounds us: from the built to the natural environment, from what is clear evidence of the past to what is recently realized. It is not important to depth single disciplines that suggest or encode mode of action (assuming specific historical era or particular geographical area). Much more interesting is to recognize that an individual perception feel exists, and above all there are questions that each operator of transformations arises when designs or achieves a design for it, especially when these projects become an integral part of the landscape.

Ultimately, the landscape no longer only applies for protection but it is expected to be exploited, fulfilling the dictates of the reform of Title V of the Italian Constitution that has distinguished the protection activity from that of enhancement: "we protect and preserve the cultural asset so that it can be offered to collective knowledge and enjoyment". Regions and Local Authorities are therefore called to organize activities aimed at creating an "integrated system" for the enhancement of the "good". Certainly, in organizing the provisions concerning the protection and enhancement of landscape assets, the National Code inevitably takes into account the European Landscape Convention and, defining the criteria for activities that can affect the landscape, it also focused the forecast of its sustainable development. "Through it the ability to minimize impacts and ensure the design quality of the works and interventions that must be carried out in areas of particular value" (Cialdea 2010).

Fundamental role is played, certainly, by the planning tool. However, for our study, the will to grasp the spirit of the engineering design of the transformations on the territory is a fundamental aim. Our project concerns Natural Infrastructural, also related to environmental risk. In our sample area, the Biferno River in Molise represents the main territorial invariant of the central area of the region, with the Matese mountain chain from which it was born. In our studies for the New Landscape Plan of the Molise Region, the Biferno River was the basic element for identifying the different homogeneous territorial areas for the entire province of Campobasso. The first area is that of the source of the river, a flat agricultural territory with numerous environmental emergencies and characterized by the presence of numerous quality agro-fed farms; the central part of the Biferno course is characterized, then, by a hilly territory with great geological-environmental difficulties and the last area is the flat area of the Lower Molise characterized by intensive agricultural production and, in the part of the river mouth, by the presence of the Industrial core of Termoli. In this context our methodology was experimented, in the implementation phase of the landscape planning, in a different way in the different contexts of the river basin to achieve the quality objectives set by the National Code, highlighting the peculiarities of the places, in correlation with the real needs of the territories and needs of the inhabitants.

The case study is particularly significant because it summarizes in itself morphological characteristics typical of internal areas but at the same time highly exploitable environmental potentialities. Therefore, it is important to respect the possibilities of environmental recovery, but also with a view to "territorial" rebalancing, aimed at local development of a territory, which as traditionally continues to have strong agricultural connotations (Cialdea 2009; Cialdea and Mastronardi 2014; Corrado 2014; Esposito 2014).

2 Landscapes in Evolution and Neutral Spaces

Territory is not just a neutral container of anything, but it is a space for urban and landscape design: actions on the territory could create new performances with new opportunities. Moreover, it could not be considered as a neutral space in which transformation actions take place, but every intervention create new interactions with complex dynamic changes, involving the landscape system (Albrechts 2003; Crosta 2010; Healey 2003; Kunzmann 2007).

All this condition becomes even more complex when the territory, as it is largely the case on the Italian territory, is subject to strong phenomena of instability. The chosen example denotes precisely such a limit situation.

In these cases, in fact, there is a multiplicity of factors: an interdisciplinary approach is clearly necessary, in order to create operational categories for territorial analysis, that are able to interpret the transformations of the landscape, historical and contemporary.

Moreover, dynamics of urban planning evolutions will be explored, because actual plans, often obsolete, are not able to create certain strategies. In the specific case of the region in question, then, the level of the vast area planning tools is absolutely missing. In Molise, in fact, the plans of the over-municipal level have never been drafted: at present there is only a preliminary study by means of territorial matrices for the province of Campobasso (Provincia di Campobasso 2007) and nothing for the other province of the region, which is Isernia. The revision of the current landscape plans, drawn up under the Galasso Law (Legge 431/1985), was carried out by our L.a.co.s.t.a. Laboratory: then we focalised our attention on these "difficult areas" analyzing transformations of the territory in relation to existing urban and regional planning tools. The Molise region is characterized by the low density, with many small towns and no cities: only Campobasso, the regional capital, Isernia, the capital of the second province and Termoli, located in the small strip overlooking the sea, have connotation of small cities. The area under study in this paper concerns a municipality in the first ring of the municipalities around Campobasso, in a part crossed by the river Biferno, the heart of our study, and by one of the few significant road infrastructures of the region, or the Fondovalle del Biferno, called "Bifernina". The region is therefore characterized by a prevalent agricultural activity, the permanence of which, however, has not succeeded in preventing the proliferation of spaces without a particular connotation, undecided, intermediate spaces, between built and not built. It is precisely on these spaces that our attention is focused, because they can be those on which it is possible to experiment new processes of landscape transformation, capable of generating processes of territorial regeneration (Musco 2009; Coyle 2011; Tejedor Bielsa 2013; Jakob 2013). Our methodological approach aimed at defining territories that will become:

(a) spaces for productive activities involvement respect therefore to the production factor of the watercourse. The work was done on two fronts: the first concerning the current situation of activities related to water and in particular with regard to agricultural productivity and the second through an overview linked to the possibilities of creating "specific productive";
(b) spaces for inhabitants involvement, carried out on two fronts, to the local inhabitants and to the tourist flows;
(c) spaces for the involvement of privileged places from a naturalistic point of view, assuming to create a connection network between them.

For the new organization of the territory in those risk areas, in order to promote territorial planning actions, a multiplicity of factors was analyzed. In summary, the territorial project must involve a complex dimension, which however change its features in a dynamic matter and implicate both the natural heritage and its fruition, obviously linked to its involvement. The river's basin is formed by different systems (infrastructural, urbanized and environmental) related each other and located in the landscape asset which is continuously in transformation: In fact, there are many elements that come into play, intertwined on the one hand to the institutional reform process underway in our country and on the other to the already mentioned Internal Areas Strategy, whose development assumes still blurred and variable contours for the different territories involved (De Nigris and Cartolano 2015). Moreover, it is specifically useful to underline the relationship with the landscape quality aims, in order to highlight the necessary link with the landscape planning activities, extending the perspective of the smart city to the vast area. It is important also to analyze the actual debate in Italy on the Law No. 56/2014, which primarily reorganizes responsibilities of Local Authorities, especially for provinces, but also it promotes cooperation between municipalities small in size. Moreover, more recently in Italy a new law was enacted (Law No. 158/2017): it specifies measures to support and enhance small municipalities, with special feature, especially related to hydro-geological instability. In this case it provides the maintenance of the territory with priority to the protection of the environment. In this way, it is possible to understand the territorial project as a territorial development project based on its position and condition, as a transcalar, multidisciplinary project, linked to a sustainable development method (Giaccaria et al. 2013).

2.1 *Spaces of Mediation Between Built and not Built in the Sample Area*

As regards to the territory involved in the river course, it is inevitable to consider the whole territory that surrounds it: the banks of the river are the areas where it is possible to identify the major land use transformations. Moreover they are also those in which there are the greatest difficulties from the point of view of the stability of the territory. Furthermore, these places have the greatest capacity for resilience. They are, then, characterized by not high population densities, but they offer particular landscapes, in which liveability can be increased. The natural function of the river is in fact important in relation to its intrinsic quality as "creator of liveable landscapes". The river is included in agriculture and forest areas, which involve the collective maintenance of the territory: the focus is, however, in its central role in the management of open spaces, which should be able to restore the relationships between urban and rural systems.

In order to reorganize the New Landscape Plan of the Molise Region, our project aimed to enhance the territorial and landscape identity: for this goal, we analyzed the territorial land and then we defined ten regional districts. For each of them, features and territorial structures were defined. This analysis, in fact, was fundamentally necessary for organizing the data collection and, in subsequent phases, for using these collected information in the context of the landscape evidence, that gradually were focused.

Now, our attention was concentrated on the central Molise, in consideration of the fact that this area is not currently covered by any landscape plans.

Moreover, Fig. 1 shows the two areas which involve the Biferno River in the central part of its course: they are the two areas of the right and the left banks.

The following sheets describe these districts (Cialdea 2014). The first one was defined as the «Biferno-Trigno Area»: it is characterized by the presence of numerous small towns, with the exception of the urban settlement of Trivento, which is the largest municipality in this area. From a production point of view, the central-northern area emerges, consisting essentially of Trivento and Montefalcone del Sannio, which together with Roccavivara make up more than 60% of the population. In the whole area the problems of soil instability are particularly emerging, which are also responsible for the demographic decline and which are also linked to the landslide to which most of the historic centers are subject. Most of the municipalities of the area, in fact, present noteworthy situations of collapse. Significant naturalistic emergencies concern both the woods and the elements of the "Morge". The territory is basically defined by a grid consisting of parallel courses of the Trigno and the Biferno rivers, between which is located the ridge that separates the two valleys. From the hydrographical point of view the presence of two different basins is found: one part falls in the Trigno basin, while the remaining part is located in the Biferno basin. The signs of the green pathways, the Ateleta-Biferno more to the north, the Celano-Foggia, which in particular crosses the Municipality of Trivento, giving rise to new territorial connotations and finally the Pescasseroli-Candela are still legible. The morphologies of the inhabited centers are inserted in this shirt, which are however characterized by the Slavic origin of many of them.

The second one is the «Biferno-Molise Centrale Area»: it is centred around the city of Campobasso, located in a central position in the region. It enjoys a system of road infrastructures higher than the average situation because it is crossed by the high-speed road «Fondovalle Biferno» and a ring around Campobasso which is connected to this artery under Montagano. The Community is rich in significant cultural heritage. In this area the valley landscape is predominant with the almost constant valley bottom amplitude, which becomes wider near Petrella and then opens towards the area characterized by a minor anthropization, where the inhabited centers located on spurs are located. The south-eastern part of the territory, which also coincides with the border of the province of Campobasso, has its own eastern limit marked by the river Fortore with the artificial basin of the Occhito Lake. The territory of this area is strongly characterized by the green pathways, named «tratturi», articulated around the city of Campobasso, with the prevalent signs of the Celano-Foggia path and the Lucera-Castel di Sangro path. The area laps the chain of the Matese and substantially is a flat "heart" surrounded by mountains and enclosed by the waterways of Biferno and Fortore. The Rocca di Oratino is situated in a privileged position on a rocky spur and allows the control of the Biferno River.

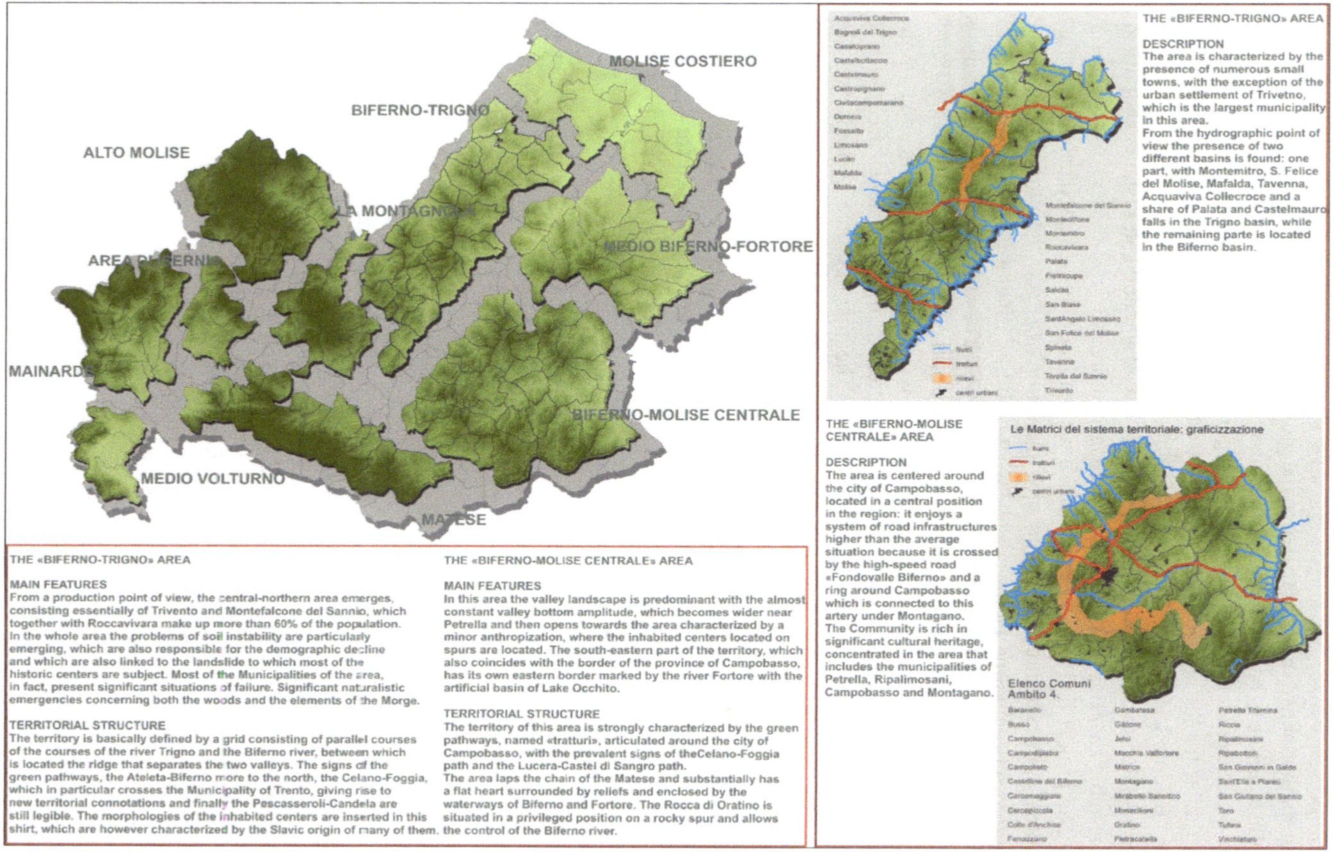

Fig. 1 The two banks of the river: diversity/limits. *Source* lab.lacosta elaboration 2015

3 Living Landscapes: Boundaries and Limits

Our methodology, oriented to define the global condition of different districts of the Molise region, is based on competitiveness and attractiveness of the places linked to the water element as vital landscapes to going towards a quality transformation. In particular for the course of the Biferno River, we defined a specific project, that is the "Biferno Laboratory" project. It is aimed at defining the environmental conditions related to actual and possible uses for environmental and sustainable tourism, renewable energies and multifunctional agriculture.

The project had a first part in which we realized the Geographical Information System, through three indicators sets: the *knowledge* set; the *production* set; the *fruition* set (including public space; sweet mobility; itineraries), as shown in Fig. 2.

This is the first part of our experimentation, which describes the territorial analysis aimed at defining the permanence and fragility of the territory under examination.

The second part of the experimentation is related to the verification of the values and detractors: territorial data, collected in the first stage, according to the five Resource Systems, were analyzed. The five resource systems are: the Physical-Environmental S., the Landscape-Visual S., the Historical-Cultural S., the Agricultural-Productive S. and the Demographic-Tourism S. In some cases the data was extrapolated from maps and they were used as shape files with polygons characterised by attributes that are easily transformable into grids; in other cases the data were in numeric form and diffusion algorithms have been elaborated for each case; in yet other cases the data was provided by the local administration (and for each case desegregations algorithms have been found for the creation of grids containing information useful for research purposes) (Cialdea 2014, 2017).

The data collection process has constantly been carried out in a dual perspective, both for the use of the database and for the data management in a mapping system. By way of example, here are the data for the Biferno area, which falls within the central area of its course, where the phenomena of landslide occurred again. Diversities will emerge on both sides of the Biferno. Basically in this picture there are some pre-imagination of scenarios.

There is therefore a confirmation of the difference in regional planning between the two shores, also from the urban point of view. In fact, on the right Biferno side, there is the regional capital of Campobasso with the ten small villages which surround it, as identified by the Master Plan of Campobasso (Beguinot 2000; Cialdea 2012). On the other side of the river, urban settlements are less consistent and they are located in a vast area oriented to agricultural activities. As well as analysed for the preparation of the Landscape Plan, it is necessary now to compare our scenarios, oriented to the landscape quality, with the current planning tools of municipalities involved in our sample area.

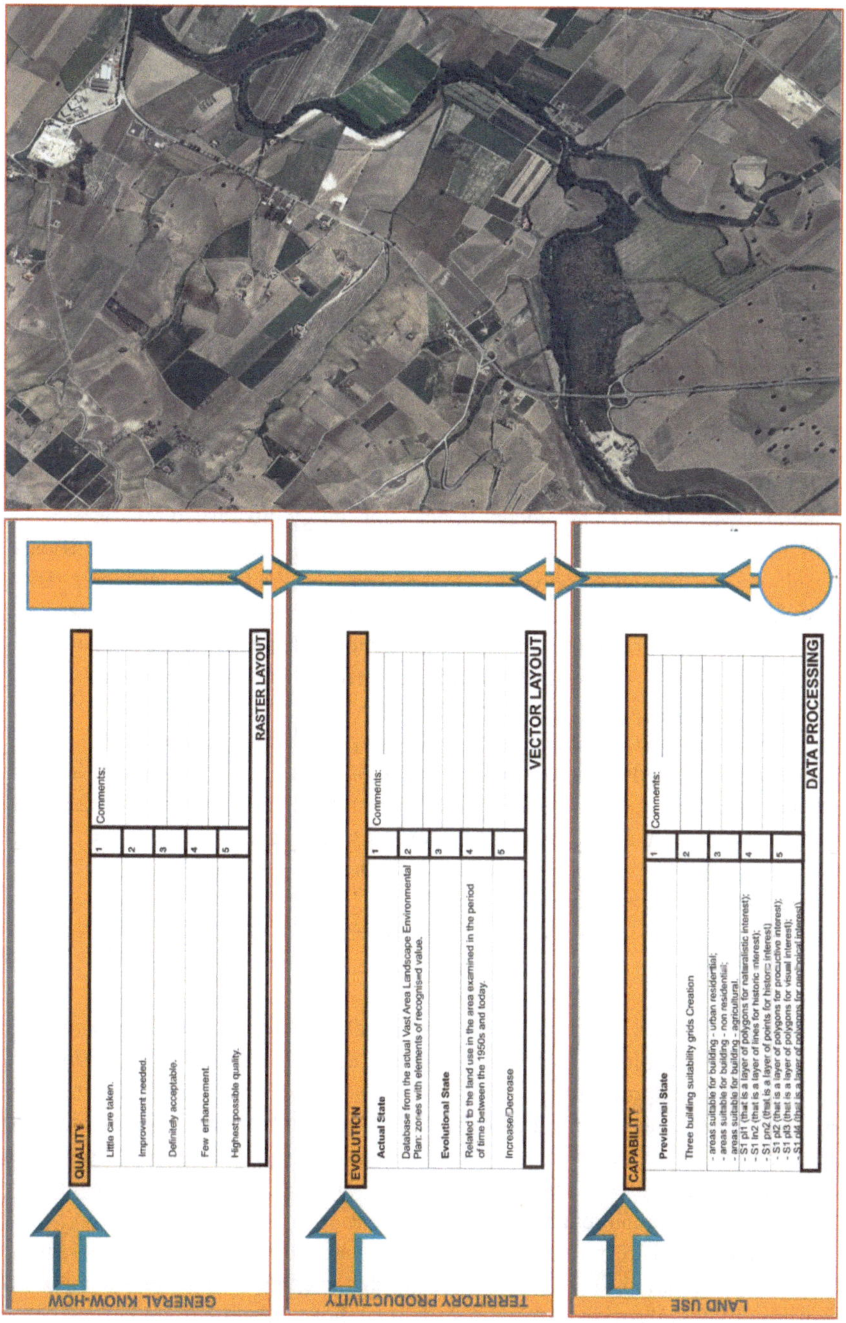

Fig. 2 The Biferno project: liveability/sustainability. *Source* lab.lacosta elaboration 2016

3.1 The Relationship with Local Planning Tools

The territorial area of the Biferno River includes, certainly, an over-municipalities district: moreover, the area under examination in this study falls within the municipality of Ripalimosani, that is one of the ten small towns around Campobasso. Therefore, we analyzed its local planning tools.

It is also necessary to remind, as already mentioned, that for the over-municipal level there is no reference: the territorial coordination plan has never been drawn up, except for a work carried out, which basically consists of an analysis of the matrices of the region.

In particular this plan highlights the socio-economic matrix (with population density and variation, territorial and social indicators), the environmental matrix (with natural and anthropic risks, correct use of resources and integration of landscape and environmental values), the historical-cultural matrix (with architectural, archaeological and historical emergencies), the Settlements matrix (with historical centers, urban systems, industrial areas and various facilities), the productive matrix (with agriculture, industries, tourism activities and other services), the infrastructural matrix (with road and railway networks and technological and energy infrastructures).

The principal aim of the Province Master Plan is the development of the Ripalimosani municipality, based on valorisation and networking of local resources. This approach leads to the definition of territorial areas, corresponding to geographical contexts whose historical-cultural, social and territorial characteristics can favour the creation of a network of relationships and concerted policies, able to bring a significant added value to the development programs. Moreover, the provincial territorial planning confers to the vegetation and woodland system priority purposes of naturalistic protection, of hydro-geological protection, of climatic and tourist-recreational function, besides productive and of scientific research. Therefore forms of utilization that may significantly alter the balance of existing wild species must be prevented, also pursuing the objective of restoring the forest heritage as a multifunctional forest ecosystem (Provincia di Campobasso 2007).

As far as the plan of the Municipality concerns, the General Modification of the Master Plan (dated 1992) is in approval (Ripalimosani 1992, 2008, 2011, 2012, 2017).

Firstly the attention is centred on the geological instability: the hydro-geological plan (adopted by resolution of the Institutional Technical Committee No. 87 of 28 October 2005, and approved by the Technical Committee No. 25 of December 16, 2004 and implemented by the Basin Authority, pursuant to Law 183/89) aims to guarantee to the territory, affected by the river basin, adequate levels of safety with respect to the hydraulic and hydrogeological phenomena. This plan consists of two sections: hydraulic alignment and slope alignment.

The Basin of the Biferno River has a draining basin surface of 1316.1 km^2. From the lake to the mouth of the river, the valley is characterized by the presence of irrigation crops with industrial destination such as sunflowers, orchards and vegetables.

Along the banks are various plant species; the most frequent are willows, poplars and robinias.

In the municipal territory there was a gypsum quarry in the "Gessiero area". This quarry was discontinued in the early 1960s. There is a curiosity about some gypsum outcrops that donate selenite samples rather impure and mostly non-transparent moreover: it is possible to find beautiful specimens of sericolite. In those areas there is an old quarry, visible from the "Bifernina Road", from which up about 50 years ago raw material was extracted for a small industry that produced plaster for construction. The gypsum quarry that has been inactive for a long time, in Ripalimosani.

In addition, the territory of Ripalimosani is affected by the presence of the following Site of Community Importance belonging to the Nature 2000 network. It is the Site code "SCI IT7222247", named "Biferno Valley from Quirino flow confluence to the Guardalfiera Lake-Rio flow" (Chlora Criteria 2015).

Furthermore, there is a part of the municipality's surface affected by the landscape constraint, declared as public interest area by the Ministerial decree dated 01/08/1977. This is a tool provided for the protection of areas of greatest landscape value, with the aim of mitigating the inclusion of buildings and infrastructures in the landscape, in order to make activities compatible with the environmental context. This restriction was introduced by the law 1497/39, integrated with the law 431/85, subsequently inserted into the Consolidated Law on the legislative provisions regarding cultural and environmental heritage Legislative Decree no. 490 of 29/10/1999, and finally confirmed by the Legislative Decree n. 42 of 22/01/2004.

Because of these reasons, in this municipality Master Plan, in relation to the described phenomena of geological and hydrogeological instability, some areas subject to the risk of flooding along the Biferno river have already been reclassified from "Zone D" (industrial area) and "Zone T" (equipped for tourism and leisure), to "Zone E3" (for controlled reforestation and/or respect of existing crops).

In addition, other actions can be pointed out:

- the complete enhancement of the existing and sometimes abandoned naturalistic and landscape, monumental and residential heritage, susceptible of a tourist attraction and of stimulating flows;
- the best use of existing plants and the development of new infrastructures suitable to correspond to the tourist vocation;
- the enhancement and qualification of events linked to existing local traditions and the planning of new cultural (theatrical and musical), recreational, sporting and cultural initiatives;
- the strengthening of relations with the emigrant associations;
- the recovery of the green pathways and of the sites linked to transhumance.

The idea of development of Ripalimosani derives and depends on being located at the border of the Campobasso territory to which it is strictly connected. There is the necessity to respond adequately to the dynamics of the neighbouring urban centre and at the same time it is necessary to enhance and preserve the landscape-environmental quality. The identified actions aim to consider the landscape as an indicator of territorial and urban quality, also through the rehabilitation of the rural building heritage with historical-testimonial value (Ripalimosani 2017).

4 Sustainability Versus Resources Utilization: The Project's Results

The "Biferno Laboratory" is related to the creation of a shared scenario for the development of the territory, describing the conditions for further strategic planning. This experience offered a training path that moves from the real knowledge of the environmental, cultural, architectural context in which it operates to the urban and landscape-environmental planning.

The analysis phase, as a fundamental moment of the planning process, lays the basis for correct and coherent actions on the territory and guarantees the creation of the critical and informative apparatus, especially in view of the identification of a unitary and shared development strategy for the Biferno river internal area.

Elaboration of project ideas and implementation of concrete examples of territorial action methods, will be intended as suggestion for new purposes on the territory, based on environmental and sustainable tourism, multifunctional agriculture and renewable energy.

With this logic, it is possible to extend the concept of smart city to the territory, with the aim of creating a smart and inclusive sustainable territory, above all oriented towards an intelligent fruition. The search for the quality of the landscape becomes, in this way, an indispensable factor for development, starting from measures to contain the exploitation of natural resources, adding value to the territory and above all linked to the intrinsic characteristics of the territory, with differentiations in its different parts.

The project's results reveal some critical issues, which should be added to those already described, deriving from official planning tools. Within the municipality of Ripalimosani some elements were found, from the comparative reading of the elements considered as values in connection with the elements considered as detractors. In conclusion, they can be summarized as follows:

1. among the environmental features, in addition to the extremely important geological criticality, the relevance of the difference between the two shores, which then consolidates the choice of two separate areas for the purposes of landscape planning;
2. the proposal for a photovoltaic system among the new insertions of energy infrastructures;
3. new purposes for productive activities.

In relation to the first factor, undoubtedly the predominant problem is the landslide phenomena. In fact, beyond the area identified in this work as a case study, another area with similar problems of failure falls again in the municipality in question. The landslide event, which had a paroxysmal evolution in 1998, severely damaged the area of the "Lama del Gallo" viaduct from branch B of the State Road n. 647 involving the prolonged closure of the road, the demolition of the unsafe batteries and the construction of a temporary by-pass.

The problem of the Molise region is always in connection with the road infrastructures: also bridges exactly along the Biferno River during the long time suffered serious crisis, as some ancient descriptions show: "The long course of the river is a cemetery of Roman bridges, Aragonese, carolini. Isolated pillars, broken arches that keep a fragment of the sixth like a rostrum. For about forty years from 1845 to 1881 the Biferno no longer had a bridge; for forty years of summer the river was ford. These steps of fortune were possible in summer and autumn: in winter they became very difficult if not impossible. Then the countries of the left bank remained cut off from the world. The Biferno had the power to upset the laws of time: the long winter with the snow that buried houses and fields became one interminable day." (Jovine 1941).

For the second factor, there are some troubles about new providing ground photovoltaic plants: the project consists in the realization of a grid-connected "ground" photovoltaic system for the production of electricity with a nominal power of 3 MWp. The plant will interest an area of about 10 ha, in the Municipality of Ripalimosani: it will work in parallel to the medium voltage electricity distribution network, totally releasing electricity to the grid (Ripalimosani 2013).

About the third factor, there are some interesting new actions for agricultural activities. The most important is the Project of the "VI.NI.CA. s.r.l. Società Agricola": it is a project that was born in 2008, where uncultivated and abandoned lands, suited for sustainable and natural agriculture they return to bear their fruits. Together with the vines, the company grows olive groves and vegetables for a total area of 220 ha in biological conversion. "The company works with conservative cultural techniques and only with the use of products natural both in the vineyard and in the cellar. The grapes are harvested manually and fermentations start thanks to indigenous yeasts. There is no pushed temperature control, yes use products for the extraction of aromas to clarify, and for keep in wine all the characteristics that the grapes gave us not we perform no filtration. Aware that working in this so our wines will be the pure expression of the territory" (www.vinica.it).

Certainly, we have seen that the city's Master Plan, following the disaster of the late nineties, has however downgraded the areas most at risk landslide, declaring the area in large part as not usable for building. But in a territorial project aimed at safeguarding but also at development, it is fundamental to read, for individual municipalities, through a matrix that examines and compares values and detractors, according to the methodology proposed by us for the New Regional Landscape Plan, illustrated here. Therefore, the form filled for the area under examination is shown below (Fig. 3).

TERRITORIAL SURVEY SCHEDULE SCHEDULE No 70/25

LOCATION

Province	Campobasso
Municipality	Ripalimosani
Locality	Rural Land
Geogr Coord.	46°90'46"N46°69'15"E
Location code	UT70
Typology Code	RU25

AREA's FEATURES

Urban Area	□
Rural Area	X
Mixed Area	□
Primarly Rural Area	□

VALUES

ENVIRONMENTAL VALUES	
SCI / SPZ	Y
Natural Reserve	N
Park	N
LANDSCAPE-VISUAL VALUES	
Restriction	Y
Residuals	Y
Forests	Y
HISTORICAL-CULTURAL VALUES	
Restriction	Y
Architectural Evidence	Y
Urban Planning Evidence	Y
AGRICULTURAL-PRODUCTIVE VALUES	
Land Use for Agricultural Activities	Y
Land Use for Agrifood Activities	Y
Land Use for Biological activities	Y

DETRACTORS

ENVIRONMENTAL DETRACTORS	
Seismic Restriction	Y
Hydro-Geological Restriction	Y
L. 445/1908	Y P
LANDSCAPE-VISUAL DETRACTORS	
Wind-Power Plants	Y P
Ground–Mounted Photovoltaic Plants S	Y P
Quarries	Y
HISTORICAL-CULTURAL DETRACTORS	
Preservation Condition	N
Public or Private Property	N
Access Facilities	N
AGRICULTURAL-PRODUCTIVE DETRACTORS	
Industrial Sites	Y
Special Districts	N
Other	N

GENERAL DESCRIPTION

Landslides of "Contrada Covatta" and "Lama del Gallo" are in the countryside of Ripalimosani, which in 1996 and in 1998 led to the closure of the important road artery to traffic because a barrage was created, then artificially drained. In addition, from the point of view of hydraulic risk, this sample area also owns interesting features, since the area immediately upstream of the narrow riverbed caused by the Contrada Covatta landslide was affected by flooding in conjunction with events of exceptional Bifemo River..

PECULIARITY

In this sample are there is no Landscape Plan.
Moreover there is the Geosite named "E10 The Covatta landslide".

LOCALIZING MAP

PHOTO

PLANNING DOCUMENTS

Municipal Level		*Regional Level*		*Notes*
P.d.F. (Buiding Plan)	X	P.T.C.P. (Province Territorial Plan)	X	There is the Landscape Restriction for a part of the municipality that includes the area in question
P.R.G. (Master Plan)	□	PAI (Idrological Asset Plan)	X	The Site of Communinity Interest "SIC IT722224 Valle Biferno from confluence Torrente Quirino to Lago di Guardalfiera-Torrente Rio" that runs along the municipal boundary is in this sample area.
Other	□	P.T.P.A.A.V. (Landscape and Environmental Protection Plan)	□	

Fig. 3 The schedule for the Covatta landslide area. *Source* lab.lacosta elaboration 2017

4.1 The Investigation Schedule

Figure 3 shows the sample area of the Biferno course, analyzed through the indicators and their elaboration aimed at achieving the aim of the "quality transformation". The schedule, named "Territorial Survey", is devoted to the identification of values and detractors. Research activities were carried out in order to obtain the recognition of historical, natural, physical and aesthetic values and their interrelationships, moreover dynamic transformations of the territory were analyzed in order to identify homogeneous areas characterized by different landscape value levels in the Molise landscape. Our work was most conspicuous for the data collection of spatial information for regional areas actually not included in Territorial Landscape Planning. The first part of our engagement, in fact, was oriented to collect these data, especially for areas around Campobasso and Isernia. Data collation, arranged as above described, was carried out in view of territorial analysis that took into account both positive and negative elements in order to analyze, subsequently, the quality of the landscape. In practice elementary data was collected and divided according to the resource systems which describe the actual state of the territory.

The second phase was carried out in order to identify the landscape quality aims, in order to allow a future articulation of the new landscape Plan, which can aim at the identification of conservation measures of connotative features of the landscape and the determination of interventions recovery and rehabilitation of severely degraded areas. Therefore it proceeded to the definition of a methodology useful for the identification of regional territory through territorial matrices. We selected two different horizons, time and spatial. For each of them are defined: preliminary analyses, main stakeholders, finally aims. As regards to the "time horizon", the first step was oriented to accord three different point of view: the *current protection*, through the transformability set by the landscape plans actually in force, *the evolution of the land use*, related to the 50s to date, and finally the *perspective of prediction*, or how the ordinary planning tools—which in Molise are basically plans at the municipal level—provide (Regione Puglia 2010; Poli 2012).

Moreover, these elements have been synthesised into two large groups, Values and Detractors. The values are elements which confer worth to the total landscape system (such as architectural, naturalistic, historical elements). They include elements of great naturalistic worth, that is the presence of particular species of flora and fauna. They are: the Important Bird Areas and the Special Protection Zones (at the sense of the Directive 79/409/EEC), the Sites of Community Importance (at the sense of the Directive 92/43/EEC "Habitat" adopted in Italy by the D.P.R. 357/1997), the Beaches free (Data are required by the Plan for Use of State Coastal Property for Tourism and Leisure of Molise Region), the Areas restricted under Law 1497/39 and under Law 431/85. What remains of areas of great natural worth and are the historical memory of these places are also included. They are Residual of dunes areas; Residual of green pathways, Residual of woods' areas. However, if values highlight landscape qualities, detractors penalize them. The detractors list was prepared after detailed land use features analyses. We undertook our analysis in ArcGis field: the detractors have been represented through graphics primitives with its geographic coordinates. Later, shapes for each detractor typology were realized.

This integration—between natural features and their awareness—has been made possible by the realization of a G.I.S. dedicated in order to integrate different types of data. The choice to realize the G.I.S. was determined by its intrinsic characteristics as an excellent management tool, for processing and querying of spatial data. After collecting all data they were collected in a geodatabase. This tool includes different types of data that have been integrated together to create and process maps through geographic analysis. The construction of a G.I.S. has the dual aim of creating the actual state maps and of creating a valuable support in the ordinary management of the phenomena. The G.I.S. has been realized with georeferenced information, while the geodatabase is composed of data in relation to the geographical location, the size, the land, detected through our schedules. This data could meet that of controlling the territory.

5 Conclusions and Remarks

The analysis carried out demonstrates the need to set up a reading methodology for the landscape which, although starting from data collection based on a homogeneous approach, in reality must always be adapted to local realities, even within the same region. The ultimate goal is to create vital landscapes. With a view to improving the quality transformation, we set up our own landscape, and landscape information and evaluation models.

The river has proved to be an extremely qualified example for the verification of the methodology aimed at the sustainability of the interventions that, on the territory that the river involves, can be realized.

Water is a perpetual element, but its performance varies in time and space: the river is an element that over time has changed its connotations, often changing a large amount. But the river is also a permeable element. And the purpose of the application of our methodology consists in the evaluation of the possible transformations from the point of view quantitative and qualitative, always with the quality transformation aim. The landslide, an element that characterizes the area in question, is certainly an element of instability, but it is also an element of strong perceptive value and it is located in the landscape with strong naturalistic connotations but in which attractive activities can be further encouraged (starting from some already successful achievements, which have been mentioned in this text). Watercourses carry out, within the territorial eco-mosaic, fundamental functions of specific ecosystem and ecological connection, as well as representing an essential functional element in order to maintain the environmental balance both physical and biological of the area, but it is also useful to implement sustainable actions.

Our work for the Landscape Plan was performed in two different scales: at the territorial level and at the local level, that is the municipal level.

At the territorial level, the indications, that are shown in the cards relating to

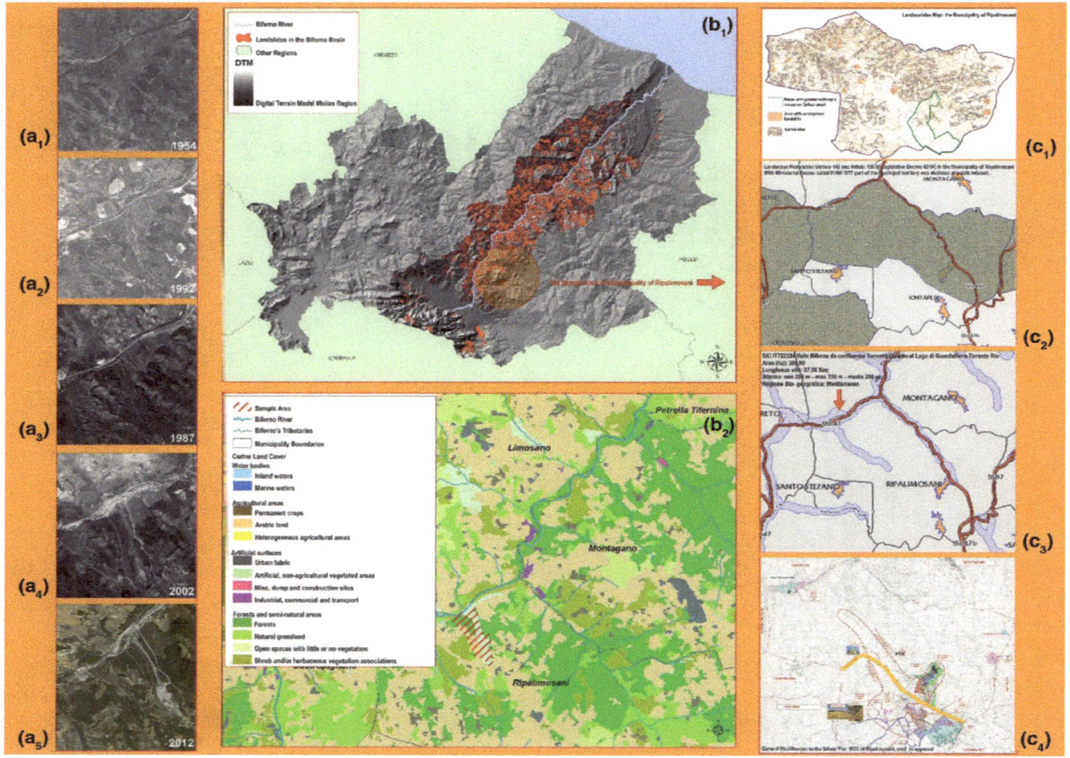

Fig. 4 The sample area of the Covatta Landslide. *Sources* $\mathbf{a_1}$–$\mathbf{a_5}$ orthophotos 1954, 1992, 1997, 2002, 2012; $\mathbf{b_1}$ digital terrain model of the molise region; $\mathbf{b_2}$ land use from Corine Land Cover 2012; $\mathbf{c_1}$–$\mathbf{c_3}$ strategical environmental assessment 2017; $\mathbf{c_4}$ master plan 2017, lab.lacosta elaboration 2017

Table 1 The landscape quality aims, general and specific, declined for the five resources systems (*source* lab.lacosta elaboration 2017)

Resources systems	General aims	Specific aims
Physical-environmental system	1. Promote the preservation of the integrity of areas of high naturalness and high ecosystem value	1.1. Safeguard geological-geomorphological systems with high integrity (especially referring to the geosite)
		1.2. Safeguard protected areas and areas of high environmental value such as those covered by the Nature 2000 Network
		1.3. Safeguard and improve environmental functionality of river systems of Molise
		1.4. Safeguard and rebuild habitats of Molise (especially for river course)
		1.5. Safeguard woods and forests of mountainous and hilly areas of Molise
		1.6. Redevelop and redesign the hill landscapes of Molise
Landscape-visual system	2. Promote improved integration of landscape and the quality of infrastructures	2.1. Define territorial and landscape quality standards in the settlement of new network infrastructure
		2.2. Define territorial and landscape quality standards in the settlement of new energy infrastructure
		2.3. Define territorial and landscape quality standards in the settlement of new productive activities
		2.4. Feasibility study for the definition of administrative attention naturalistic qualification of existing infrastructure at sensitive areas

(continued)

Table 1 (continued)

Resources systems	General aims	Specific aims
Historical-cultural system	3. Promote the preservation of cultural values	3.1. Preserve cultural value and witnesses of settlements and historical settlement
		3.2. Preserve cultural value of traditional rural buildings
		3.3. Preserve the visible cattle-tacks residual
		3.4. Redevelop the historic rural landscapes
Agricultural-productive system	4. Promote the conservation of agricultural landscapes	4.1. Develop the agricultural landscape of Molise, recognize and promote its social functions
		4.2. Preserve open landscapes of the hill territories
		4.3. Redevelop the agricultural landscape of Molise
		4.4. Promote new initiatives in order to valorize the inner landscape
Demographic-touristic system	5. Promote the improvement of the quality of the settlements	5.1. Improve quality of urban settlements and their environmental performance, for greater well-being of the population
		5.2. Redevelop degraded contemporary urbanization landscapes
		5.3. Improve urban quality of and touristic settlements
		5.4. Improve urban quality of agricultural and productive settlements
		5.5. Improve soft mobility quality (walking, cycling, trekking on horse) and its interconnection with the traditional mobility

the ten areas identified by the Landscape Plan, are intended to address and propose planning actions.

At local level, the municipal plans should act in accordance with the landscape plan suggestion: within the areas mentioned above, different situations are identified (in relation to the features and values of the natural environment and human actions) that require specific norms and guidelines.

In particular, for the area under study, a Landscape-Environmental Regeneration Plan will be recommended (thus including both the interventions from the environmental point of view and those from the point of view of the landscape quality improvement.

The greatest attention was paid to the variations foreseen by the General Modification of the current Master Plan (see Fig. 4 and Table 1). In the Fig. 4 evolution of the landscape in our sample area is shown: orthophotos from 1954 to ourday (a1–a5) and pictures b1 and b2 revealed the current land use, fruition and enhancement of resources in connection with the orographic status and the strategic planning tools (c1–c3).

In the Table 1 the landscape quality aims, general and specific, were declined for our five resources systems: the proposed actions are organized for the ordinary maintenance and management of river territories in connection with the land uses otpimization.

The overlap of the current and forecast condition also allowed us to have an immediate (and quantifiable) perception of dangerous, intense, extensive and persistent modifications, especially where there is neglect and poor preservation.

The conclusion could be that even the municipal administrations promote, as is already happening in some other regions, an active tool for the protection of the landscape within their municipal planning tools, in compliance with what is defined by the Regional Landscape Plan. This tool must naturally understand how the individual management tools for parts of their territories already foresee, but further implement them in a more specific logic for urban planning.

References

Albrechts L (2003) Planning and power: towards an emancipatory planning approach. Env Plan C 21:905–924

Beguinot C (2000) Relation in attachment to D.C.C. No. 82 December 11 2000 "Variante Generale al Piano Regolatore Generale di Campobasso-Adozione, art.8 della L. 1150/42"

Chlora s.a.s. Criteria (2015) Management Plan of the SCI IT7222247 "Valle Biferno da confluenza Torrente Quirino al Lago Guardalfiera - Torrente Rio", Rural Development Program 2007/2013 Measure 3.2.3, action to "supporting the planning of the management of the recommended sites in the Natura 2000 areas"

Cialdea D (2009) Valorisation strategies for Molise coastal area: archaeological sites and settings of environmental value, Chap. 13. In: Nijkamp P, Fusco Girard L (eds) Cultural tourism and sustainable local development, England. Ashgate Puplishing, Ltd., Farnham, pp 181–200

Cialdea D (2010) Territori antichi e nuovi scenari. Prolusione dell'Inaugurazione dell'A.A. 2010–2011 Università degli Studi del Molise, pp 1–48

Cialdea D (2012) A new landscape planning approach in areas with a strongly rural matrix. In: Campagna M, De Montis A, Isola F, Lai S, Pira C, Zoppi C (eds) Planning support tools: policy analysis, implementation and evaluation, vol. 5. Franco Angeli, Milano, pp 453–464

Cialdea D (2014) Ricerca per le attività di redazione del Nuovo Piano Paesaggistico della Regione Molise: Relazione generale

Cialdea D (2017) New performances in urban nature. Recovery examples for the city crossing green pathways. In: Proceedings of the international conference on changing cities III: spatial, design, landscape & socio-economic dimensions Syros, Delos, Mykonos Islands, Greece, 26–30 June 2017, pp 79–93

Cialdea D, Mastronardi LA (2014) Multi-criteria methodology oriented to the preservation of the coastal areas landscape. In: Batzias F, Mastorakis N, Guarnaccia C (eds) Proceedings of the 5th international conference on development, Energy, Environment, Economics (DEEE '14), WSEAS Press, pp 237–243

CoE (2000) European landscape convention. Florence

CoE (2008) Guidelines for the implementation of the European Landscape Convention, Recommendation CM/Rec

Corrado F (2014) Nuove forme dell'abitare nelle terre alte e produzione di territorio in Atti della XVII Conferenza nazionale SIU, L'urbanistica italiana nel mondo, Milano 15-16 maggio 2014. Planum Publisher, Roma-Milano, pp 941–946

Coyle S (2011) Sustainable and resilient communities. Wiley, New Jersey

Crosta PL (2010) Pratiche. Il territorio è l'uso che se ne fa. Franco Angeli, Milan

Decreto Legislativo 22 gennaio 2004, n. 42 recante il "Codice dei beni culturali e del paesaggio" ai sensi dell'articolo 10 della legge 6 luglio 2002, n. 137 e successive modifiche ed integrazioni. G.U. n. 45 del 24/02/2004

De Nigris A, Cartolano A (2015) Smart city smart land. Approccio Strategico ai Processi di Trasformazione dei Territori: Le possibili forme di governance, XXXVI Conferenza Italiana Di Scienze Regionali

Esposito F (2014) Ingegneria&Paesaggio in Liguria. Una proposta per una possibile integrazione in Atti della XVII Conferenza nazionale SIU, L'urbanistica italiana nel mondo, Milano 15–16 maggio 2014. Planum Publisher, Roma-Milano, pp 962–969

Giaccaria P, Rota FS, Salone C (eds) (2013) Practicing territoriality. Reflections on policies for the green economy, agro-industry and culture in Piedmont. Carrocci Editore, Rome

Healey P (2003) Collaborative planning in perspective. Plan Theor 2:101–123

Ingaramo R, Voghera A (eds) (2016) Topics and methods for urban and landscape design. From the river to the project. Springer International Publishing AG, Berlin

Jakob D (2013) The eventification of place: urban development and experience consumption in Berlin and New York City. Eur Urban Reg Stud 20(4):447–459

Jovine F (1941) Journey to Molise, Isernia, Libreria Editrice Marinelli

Legge 8 agosto 1985, n. 431, Conversione in legge, con modificazioni, del decreto-legge 27 giugno 1985 n. 312, recante disposizioni urgenti per la tutela delle zone di particolare interesse ambientale. G.U. n. 197 del 22/08/1985

Legge aprile 2014, n. 56 "Disposizioni sulle Città Metropolitane, sulle Province, sulle Unioni e Fusioni di Comuni 14G00069" (GU n.81 del 7-4-2014)

Legge 6 ottobre 2017, n. 158 "Misure per il sostegno e la valorizzazione dei piccoli comuni, nonche' disposizioni per la riqualificazione e il recupero dei centri storici dei medesimi comuni. 17G00171" (GU n.256 del 2-11-2017)

Kunzmann KR (2007) Territoire, Territorialité, Territorialisation. Conference. Institut de Géographie Alpine, Grenoble, 7–8 June 2007

Musco F (2009) Rigenerazione urbana e sostenibilità. Franco Angeli, Milano

Poli D (ed) (2012) Regole e progetti per il paesaggio. Verso il nuovo piano paesaggistico della Toscana, Firenze University Press

Provincia di Campobasso (2007) Piano Territoriale di Coordinamento (PTCP) di Campobasso, Approvazione Atto deliberativo D.C.P. n. 45 del 19 giugno

Regione Molise (2011) Convenzione per l'attribuzione del contributo di ricerca per le attività di redazione del Nuovo Piano Paesaggistico Regionale del Molise tra La Regione Molise e il Laboratorio L.A.CO.S.T.A dell'Universita degli Studi del Molise, Responsabile scientifico prof. Donatella Cialdea, Responsabile del procedimento arch. Marcello Vitiello, Dirigente del Servizio Beni Ambientali, Statistico e Cartografico della Regione Molise, stipulata in data 28/02/2011
Regione Molise (2015) The European agricultural fund for rural development-europe investing in rural areas, Italy–Rural Development Programme (Regional)—Molise 2014–2020
Regione Puglia (2010) Piano Paesaggistico Territoriale Regionale, Lo scenario Strategico, www.regione.puglia.it
Ripalimosani (1992) Chronological sequence of the urban planning tools: Master Plan in force, approved by resolution G. R. n ° 5805 of 21.12.1992
Ripalimosani (2008) Partial modification to the Master Plan for the industrial zone, approved by resolution G. R. No. 325 of 31.03.2008
Ripalimosani (2011) Partial modification to the Master Plan for the zone T "c.da Selva", approved by resolution GR n ° 359 of 20.05.2011
Ripalimosani (2012) Partial modification to the Master Plan for the "Fondovalle Biferno area", approved by resolution GR n ° 302 of 08.05.2012
Ripalimosani (2013) Il Piano d'Azione per l'Energia Sostenibile del Comune di Ripalimosani, approvato con delibera del Consiglio Comunale n. 11 del 09/03/2013
Ripalimosani (2017) General Modification of the whole Master Plan proposed in 2017, not yet approved; Ripalimosani (2017) Environmental preliminary report for the general modification version of Master Plan 2017
SNAI (2014) Strategia Nazionale Aree Interne in relazione all'approccio integrato allo sviluppo territoriale di cui all'art 15 comma 2 lettera a) del reg. (UE) 1303/2013. Italy, National Internal Areas Strategy (SNAI) in relation to the integrated approach to territorial development set out in Article 15 paragraph 2 letter a) of EU Regulation. 1303/2013
Socco C (1999) Italian landscapes, for the government of transformations, Fondazione Benetton, Castelfranco Veneto, 26–29 May 1999
Tejedor Bielsa J (2013) Nuevos Instrumentos de Planificación y Gestión de la rehabilitación y la regeneración urbana. Monografías de la Revista Aragonesa de Administración Pública XV:27–72
VI.NI.CA. s.r.l. Società Agricola, www.vinica.it

Reducing Urban Entropy Employing Nature-Based Solutions: The Case of Urban Storm Water Management

Raffaele Pelorosso, Federica Gobattoni and Antonio Leone

Abstract This paper presents the theory behind the concept of low-entropy cities based on the second law of thermodynamics. This concept aims to provide a bridge among different approaches on city sustainability studies, highlighting the links between natural processes and the socio-ecological complexity of urban systems. A practical low-entropy application is then proposed for urban storm water management, examining the planning of nature-based solutions with the support of a modelling approach. A further novelty of this work is the attempt to combine entropy with resilience assessment for urban green infrastructure planning.

1 Introduction

In a previous work (Leone et al. 2016), the authors highlight the importance of taking the laws of thermodynamics into consideration when planning actions to improve the sustainability of cities and landscapes. In particular, a SLT approach, also defined as "second-law thinking", is proposed as a planning paradigm aimed at increasing exergy (i.e. the energy component able to produce work) and, consequently, at reducing the production of wastes (entropy) which impede the functionality of the socio-ecological system. In this work, we explore another aspect of SLT, the entropy concept. Entropy is a measure of the disorder, or waste of the city, and as such can be considered an indicator of the diversified impacts of the urban development on the biosphere.

The entropy release of a city today is excessive because it overcomes Earth's natural capacity of regeneration and threatens to destabilise the urban (human) civilization itself, which is causing it.

Indeed, a link has clearly been established between consumption processes (also in terms of soil sealing, and relative waste productions) and social-environmental

R. Pelorosso (✉) · F. Gobattoni · A. Leone
DAFNE, Tuscia University, Viterbo, Italy
e-mail: pelorosso@unitus.it

R. Papa et al. (eds.), *Smart Planning: Sustainability and Mobility in the Age of Change*, Green Energy and Technology,
https://doi.org/10.1007/978-3-319-77682-8_3

problems (e.g. increasing soil erosion, decreasing biodiversity, climate change and migration), which are all consequences of the natural response by Gaia to the human imposed constrictions on the irreversible growth of global entropy (Costanza et al. 2012; Fistola 2011; Gobattoni et al. 2011; Karnani and Annila 2009; Lovelock and Margulis 1974).

Recently, a new thermodynamic concept of cities has been proposed to support a more systematic evaluation of the urban system sustainability (Pelorosso et al. 2016a). In particular, Pelorosso et al. (2017) propose a low-entropy urban green infrastructure (UGI) planning strategy and several entropy indicators to face the main environmental, social and economic phenomena affecting urban systems. The low-entropy city approach can lead to a reduction in consumption, but more importantly, it aims to investigate what is consumed, how energy and matter are used and where waste might induce socially and environmentally negative consequences.

Nature-based solutions (NBSs) are engineered green/ecological systems inspired or supported by, or copied from, Nature (EU 2015). NBSs are part of UGI, examples include Sustainable Urban Drainage Systems (SUDS) or urban BMPs to control pollution, runoff and, in general, to ensure a sustainable urban water management. These include green roofs, pervious surfaces, constructed wetlands, detention basins, infiltration basins and filter drains (Pelorosso et al. 2013; Recanatesi et al. 2017). Such NBSs aim to regulate the water cycle through their adsorption, stoking and infiltration capacities and then they represent potential sustainable solutions for the reduction of city entropy.

The objective of this contribution is to present a first application of the low-entropy city concept in the context of UGI planning aimed at optimizing the storm water cycle and thus reducing its negative impacts on urban and extra-urban systems. Accordingly, some new indicators are used to evaluate the entropy of an urban system in terms of water quantity and quality parameters in a base scenario and a NBS scenario. The latter is proposed with the aim of increasing the system's hydrological resilience (Pelorosso et al. 2018). The variation in entropy indicator values shows the operational potential of the low-entropy concept for storm water management in urban contexts as well as being the first proposed integration of entropy and resilience concepts in UGI planning.

2 The Low-Entropy City

Every single living organism of the landscape, Man included, uses and exchanges energy and matter (e.g. fossil fuels, solar radiation, organic matter, food, minerals, soil) struggling to develop highly (social-ecological) ordered, lower entropy structures to increase the total dissipation of energy and maximize the global production of entropy (at Earth scale) following the Second Law of Thermodynamics (SLT). Man, more than other species, plays a fundamental role in the evolution process of the global system due to our ability to modify the biosphere (and consequently, the landscape) drastically in order to reach our objectives (Norra 2014; Pasimeni et al. 2014).

Cities can be seen as open systems and, according to the SLT, they can increase their socio-ecological structure and complexity (negentropy) only by increasing the disorder and random-ness (entropy) in their host system, the biosphere.

Thus, low-entropy cities are urban systems with lower entropy release into the biosphere adopting strategies that mimic, as far as possible, Nature and the structure and functioning of ecosystems according to the SLT and the circular economy of Nature (Ho 2013). A low-entropy city will therefore evolve and grow enhancing its socio-ecological and structural complexity (reducing internal entropy) by adding and optimizing functional elements and synapses among them while wastes (exported entropy to the biosphere) are minimized. Entropy exported to the biosphere is represented by direct and indirect city wastes. Direct wastes derive from urban metabolic cycles of socio-ecological and technical systems, indirect wastes are produced by unsustainable extraction of resources in source countries and ecosystems in order to sustain the same urban system.

In this work a first application of this concept is presented in the context of storm water management and NBS planning.

3 Nature-Based Solutions and Second Law of Thermodynamics

Cities are human-modified complex ecosystems (Bai 2016) and associations between natural and urban processes are possible.

In nature, an ecosystem such as a forest, has no useless features because each component has a role in the functioning of the whole system. An open space within a dense forest created by tree falls can host colonizing plant species, can be a habitat for wild fauna that spread seeds or pollinate flowers allowing a complex ecosystem to evolve and regenerate maintaining sufficient levels of identity and essential resources (e.g. soil fertility). The bark on the sunny side of a tree trunk may host a most beautiful orchid, while the shady side supports an incredible lichen community. All these organisms appear and grow in a suitable habitat because they have evolved to use the free available energy present in that site and those conditions. In natural ecosystems, very little energy is wasted but instead it is stored (e.g. as biomass) and transformed into different forms that can be used by other organisms for the benefit of the whole system. Similarity and in a scalar and fractal manner, a cell plays its role in an organism, a cell sub-component within that cell. All the components of an organism (from unicellular to an entire ecosystem) carry out a role in the evolution of the upper system through cooperation and competition mechanisms that often can be observed and understood only from the highest levels of the system (Capra and Luisi 2014).

Good farmers apply rotational crop management, recycle and reuse everything in their productivity systems (Ho 2013; Leone et al. 2014). Entrepreneurs restructure, close or sell inefficient society components. Engineers reduce heat loss and employ used heat to support other industrial processes, thus increasing global efficiency. Like these examples, in a low-entropy city view, urban planners should aim to pursue social-economic objectives employing all available and free forms of energy, defining the best employment or, at least, the main green strategies for different underused and unused urban spaces that represent an opportunity for transformation and urban regeneration (Gobattoni et al. 2016; Pelorosso et al. 2017).

In a SLT view, NBSs providing several UESs (e.g. climate or water regulation) are able not only to use the incoming solar energy and rainfall, stocking it in biomass or in the soil layer, but also to reduce the dispersion of entropy and low-value forms of energy such as reflected radiation, radiant temperature or water runoff and associated pollutants. Clearly, following the low-entropy concept, NBSs should be planned where excessive or unused energy, water and wastes are present. Indeed, NBSs must be adapted to local conditions to be energy and resource-efficient and resilient to change (EU 2015).

4 Study Case

The considered study area has an extension of about 330 ha and corresponds to the second municipality of the Bari Metropolitan Area (Fig. 1) where the urban drainage network is inadequate to manage storm water runoff efficiently. The combined sewer system, during heavy precipitation events, is not capable of managing all the volumes. Consequently the overloads affect the urban system, flowing directly to the seafront, contaminating the adjacent public beaches with obvious consequences on health and fruition.

5 Methods

In this work, some new indicators are used to evaluate the entropy of the urban system in terms of water quantity and quality regulating capacity. A base (status quo) scenario has been modelled and subsequently compared with a Nature-based scenario, proposed to increase the hydrological resilience (Pelorosso et al. 2018). The difference in entropy indicators shows a clear potential for implementation of the low-entropy approach for the storm-water management in urban contexts.

Few applications of the entropy concept have been presented in a context of spatial urban planning (Balocco and Grazzini 2000; Fistola and La Rocca 2014) and, to our knowledge, none within urban storm water management. Pelorosso et al. (2017) propose new kinds of entropy indicators intended to be easily employable by urban planners with the support of a modelling/assessment approach similar to

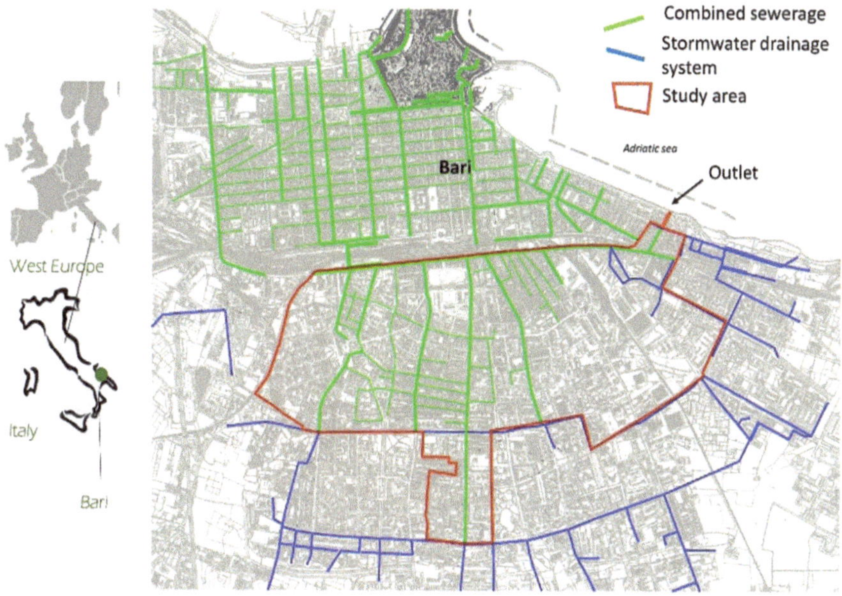

Fig. 1 Study area

Table 1 Entropy indicators and NBSs for urban storm water management (modified from Pelorosso et al. 2017)

Low-entropy nature-based solutions	Diffuse NBSs (green roofs, permeable pavements and bio-retention basins)	End-of-pipe NBSs (phytoremediation plants, wetlands)
Internal entropy indicators		
Ecological complexity	n° and typology of NBS	n° and typology of species, local freshwater ecosystems health status
Social complexity	People and water enterprises	People and water enterprises
Structural/physical complexity	% infiltration or stored water	Re-used water
External entropy indicators		
Biosphere/regional complexity	Outlet flooding, global runoff	Health status of receiving freshwater and sea ecosystems, CO_2 emission

that presented herein. Table 1 reports the entropy indicators proposed by Pelorosso et al. (2017) to evaluate a NBS scenario in terms of urban storm water management efficiency following a SLT view.

The proposed indicators have an explicit spatial character to allow for practical decisions on urban planning. They are classified in two groups: internal and external entropy indicators. Internal urban system complexity (internal entropy) can be increased by NBSs and it is divided into ecological, social and structural/physical complexity. Internal entropy indicators can therefore help to measure UGI socio-ecological cycles within urban systems (see Sect. 2). External entropy indicators measure the impact of urban systems on external areas. The identification of the borders of open systems such as cities is a hard issue. Nevertheless, such external areas can be identified at local scale, with opportune justifications, for a real pragmatic and operative planning (Pelorosso et al. 2017).

In this work, we propose more detailed entropy indicators adapted to the specific study case. Specifically, only structural/physical complexity is presented leaving further assessments to a future research development. External indicators are referred to the outlets of each sub-basin and at the outlet of the whole urban catchment that converges to the treatment plant. It is worth mentioning that water amount can exceed the overflow weir at the outlet of the study area during critical rainfall events. At such times, the entropy (waste) is not managed by the treatment plant and is discharged into the sea by the drainage pipe directly.

All the following entropy indicators are derived from the outputs obtained through the implementation of the US-EPA managerial model called Storm Water Management Model (SWMM, release 5.1.012), which is a well-recognized tool for dealing with storm water management issues (Zhou 2014). The simulated critical rainfall event has a three hour duration (equal to the concentration time of the urban catchment) and a return period of five years (as requested by Italian technical legislation for urban drainage systems) (Pelorosso et al. 2016b).

Two internal entropy indicators, E^i, have been proposed to evaluate the efficiency of the actual land use assemblage in terms of SLT

$$E^i_{infiltration\_j} = \frac{infiltration\,(\text{mm})_j}{rainfall\,(\text{mm})_j}. \tag{1}$$

$E^i_{infiltration\_j}$ describes the amount of water infiltrated into a sub-basin j with respect to the total amount of rainfall in the sub-basin j. It can be used to determine the efficiency of the current land use in terms of groundwater recharge for the sub-basin j. The hypothesized groundwater recharge would allow local use of water and the maintenance of a healthy status for green areas (e.g. urban trees, parks)

$$E^i_{storage\_j} = \frac{storage\,(\text{mm})_j}{rainfall\,(\text{mm})_j}. \tag{2}$$

$E^i_{storage\_j}$ describes the amount of water stocked in a sub-basin j with respect to the total amount of rainfall in the sub-basin j. It is related to the efficiency in water reuse for UES provision (e.g. biodiversity support or irrigation for local food or non-food production).

Both indicators can be used to design more liveable environments with positive potential outcomes in social and cultural terms, such as the possibility of creation of green meeting points, urban gardening etc.

The other proposed external entropy indicator, E^e, considers quantitative and qualitative water parameters:

$$E^e_{overflow\_j} = \frac{overflow\,(\mathrm{L})_j}{inflow\,(\mathrm{L})_j}. \tag{3}$$

$E^e_{overflow\_j}$ takes into consideration the efficiency of the drainage system in managing water runoff. The measurement is realised in the node of the drainage system related with sub-basin j. In this case, it describes, for each node j, the amount of combined sewerage overflow during the simulated rainfall event divided by the total inflow into the same node due to sub-basin runoff and the current pipe load

$$E^e_{sewagedischarge\_s} = \frac{Total\,volume\,(\mathrm{L})_s}{rainfall\,(\mathrm{L})_s}. \tag{4}$$

$E^e_{sewagedischarge\_s}$ describes the efficiency of the green and grey infrastructure in managing the total amount of rainfall at whole system level. Total volume is the amount of discharge (L) at the final outlet; rainfall is the total amount (L) of rainfall in the whole system s

$$E^e_{TSS\_j} = \frac{TSS\,washoff\,(\mathrm{kg})_j}{area\,\left(\mathrm{m}^2\right)_j}. \tag{5}$$

$E^e_{TSS\_j}$ describes the Total Suspended Solid (TSS) produced in a sub-basin j by washoff related to the total area of sub-basin j producing the TSS. *TSS* is listed as a potential pollutant by the Handbook of Urban Runoff Pollution, Prevention and Control Planning (US EPA 1993) and is widely recognized as a conventional contaminant in the international scientific literature (Borris et al. 2014; Hilliges et al. 2017; Revitt et al. 2014). TSS may have negative impacts on natural Habitat and on aesthetic and recreational value, may cause and/or increase erosion phenomena and transport pollutants (US EPA 1993)

$$E^e_{TSS\_overflow\_j} = \frac{TSS\,overflow\,(\mathrm{kg})_j}{TSS\,inflow\,(\mathrm{kg})_j}. \tag{6}$$

$E^e_{TSS\_overflow\_j}$ considers the amount of TSS into the overflow of each node. It describes the efficiency of the drainage system in managing the TSS coming from a sub-basin j plus the current TSS load in the upstream pipes

$$E^e_{TSS\_s} = \frac{TSS\,(\mathrm{kg})_s}{area\,\left(\mathrm{m}^2\right)_s}. \tag{7}$$

$E^e_{TSS\_s}$ describes the efficiency of the green and grey infrastructure in managing the total amount of TSS at whole system level s.

It is worth noting that indices (1, 2, 3, 4, and 6) are unit-less, while indices (5) and (7) are expressed in (kg/m^2). All these entropy indicators give us a measure of the system's efficiency in terms of quantitative and qualitative water parameters but, most importantly, they provide us with a measure of the spatial distribution of wastes (entropy) with respect to certain fixed variables relating to the context of the study case (specific morphological characteristics of the basin and sub-basin, green and grey infrastructure) and the simulated rainfall event. Following a SLT view, planners and designers should then analyse the proposed entropy indicators in order to increase and optimize the local reuse of water (as described by the internal entropy indicators 1 and 2) and reduce wastes or even transform them into a resource for other metabolic activities (as described by external entropy indicators, 3, 4, 5, 6, 7). Finally, a composite indicator of external entropy at sub-basin scale is calculated as the sum of standardized entropy indicators $E^{e}_{overflow\_j}$, $E^{e}_{TSS\_j}$ and $E^{e}_{TSS\_overflow\_j}$ obtained by dividing each one by the maximum value registered in the base scenario. This index of external entropy was then reclassified in three classes and mapped to add further judgment criteria about the spatial distribution of critical points amongst the different sub-basins.

6 Results

Figure 2 shows the spatial distribution of the entropy indicators calculated at sub-basin scale. In map 2a, b an inverse colour scale has been used since high values of the indicators represent more internal complexity (i.e. water infiltration and storage) and then more possibilities for water reuse and an increase in system complexity. The distribution of these indices shows a reduced functionality of the system for sustainable storm-water management. The external entropy indicators show several critical sub-basins (see also the composite indicators of Fig. 2f), mostly in the south-central part of the study area, with particular reference to the sub-basins 40, 37, 17, 31, 22 and 33. At the final outlet, entropy indicators at system level report: (a) a water volume discharge of around half of the total rainfall volume, (b) a mean TSS of 2.3 kg/ha produced by the whole basin (Table 2).

The variation in entropy in the NBS scenario can be evaluated at system (Table 2) and sub-basin scale (Table 3). Internal entropy indices show a slight increase in values with respect to the base scenario where the infiltration and storage of stormwater processes were proximate to zero due to the high imperviousness of the urban district. A significant reduction in external entropy indices is shown in the sub-basins where the introduction of NBSs and the grey infrastructure re-design are proposed. Moreover, two sub-basins (nos. 39 and 41) display an entropy reduction above all

Table 2 External entropy indicators at system level

Entropy indicator	Base scenario	NBS scenario	Δ%
$E_{sewagedischarge\_s}$ (Eq. 4)	0.518	0.513	−0.965
E_{TSS_s} (Eq. 7)	2.328	2.283	−1.968

Base scenario and NBS scenario

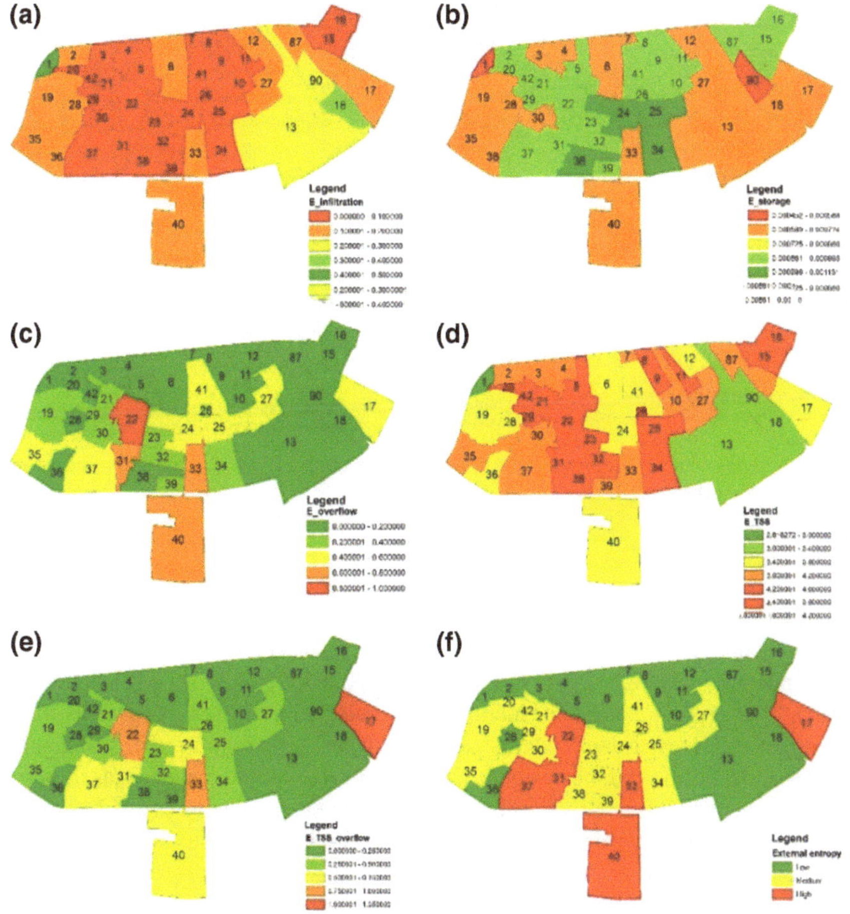

Fig. 2 Indicators of internal entropy **a**, **b**, external entropy **c**–**e** and composite external entropy (**f**) at sub-basin scale

in terms of overflow as they are downstream to the proposed interventions and they receive a smaller amount of sewerage inflow.

7 Discussion and Conclusions

The strength of thermodynamics is its ability to describe whole systems and aggregate properties, on macro scales, that emerge from complex micro-scale processes (Bristow and Kennedy 2015). Despite the widespread scientific awareness of thermodynamics as a key for interpreting evolution mechanisms, practical ways to employ it in land use decision making, landscape ecology and urban planning are challeng-

Table 3 Variation in entropy indicators at sub-basin scale

Sub-basin	Proposed intervention	Δ $E_{infiltration}$	Δ $E_{storage}$	Δ% $E_{overflow}$	Δ% E_{TSS}	Δ% $E_{TSS\_overflow}$
17	Infiltration basin	0.05	0.18	−23.88	−50.94	−60.22
24	Extensive green roof	0.20	0.07	−46.13	−21.73	−51.77
37	Permeable pavement	0.56	0.00	−33.72	−59.07	−34.60
39		0.00	0.00	−100.00	0.00	0.00
40	Grey infras-tructure project	0.00	0.00	−18.25	0.00	−18.26
41		0.20	0.07	−12.83	−21.21	−12.68

Only sub-basins with significant variation in entropy are reported. $E_{infiltration}$ and $E_{storage}$ variations are expressed in absolute values

ing. In practice, entropy and the second law of thermodynamics tend to be applied only metaphorically, while the required equations and parameters are often abstract quantities and are not usually available to and manageable by land managers and urban planners (Cushman 2015; Filchakova et al. 2007; Pelorosso et al. 2011; Leone 2014). There is also a need for hierarchical levels and multi-scale processes to be investigated in greater depth than is generally the case. In contrast, real sustainable futures and cities need to be designed under the lens of thermodynamics as they are the fundamental laws of universal evolution (Leone et al. 2016; Rees 2012).

The proposed entropy indicators represent new tools towards sustainable land-use decision making. This work shows how a quantitative modelling assessment approach could be successfully applied in urban planning practice while dealing with green infrastructure and grey infrastructure design. However, in this work, only the entropy release associated to hydrological processes is studied and further research is therefore necessary both to define entropy more appropriately in ecological/environmental terms (e.g. impact on urban climate, biodiversity) and to investigate possible implications on social aspects (Pelorosso et al. 2017).

The integration of the SLT into the planning of NBSs will require a transdisciplinary approach with a proactive contribution from different research fields and experts. Proposed assessment methodology and entropy indicators can be used to evaluate different NBS scenarios and/or to guide the design process in building low-entropy cities with multifunctional UGI components. An example of this multi-perspective view-point implementation is presented in another work (Pelorosso et al. 2018), where an integrated climate resilience and vulnerability assessment allows the best NBSs to be allocated to the most suitable areas. The choice and the spatial localization of NBSs would therefore focus on the optimization of UGI multifunctionality, considering SLT (i.e. available and recyclable water or intercepted waste), resilience (i.e. adaptive and recovery capacity) and a UES viewpoint (e.g. number and typology of beneficiaries). This work, using the scenario NBS coming from (Pelorosso et al. 2018) can thus be considered as a first step towards the integration of the resilience and entropy concepts into a process of sustainable planning of urban systems.

References

Bai X (2016) Eight energy and material flow characteristics of urban ecosystems. Ambio 45(7):819–830

Balocco C, Grazzini G (2000) Thermodynamic parameters for energy sustainability of urban areas. Sol Energy 69(4):351–356

Borris M, Viklander M, Gustafsson AM, Marsalek J (2014) Modelling the effects of changes in rainfall event characteristics on TSS loads in urban runoff. Hydrol Process 28(4):1787–1796

Bristow D, Kennedy C (2015) Why do cities grow? Insights from nonequilibrium thermodynamics at the urban and global scales. J Ind Ecol 19(2):211–221

Capra F, Luisi PL (2014) Vita E Natura: Una Visione Sistemica. Aboca

Costanza R et al (2012) Sustainability or collapse: what can we learn from integrating the history of humans and the rest of nature. Ambio 36(7):522–527

Cushman SA (2015) Thermodynamics in landscape ecology: the importance of integrating measurement and modeling of landscape entropy. Landscape Ecol 30(1):7–10

EU (2015) Towards an EU research and innovation policy agenda for nature-based solutions & re-naturing cities. Final Report of the Horizon 2020 Expert Group on 'Nature-Based Solutions and Re-Naturing Cities'

Filchakova N, Robinson D, Scartezzini JL (2007) Quo Vadis thermodynamics and the city: a critical review of applications of thermodynamic methods to urban systems. Int J Ecodyn 2(4):222–230

Fistola R (2011) The unsustainable city. Urban entropy and social capital: the needing of a new urban planning. Procedia Eng 21:976–984

Fistola R, La Rocca RA (2014) The sustainable city and the smart city: measuring urban entropy first. In: The sustainable city, vol IX. WIT Press, p 537

Gobattoni F, Pelorosso R, Leone A (2016) Strategie Di Riqualificazione Urbana Clima-Adattiva: Nature-Based Solutions per Città Più Resilienti. Planum 101–110

Gobattoni F, Pelorosso R, Lauro G, Leone A, Monaco R (2011) A Procedure for mathematical analysis of landscape evolution and equilibrium scenarios assessment. Landscape Urban Plan 103(3–4):289–302

Hilliges R, Endres M, Tiffert A, Brenner E, Marks T (2017) Characterization of road runoff with regard to seasonal variations, particle size distribution and the correlation of fine particles and pollutants. Water Sci Technol 75(5):1169–1176

Ho M-W (2013) Circular thermodynamics of organisms and sustainable systems. Systems 1(3):30–49

Karnani M, Annila A (2009) Gaia again. Bio Syst 95(1):82–87

Leone A (2014) Planificación, Fisica de La Complejidad, Entropía. Cerrar La Brecha Entre Los Principios Teoricos Y La Praxis. Plurimondi VII(15):141–167

Leone A, Gobattoni F, Pelorosso R (2014) Sustainability and planning: thinking and acting according to thermodinamics laws. J Land Use Mobility Environ (special issue eighth international conference INPUT)

Leone A, Gobattoni F, Pelorosso R (2016) Energy supply, thermodynamics and territorial processes as a new paradigm of sustainability in planning science and practice. In: Papa R, Fistola R (eds) Smart energy in the smart city. Urban planning for a sustainable future. Springer International Publishing, Berlin pp 83–101

Lovelock James E, Margulis Lynn (1974) Atmospheric homeostasis by and for the biosphere: the Gaia hypothesis. Tellus A 26(1–2):1–2
Norra Stefan (2014) The biosphere in times of global urbanization. J Geochem Explor 147:52–57
Pasimeni MR et al (2014) Scales, strategies and actions for effective energy planning: a review. Energy Policy 65:165–174
Pelorosso R, Gobattoni F, Lopez N, Leone A (2013) Verde Urbano E Processi Ambientali: Per Una Progettazione Di Paesaggio Multifunzionale. J Land Use Mobility Environ 6(1):95–111
Pelorosso R, Gobattoni F, Leone A (2016a). Nature-based cities: a low-entropy approach to sustainability. In: Ecosummit conference. Montpellier, France
Pelorosso R, Gobattoni F, Lopez N, Leone A (2016b) Verde Urbano E Regolazione Delle Acque Meteoriche: L'approccio Modellistico Come Base per Nuovi Standard Urbanistici. Sentieri urbani 19:71–77
Pelorosso R, Gobattoni F, Leone A (2017) Low-entropy cities: a thermodynamic approach to reconnect urban systems with nature. Landscape and Urban Planning 168:22–30
Pelorosso R, Gobattoni F, Leone A (2018) Increasing hydrological resilience employing nature-based solutions: a modelling approach to support spatial planning. In: Papa R, Fistola R (eds) Smart Planning: sustainability and mobility in the age of change. Springer
Recanatesi F, Petroselli A, Ripa MN, Leone A (2017) Assessment of stormwater runoff management practices and BMPs under soil sealing: a study case in a peri-urban watershed of the Metropolitan Area of Rome (Italy). J Environ Manag 201:6–18
Rees W (2012) Cities as dissipative structures: global change and the vulnerability of urban civilization. In: Weinstein MP, Turner RE (eds) Sustainability science: the emerging paradigm and the urban environment. Springer, London
Revitt DM, Lundy L, Coulon F, Fairley M (2014) The sources, impact and management of car park runoff pollution: a review. J Environ Manage 146:552–567
US EPA (1993) Handbook: urban runoff pollution prevention and control planning. p 175
Zhou Q (2014) A review of sustainable urban drainage systems considering the climate change and urbanization impacts. Water 6(4):976–992

Cities and Energy Consumption: Strategies for an Energy Saving Planning

Carmela Gargiulo and Laura Russo

Abstract A growing body of international researchers has been studying the complex relationship between cities and energy consumption so to support local policy makers' decisions and foster the transition towards a low-carbon future. However, despite the great interest of the literature for this topic, a consistent number of interactions between urban features and energy use at urban scale still lacks consensus. Therefore, this research aims to identify the urban factors that significantly affect a city's energy and carbon footprint, thus supporting policy-makers in the definition of effective strategies and policies that can be implemented on an urban scale to reduce energy consumption and resulting CO_2 emissions. Using a holistic approach rather than a sectorial one, we consider together a comprehensive set of urban factors—-physical, functional, geographical, and socio-economic—describing the complexity and multidimensionality of cities for measuring their impacts on CO_2 emissions. The results of the statistical analyses show that the two main categories of urban factors directly affecting CO_2 emissions per capita are the geographical and physical features, whereas the functional and socio-economic characteristics of urban areas have an indirect effect on CO_2 emissions. In other words, the climate condition of a city and its physical structure (both in terms of urban density and buildings characteristics) are in large part responsible for the use of energy and the resulting CO_2 emissions within the urban perimeter. Given that the geographical factors of cities cannot be changed by human intervention, the key role of urban policies and spatial planning in addressing energy and environmental issues becomes of strategic importance for addressing climate change.

C. Gargiulo (✉) · L. Russo
Department of Civil, Architectural and Environmental Engineering, University of Naples Federico II, Naples, Italy
e-mail: gargiulo@unina.it

R. Papa et al. (eds.), *Smart Planning: Sustainability and Mobility in the Age of Change*, Green Energy and Technology,
https://doi.org/10.1007/978-3-319-77682-8_4

1 Introduction

According to IEA (2016), urban areas consume about two-thirds of primary energy demand and produce over 70 per cent of global carbon dioxide emissions (CO_2). Consequently "cities are the heart of the decarbonisation effort" (IEA 2016) and can be the solution to climate change.

In order to support local policy makers' decisions and foster the transition towards a low-carbon future, a growing body of international researchers has been studying the complex and multidimensional relationship between cities and energy consumption.

Urban planning policies, indeed, can effectively improve energy saving in cities and reduce urban emissions only if the interactions between urban factors and energy use are investigated and are found to be significant. However, despite the great interest of the literature for this topic, a consistent number of interactions between urban features and energy use at urban scale still lacks consensus. Therefore, this research aimed to investigate the relationship between cities and energy consumption to identify the urban factors that significantly affect a city's energy and carbon footprint, thus supporting policy-makers in the definition of effective strategies and policies that can be implemented at an urban scale to reduce energy consumption and resulting CO_2 emissions.

This paper includes five sections. Section 2 provides a review of recent studies on the relationship between cities and energy consumption. The review highlights the knowledge gap between what is known and what is still under debate. Section 3 focuses on the methodology for developing the statistical models performed to investigate the relationship between urban features and energy consumption. More specifically, it presents the set of physical, functional, geographical, socio-economic and energy variables selected for the model, as well as the data collection procedure. Section 4 shows the results of the statistical analyses performed on the dataset. The results are carefully interpreted and discussed considering previous findings found in the scientific literature in order to highlight two types of significant relationships: (1) the relationships between the different urban factors, which may indirectly affect CO_2 emissions; (2) the direct relationships between urban factors and CO_2 emissions. Section 5, lastly, provides some concluding remarks that could be helpful for supporting policy makers in the definition of effective strategies to be implemented at an urban scale to reduce energy consumption and resulting CO_2 emissions. Furthermore, this Section highlights the main limitation of this work and outlines potential directions for future research on this topic.

2 Literature Review[1]

The review of the literature combined interdisciplinary researches that investigate the multidimensional relationship between cities and energy consumption. Using a holistic perspective, the critical review of these contributions revealed that different studies have considered different categories of urban features influencing energy consumption and CO_2 emissions. We have classified and summarized these features into four groups, each including a different number of variables (Fig. 1): (1) physical features; (2) functional features; (3) geographical features; (4) socio-economic features. Giving that there is no single way of identifying different categories (Stead and Marshall 2001), this classification is based on the General System Theory (von Bertalanffy 1969) applied to the urban phenomenon (Gargiulo and Papa 1993). In particular, according to the systemic principles, cities can be defined "as sets of elements or components tied together through sets of interactions" (Batty 2008) and an urban system can be represented as a set of four subsystems: physical subsystem; functional subsystem; geomorphological subsystem; anthropic subsystem (Papa et al. 1995). The four categories of urban features previously introduced reflect the aforementioned four urban subsystems.

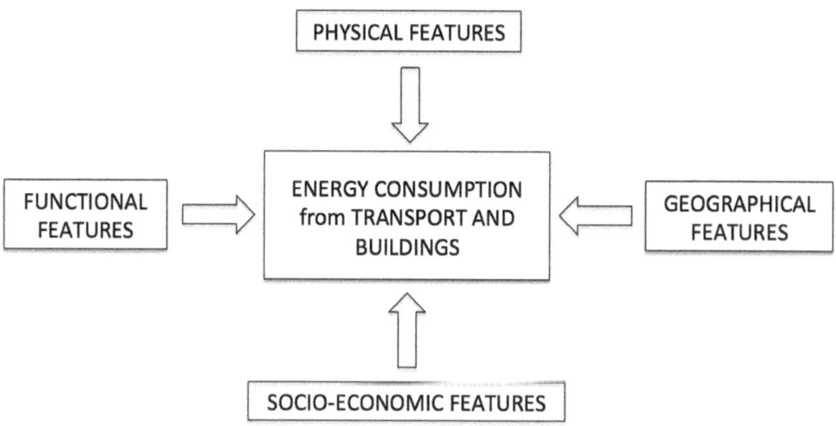

Fig. 1 Theoretical approach

[1]This section summarizes the results of the literature review on the relationship between cities and energy consumption on a city scale discussed in detail in a previous article (Gargiulo and Russo 2017).

2.1 Physical Features and Energy Consumption

Two main groups can be recognized in the debate on the relationship between urban form and energy consumption: those who support the compact city (Banister et al. 1997; Clark 2013; Ewing and Rong 2008; Marshall 2008; Newman and Kenworthy 1989) and those who question the magnitude of its environmental benefits (Baur et al. 2013; Brownstone and Golob 2009; Echenique et al. 2012; Mindali et al. 2004). While compact development advocates support the idea that people living in dense urban settlements are less automobile dependent, tend to live in multifamily houses, and thus consume less energy than do residents in sprawl areas, critics suggest that the energy savings associated with the intensification of land use are too small to be considered significant, and they may be associated with negative externalities such as congestion, higher housing price, and less availability of green areas.

2.2 Functional Features and Energy Consumption

There are relatively few studies that investigate the impacts of urban functional features on energy consumption. Although some results may appear contradictory, the general argument that emerges is that the positive effect of mixed-use development on energy saving from transport is not significant by it self, but becomes significant when combined with high density and supply of transit services (Banister et al. 1997; Holden and Norland 2005).

2.3 Geographical Features and Energy Consumption

The relationship between geographical features and energy consumption has been interpreted by the literature as that between climate—specifically HDDs—and energy consumption from buildings. In this context, it is widely argued that an increase in HDDs is associated with an increase in CO_2 emissions from heating (Creutzig et al. 2015; Ewing and Rong 2008). As far as the geographical location of cities is concerned, only one research finds that the proximity to the ocean does not affect energy consumption (Creutzig et al. 2015). Future research should further investigate the importance of these aspects as well as that of urban topography with respect to energy consumption.

2.4 *Socio-economic Features and Energy Consumption*

It is widely recognized that social and economic factors affect energy consumption. However, while there is great consensus about the relationships between economic variables—income (Battarra et al. 2016; Brownstone and Golob 2009; Clark 2013; Ewing and Rong 2008; Holden and Norland 2005; Kennedy et al. 2009; Newman and Kenworthy 1989), fuel price (Creutzig et al. 2015; Ewing and Rong 2008; Newman and Kenworthy 1989), and car ownership (Banister et al. 1997; Mindali et al. 2004)—and energy consumption, there is far less agreement about the way social characteristics, such as demographic growth, household composition, education, and race may influence energy use.

3 Methodology

The review of the scientific literature described above represents the starting point for the selection of the variables to be included in the statistical model in order to assess the relationship between urban characteristics and energy consumption on a city scale. The final dataset chosen for this analysis includes 18 variables describing the urban system and 5 variables measuring CO_2 emissions from energy use (Table 1). These data were collected for a sample of 73 Italian provincial capitals uniformly distributed across the country (Fig. 2), which account for 12.340.751 inhabitants, corresponding to the 21% of the total Italian population (recorded in 2011).

3.1 *Urban Variables*

Following the general system theory applied to the urban phenomenon (see Sect. 2), the 18 urban variables were classified into four categories, which reflect the four subsystems of a city: (1) physical features; (2) functional features; (3) geographical features; (4) socio-economic features.

(1) This category of urban features describes the physical component of a city, i.e. the spaces that people live in. (2) The functional features describe the functional subsystem of a city, which refers to the frequency and variety of activities occurring within the urban perimeter.

The choice of including in the dataset four variables identifying the economic base of a city comes from the consideration that urban economics may be a factor influencing the energy and carbon footprint of an urban settlement because different activities produce and consume different amount of energy (Habitat—UN 2011). Nevertheless, the relationship between the economic base of a territory and CO_2 emissions have not been explored yet. Thus, including these variables in this analysis represents one small step forward in the scientific debate on the relationship between

Table 1 List of the urban and energy variables selected for the analysis

Category	Variable	Description	Source
Physical features	Housing density	Housing units per square kilometer	ISTAT (2011)
	House size	Average floor area per capita	ISTAT (2011)
	House age	% of buildings built after 1980	ISTAT (2011)
	House material	% of masonry buildings	ISTAT (2011)
	Green areas	% of green areas	ISTAT (2011)
Functional features	Land use mix	Jobs per square kilometer	ISTAT (2011)
	Functional specialization	Concentration of manufacturing activities	ISTAT (2011)
		Concentration of commercial activities	ISTAT (2011)
		Concentration of public activities	ISTAT (2011)
		Concentration of touristic activities	ISTAT (2011)
Geographical features	Degree days	°C per day/year[a]	DPR 412/09
	Coastal location	Binary variable (0 = coastal; 1 = inland)	ISTAT (2009)
	Topography	Binary variable (0 = non mountain; 1 = mountain)	ISTAT (2009)
Socio-economic features	Income	Average income per capita	Ministry of Economy and Finance (2012)
	Car ownership	Number of cars per 1.000 inhabitants	ISTAT (2011)
	Household composition	% of children (<15 years old)	ISTAT (2011)
	Education	Graduates per 1.000 inhabitants	ISTAT (2011)
	Ethnicity	% of foreign residents	ISTAT (2011)
Energy consumption	CO_2 emissions	Residential CO_2 emissions per capita	Sustainable Energy Action Plans (2005)

(continued)

Table 1 (continued)

Category	Variable	Description	Source
		Transport CO_2 emissions per capita	Sustainable Energy Action Plans (2005)
		Tertiary CO_2 emissions per capita	Sustainable Energy Action Plans (2005)
		Municipal CO_2 emissions per capita	Sustainable Energy Action Plans (2005)
		Total CO_2 emissions per capita	Sustainable Energy Action Plans (2005)

Calculated as the sum of daily positive differences between a temperature of reference of 20 °C (for Italy) and the average daily outdoor temperature, extended to the whole year

cities and energy consumption. (3) The geographical features describe the climatic and topographic characteristics of Italian capital cities. (4) The five socio-economic variables included in the dataset describe the main characteristics of the population living in the 73 Italian capital cities that may affect energy consumption and CO_2 emissions on an urban scale. Table 2 shows some descriptive statistics for the sample of this analysis.

3.2 *Energy Variables*

Data on the energy and carbon footprint of the Italian cities considered in this research were collected using the Sustainable Energy Action Plans (SEAP)[2]. More specifically, each SEAP provided data on the amount of CO_2 emissions due to energy consumption by sector. The sectors considered were the following: transport and residential buildings, tertiary buildings and municipal buildings. Table 3 provides some descriptive statistics on CO_2 emissions by sector for the sample of 73 Italian cities.

The SEAP quantifies the emissions that occurred in the baseline year, which is the year against which the achievements of the emission reductions in 2020 shall be compared. Different cities chose different baseline years, mainly depending on data availability. Therefore, we identified the most frequent baseline year in the sample, which is 2005 (corresponding to 35 out of 73 cities, equal to 48% of the sample), and transformed the other data. In order to report all emissions data to the same baseline

[2]The European Commission (2010) has launched the Covenant of Mayors initiative in 2008, after the adoption of the 2020 EU Climate and Energy Package. This initiative aims to "endorse and support the efforts deployed by local authorities in the implementation of sustainable energy policies". The Covenant of Mayors successfully managed to involve a great number of local and regional authorities, which committed to (1) prepare a Baseline Emission Inventory (BEI) and (2) develop and implement a Sustainable Energy Action Plan (SEAP) within the year following their formal adhesion to the initiative.

Table 2 Descriptive statistics on urban characteristics for the sample of 73 Italian capital cities

Variable	Minimum	Maximum	Average	Standard deviation	Unit of measurement
Housing density	28.36	3214.38	579.96	623.96	Housing units/km^2
House size	31.67	49.71	42.05	4.01	m^2/inhabitants
House age	2.61	53.00	22.64	10.49	%
House material	23.43	91.19	53.40	14.13	%
Green areas	0.09	30.70	3.97	5.52	%
Land use mix	15.39	2496.70	422.98	467.74	Jobs/km^2
Concentration of manufacturing activities	0.21	2.11	0.96	0.44	Location quotient
Concentration of commercial activities	0.48	1.65	1.02	0.21	Location quotient
Concentration of public activities	0.33	1.97	1.03	0.39	Location quotient
Concentration of touristic activities	0.09	8.67	1.08	1.12	Location quotient
Degree days	707	3001	1860.33	599.69	°C per day/year
Coastal location	0.00	1.00	0.44	0.50	Binary variable
Topography	0.00	1.00	0.32	0.47	Binary variable
Income	12,183	26,744	20,411	2895	€
Car ownership	412.19	745.12	614.21	62.06	Car/1ooo person
Household composition	9.92	17.50	12.87	1.39	%
Education	55.50	242.21	150.51	38.64	Grad./1ooo person
Ethnicity	0.50	15.43	7.23	4.18	%

Fig. 2 Map of the 73 Italian provincial capitals included in the sample

year, we used time series data on greenhouse gas (GHG) emissions provided by the ISTAT for Italian regions. More specifically, for each region, the ISTAT offers data on GHG emissions for the following years: 1990, 1995, 2000, 2005, 2010. Based on these data, we calculated the annual change in emissions and we used this value for estimating the CO_2 emissions at 2005 for the 52% of the sample with a different baseline year.[3]

[3]The following example better illustrates the procedure here adopted. Novara is a city in Piedmont region and the baseline year of its BEI was 1998; data by the ISTAT for Piedmont show that GHG emissions were 8.8 t per capita in 1995 and 9.8 t per capita in 2005, corresponding to a 10.4%

Table 3 Descriptive statistics on CO_2 emissions by sector for the sample of 73 Italian capital cities

Variable	Minimum	Maximum	Average	Standard deviation	Unit of measurement
Residential CO_2 emissions	0.59	3.57	1.85	0.70	t per capita
Transport CO_2 emissions	0.07	4.58	1.78	0.89	t per capita
Tertiary CO_2 emissions	0.00	2.35	1.13	0.53	t per capita
Municipal CO_2 emissions	0.00	0.40	0.11	0.08	t per capita
Total CO_2 emissions	1.95	8.77	4.87	1.47	t per capita

4 Results

The dataset selected and described in the previous Section was explored and analyzed using different statistical methods—exploratory data analysis, correlation analysis, regression analysis—each of which provided useful insights into the complex relationship between cities and their carbon footprint. The statistical software SPSS 20 was used for these analyses.

Exploratory data analysis (EDA) was performed in order to identify potential outliers and evaluate the distribution of the data, thus gaining a better knowledge of the research dataset and sample. After EDA, the data were analyzed using a correlation analysis, which enabled the measurement of the association between the eighteen urban variables in order to identify redundant information and, most importantly, significant interconnections amongst these factors. Lastly, multiple linear regression analysis was performed to estimate the relationship between CO_2 emissions per capita and a number of determinants selected from the initial set of eighteen urban variables. Models were estimated separately for three categories of CO_2 emissions (i.e. residential, transport, and total CO_2 emissions per capita).

The results provided by these statistical analyses are discussed here considering previous findings found in the scientific literature discussed in Sect. 2 (for more details see Gargiulo and Russo 2017). In particular, in order to address the research goal, the results are described highlighting the two main types of relationships affecting energy consumption and CO_2 emissions on a city scale: namely, (1) the primary relationships between the urban features and CO_2 emissions, which directly affect a city's energy and carbon footprint; and (2) the secondary relationships among the

increase during the ten years of reference, i.e. 1.04% annual increase, which corresponds to a 7.32% growth between 1998 and 2005.

different urban features, which indirectly (but significantly) affect a city's energy and carbon footprint.

4.1 Correlation Analysis

Correlation analysis was used to determine the strength and direction of the linear relationships between each pair of both urban and energy variables. Before performing the correlation analysis, the four variables positively skewed were log transformed to normalize their distribution and calculate Pearson's correlation coefficients. Table 4 reports the correlation coefficients for the twenty-three variables, and the values above 0.65 are in bold and red color. The threshold of 0.65 was later used to identify the best subset of urban variables to be included in the regression model to avoid multicollinearity issues.

The correlation analysis allowed the assessment of the relationships among the different physical, functional, geographical and socio-economic characteristics, which may indirectly affect energy consumption and CO_2 emissions on a city scale, and therefore represent a crucial result of this research.

With respect to this type of relationships, a number of significant associations ($r > 0.65$) emerge. Housing density is highly positively correlated with the concentration of jobs per square kilometer ($r = 0.98$) and is negatively correlated with the percentage of buildings built after 1980 ($r = -0.66$); subsequently the concentration of jobs per square kilometer is also negatively correlated with the percentage of buildings built after 1980 ($r = -0.66$). In other words, densely built cities have also higher land use mix, as well as a higher concentration of older buildings. Another significant association is that between house size (i.e. average floor area per capita) and household composition (i.e. percentage of children), which are negatively correlated ($r = -0.63$), meaning that a greater concentration of children corresponds to a lower average floor area per person. Moreover, income is positively correlated with the share of graduates ($r = 0.73$) and foreigners ($r = 0.69$), as well as with the number of jobs per square kilometer ($r = 0.67$): richer cities have higher levels of education, higher concentration of foreign residents and higher concentration of jobs. With respect to the geographical characteristics or urban areas, not surprisingly, degree-days are negatively correlated to coastal location ($r = -0.69$), i.e. colder cities tend to be inland cities and vice versa, and, less predictably, are positively correlated with the percentage of foreign residents ($r = 0.69$), i.e. warmer cities attract less foreigners than colder ones. The last strong correlation is that between manufacturing cities and those with a high concentration of public activities ($r = -0.79$), thus meaning that cities are likely to be either specialized in manufacturing or in public activities, not both.

A number of similarities and differences emerge when these results are compared with those found in the scientific literature. More specifically, given the equivalence

Table 4 Correlation analysis, Pearson's correlation coefficients

	Housing density (ln)	Average floor area per capita	Percentage of buildings built after 1980	Percentage of masonry buildings	Percentage of green areas (ln)	Jobs per kmq (ln)	Concentration of manufacturing	Concentration of commercial	Concentration of public activities	Concentration of touristic activities (ln)	Degree days	Coastal location	Topography	Average income per capita	Cars per 1.000 inhabitants	Percentage of children	Graduates per 1.000 inhabitants	Percentage of foreign residents
Housing density (ln)	1.00																	
Average floor area per capita	0.00	1.00																
Percentage of buildings built after 1980	-0.66**	-0.12	1.00															
Percentage of masonry buildings	-0.12	0.35**	-0.11	1.00														
Percentage of green areas (ln)	0.61**	0.09	-0.45**	-0.14	1.00													
Jobs per kmq (ln)	0.98**	0.07	-0.66**	-0.10	0.63**	1.00												
Concentration of manufacturing	0.05	0.18	-0.18	0.02	0.13	0.11	1.00											
Concentration of commercial	-0.16	0.03	0.22	-0.06	-0.08	-0.16	-0.11	1.00										
Concentration of public activities	0.07	-0.18	0.01	-0.05	-0.06	0.01	-0.79**	-0.46**	1.00									
Concentration of touristic activities (ln)	-0.04	0.04	0.08	0.22	0.01	-0.04	-0.22	0.17	-0.07	1.00								
Degree days	0.09	0.46**	-0.21	0.08	0.36**	0.18	0.42**	-0.04	-0.38**	0.03	1.00							
Coastal location	0.06	-0.36**	0.01	0.01	-0.18	-0.01	-0.30*	0.05	0.22	0.21	-0.69**	1.00						
Topography	-0.32**	-0.17	0.39**	-0.04	-0.31**	-0.36**	-0.29*	0.05	0.23	0.05	-0.08	-0.12	1.00					
Average income per capita	0.59**	0.52**	-0.42**	0.12	0.61**	0.67**	0.08	-0.13	0.00	-0.06	0.53**	-0.29*	-0.24*	1.00				
Cars per 1.000 inhabitants	-0.41**	0.10	0.50**	-0.10	-0.24*	-0.39**	-0.19	0.04	0.13	0.05	-0.08	-0.12	0.24*	-0.14	1.00			
Percentage of children	-0.15	-0.67**	0.31**	-0.28*	-0.15	-0.17	0.11	0.13	-0.18	-0.01	-0.23*	0.12	0.13	-0.46**	0.05	1.00		
Graduates per 1.000 inhabitants	0.40**	0.48**	-0.30*	0.00	0.42**	0.46**	-0.23*	-0.19	0.31**	0.04	0.31**	-0.24*	-0.16	0.73**	0.12	-0.58**	1.00	
Percentage of foreign residents	0.42**	0.50**	-0.46**	0.23*	0.45**	0.50**	0.47**	-0.05	-0.43**	0.07	0.69**	-0.37**	-0.33**	0.69**	-0.25*	-0.19	0.31**	1.00

of housing and population density[4], the relationship between population density and heating degree days (Ewing and Rong 2008) does not come out as a significant result here (r = 0.09), and that between population density and house size (Ewing and Rong 2008; Lee and Lee 2014) also turns out to be not significant here (r = 0.00).

Even opposite to those found in the literature are the results on the relationship between density and income: while Brownstone and Golob (2009) find a negative association for a sample of California households, the correlation found for the seventy-three Italian cities is positive (r = 0.59). This difference does not have to surprise if we consider the substantial differences of the two contexts in both socio-economic development and historical background.

Only two similarities can be found comparing the results presented above and those found in the literature; namely, the relationship between density and land-use mix (Chen et al. 2008) and the relationship between house size and household composition (Ewing and Rong 2008).

The few similarities and many differences in the results substantiate the argument that researches on the relationship between cities and energy consumption should pay more attention on these aspects rather than mainly focus on the direct relationships between urban features and energy use. The relationship between cities and energy consumption is complex and multidimensional because of the complexity and multidimensionality of cities, therefore in order to better understand this complexity is necessary to consider both the direct relationships between the different urban features and energy consumption and the relationships among the urban features, which may indirectly but significantly affect energy consumption and CO_2 emissions.

4.2 Regression Analysis

Three regression models (OLS) were estimated in order to measure the direct relationships between urban and energy features. In these three models, the dependent variables are three of the five categories of CO_2 emissions—residential, transport and total—and the eleven independent variables are housing density, house material, green areas, concentration of manufacturing activities, concentration of commercial activities, concentration of touristic activities, degree-days, topography, income, car ownership and household composition.

The selection of the eleven independent variables was based on the results of the correlation analysis: variables with a Pearson's correlation coefficient greater than 0.65 were discarded (i.e. average floor area per capita; percentage of buildings built after 1980; jobs per square kilometer; concentration of public activities; coastal location; graduates per 1.000 inhabitants; percentage of foreign residents).

[4]For the sample of 73 Italian cities, the Pearson's correlation coefficient between housing density and population density is 0.99.

Table 5 OLS results for residential CO_2 emissions

	Residential CO_2 emissions per capita		
	Coefficient	*p*-value	VIF
Degree-days	0.753	0.000	1.000
Adjusted R^2	0.420		

Both dependent and independent variables included in the three multiple regression analyses are in natural log form[5], with the exclusion of the dummy variable "Topography". Therefore, the regression coefficients can be interpreted "as the average percentage change in the dependent variable corresponding to a percentage change in the independent variable" (Creutzig et al. 2015). The predictors initially included in each of the three regression models were then reduced by applying the backward stepwise method[6].

The results for the regression model with residential CO_2 emissions per capita as dependent variable (Table 5) show that emissions from buildings change considerably with changes in climate conditions: every 1% increase in degree-days corresponds to a statistically significant 0.75% increase of residential emissions. This result is in line with previous findings (Creutzig et al. 2015; Ewing and Rong 2008). Furthermore, the model also shows that only one determinant explains two-fifth of the variance in residential emissions per capita. The other independent variables, including those describing the physical subsystem of a city, are not statistically significant, and thus removed from the final model (backward elimination procedure).

The results for the regression model with transport CO_2 emissions per capita as dependent variable (Table 6) show that emissions from transportation moderately depend on the geographical and functional characteristics of urban settlements. Emissions are higher in colder cities with a concentration of touristic activities. Moreover, valley cities (topography = 0) emit 41% more for transportation than mountain cities (topography = 1).

When residential and transportation emissions are considered together with emissions from municipal and tertiary buildings, results become even more interesting. Table 7 shows that total CO_2 emissions per capita decrease with increasing housing density, increase with increasing house age, degree days, the concentration of green areas and that of commercial activities, and that mountain cities emit less than valley ones.

The geographical features are the most important factors: every 1% increase in degree days corresponds to a statistically significant 0.39% increase of total emissions; valley cities emit 0.13% more CO_2 emissions per capita than mountain ones.

[5]See paragraph 3.4.3.2 for more details on the loglinear model.

[6]"*Backward elimination of variables chooses the subset models by starting with the full model and then eliminating at each step the one variable whose deletion will cause the residual sum of squares to increase the least. This will be the variable in the current subset model that has the smallest partial sum of squares*" (Rawlings et al. 2001).

Table 6 OLS results for transport CO_2 emissions

	Transport CO_2 emissions per capita		
	Coefficient	*p*-value	VIF
Degree days	0.391	0.066	1.012
Concentration of touristic activities	0.228	0.015	1.030
Topography	−0.410	0.027	1.019
Adjusted R^2	0.140		

Table 7 OLS results for total CO_2 emissions

	Total CO_2 emissions per capita		
	Coefficient	*p*-value	VIF
Housing density	−0.128	0.002	2.265
% of masonry buildings	0.234	0.037	1.061
% of green areas	0.092	0.006	2.493
Concentration of commercial activities	0.243	0.079	1.020
Degree-days	0.391	0.000	1.236
Topography	−0.132	0.046	1.151
Adjusted R^2	0.458		

With respect to the physical variables, every 1% increase in housing density corresponds to a statistically significant 0.13% decrease of total emissions. This result substantiates the argument that compact cities consume less energy and emit less CO_2 than sprawl ones (Bereitschaft and Debbage 2013; Clark 2013; Creutzig et al. 2015; Ewing and Rong 2008; Makido et al. 2012). Similarly, every 1% increase in the percentage of masonry buildings corresponds to a 0.23% increase of total emissions, thus meaning that house material significantly affects the energy performance of buildings when residential, tertiary and municipal buildings are considered together. If tertiary and municipal buildings are excluded, indeed, this correlation is not found significant (Table 5). Furthermore, if we look at the matrix of correlation coefficient (Table 4), tertiary CO_2 emissions and the percentage of masonry buildings have a correlation coefficient of 0.49, which suggests that the different construction materials have a significant influence on the energy use of tertiary buildings.

Another interesting finding is that every 1% increase in the percentage of green areas corresponds to a 0.09% increase of total CO_2 emissions. This result is particularly interesting because, up to now, green areas have always been considered for their microclimatic benefits. Vegetation, indeed, can effectively contribute to mitigate the urban heat island phenomenon (Dimoudi and Nikolopoulou 2003; Gargiulo et al. 2016, 2017; Oliviera et al. 2011; Zoulia et al. 2009), i.e. the increase in urban temperatures compared to the surrounding rural areas. However, the studies on the

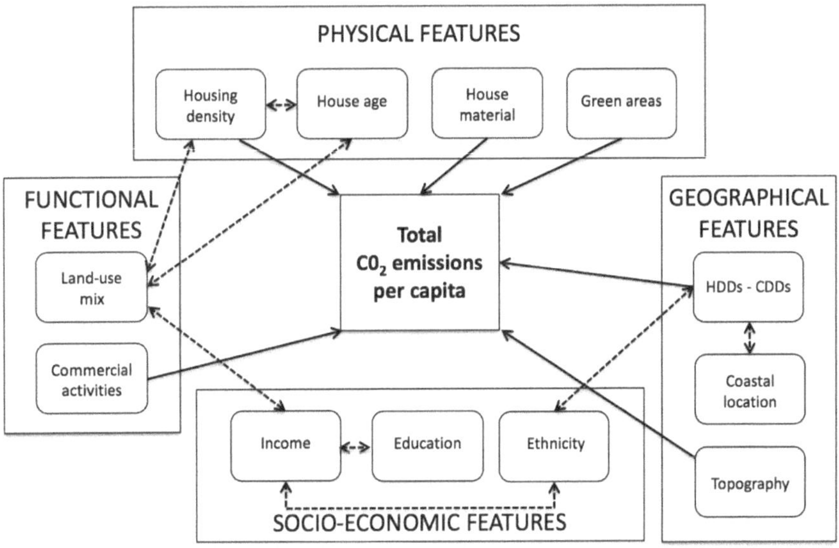

Fig. 3 Direct and indirect relationships between the urban features and total CO_2 emissions

positive effects of green spaces are limited to the cooling effect during summer time. Therefore, they did not consider the potential negative impacts of vegetation during winter time, which may increase energy use for space heating by increasing winter air temperatures. This negative effect can explain the positive correlation between the density of green areas and total CO_2 emissions that is shown in Table 7.

These results are show that two categories of urban features mostly affect CO_2 emissions on a city scale: namely, physical and geographical features. Although the geographical characteristics of urban areas cannot be modified by human actions, the physical attributes can be transformed through land-use planning decisions, and thus are of main interest for planners and policy makers.

4.3 *Discussion*

The complex relationship between cities and energy consumption/CO_2 emissions involves two main types of interactions: (1) the direct relationships between urban and energy variables; and (2) the relationships among the different urban features, which indirectly but significantly affect the carbon and energy footprint of cities.

Based on this conceptual framework, the main results of this research are summarized in Fig. 3, which includes both types of interactions: solid arrows represent the relationships between the urban features and total CO_2 emissions (found through the regression analysis); dashed arrows represent the relationships amongst the four groups of urban features (found through the correlation analysis).

When considering both direct and indirect effects, all the four groups of urban variables affect total CO_2 emissions per capita. More specifically, three physical variables (i.e. housing density, house material and green areas), one functional variable (i.e. the concentration of commercial activities), and two geographical variables (i.e. degree-days and topography) have a direct effect on CO_2 emissions: lower density of dwelling units and lower air temperatures, as well as valley topography and higher concentrations of masonry buildings, green areas and commercial activities increase CO_2 emissions per capita.

On the other hand, three socio economic variables (i.e. income, education and ethnicity) and one functional variable (i.e. land-use mix) indirectly affect CO_2 emissions through the mediators of other urban features.

In particular, a higher level of education and a higher share of foreign residents are both associated with higher income that, in turn, is associated with higher land-use mix that corresponds to higher housing density, which reduces CO_2 emissions per capita. However, at the same time, a higher concentration of foreigners also corresponds to higher degree-days, which increases CO_2 emissions per capita.

5 Conclusions

As a factor of economic growth, technological progress and social change, energy can be considered the driving force behind urban development. Nevertheless, using and producing energy is responsible for about 65% of global greenhouse gas (GHG) emissions (Marrero 2010), which, in turn, are responsible for climate change. The dramatic consequences of climate change—such as the warming of the atmosphere and the oceans, the intensification of extreme weather events, and the issue of food security, to name a few—are expected to increase in the coming years, especially if no action is taken.

Therefore, in order to prevent climate change and preserve our planet for future generations, the European Commission has recently proposed new ambitious energy goals to cut its GHG emissions substantially and turn Europe into a highly energy efficient and low-carbon economy. In particular, key EU targets for 2030 require EU countries to increase energy efficiency by 27%, and cut GHG emissions by 40% compared with 1990. Thus, European countries have to consider energy inside the urban planning process (Papa and Fistola 2016), in order to pursue together social, economic and environmental goals.

In this context, urban areas should be at the center of these sustainable policies, because "urban energy systems provide significant opportunities for increased efficiency in delivering transport and building services" (IEA 2016). Cities, indeed, consume up to 75% of global energy and account for 78% of carbon emissions (CO_2) produced by human activities (Habitat-UN 2011; IEA 2016). Thus, urban areas play a key role in addressing climate change.

Based on these considerations, this research aimed to identify the urban factors that significantly affect a city's energy and carbon footprint, thus supporting policy-makers in the definition of effective strategies and policies that can be implemented at urban scale to reduce energy consumption and resulting CO_2 emissions.

Two main innovations were introduced in this research. The first innovation concerns the approach: we used a holistic approach rather than a sectorial one, thus considering at the same time a comprehensive set of urban factors—physical, functional, geographical, and socio-economic—describing the complexity and multidimensionality of cities for measuring their impacts on CO_2 emissions. Second, we didn't limit the analysis to the direct relationships between the urban features and energy consumption, but we also investigated the relationships among the different urban features, which indirectly but significantly affect energy consumption on an urban scale.

This integrated approach allowed the identification of the existing trade-offs between different urban features and energy saving, providing a broader and more complete picture on such a complex topic.

The first step in this research was to review the scientific literature on the relationship between cities and energy consumption over the past twenty years. This review allowed the identification of the urban and energy variables to be included in the statistical models later developed to investigate this relationship. In particular, a set of eighteen urban variables and five energy variables was collected for a sample of seventy-three Italian capital cities, uniformly distributed across the country. The complete dataset was then explored and analyzed using different statistical methods, each of which provided useful insights into the complex relationship between cities and their carbon footprint.

The results of the regressions analysis show that three physical features—housing density, house material, green areas—and two geographical features—degree-days and topography—significantly affect total CO_2 emissions per capita ($R2 = 0.46$, p value < 0.05): with each 1% increase in housing density, total CO_2 emissions decrease by 0.13%; every 1% increase in the percentage of masonry building respect to concrete ones corresponds to a 0.23% increase in emissions; every 1% increase in the density of green spaces corresponds to a 0.09% increase in emissions; every 1% increase in degree-days corresponds to a 0.39% increase in emissions; and total CO_2 emissions decrease by 0.13% when passing from valley to mountain cities. The significant effects of housing density and degree-days on CO_2 emissions substantiate previous findings (Bereitschaft and Debbage 2013; Clark 2013; Creutzig et al. 2015; Ewing and Rong 2008; Gargiulo and Lombardi 2016; Makido et al. 2012). On the other hand, results on the influence on CO_2 emissions of construction materials, green areas and topography have not been found in the literature so far, and therefore require further investigation for being validated. Green areas, in particular, have always been considered a positive element within the urban context because of their capability of reducing air temperature during summer time; the potential negative effects of green spaces on urban temperature during the winter have not been investigated so far.

The results of the correlation analysis show that more compact cities have a higher density of jobs that, in turn, corresponds to a higher average income, a higher number of graduates and a higher share of foreign residents. These findings partially contradict previous results found in the literature. More specifically, Brownstone and Golob (2009) as well as Ewing and Rong (2008) find that density and income are negatively correlated in the US: richer people are more likely to live in sprawl counties. Within the Italian context, in contrast, the association between density and income is positive: higher levels of income are concentrated in urban settlements with a higher share of jobs and dwelling units. This difference can be explained considering the dissimilarities between North American and Italian cities in terms of urban development due to the very different historical backgrounds and economic growth paths. These considerations highlight the importance of sample homogeneity for the investigation of the relationship between cities and energy consumption, because the physical, functional and socio-economic characteristics of urban areas may significantly differ among cities of different countries. Therefore, when the relationship between cities and energy consumption is investigated at global scale (i.e. considering a sample of cities of different countries), cities should be clustered in order to account for historical and socio-economic differences, which might confound the final results.

Integrating both correlation and regression analysis results (Fig. 3), this research shows that the two main categories of urban factors directly affecting CO_2 emissions per capita are the geographical and physical features, whereas the functional and socio-economic characteristics of urban areas have an indirect effect on CO_2 emissions.

In other words, the climate condition of a city and its physical structure (both in terms of urban density and buildings characteristics) are in large part responsible for the use of energy and the resulting CO_2 emissions within the urban perimeter. Given that the geographical factors of cities cannot be changed by human intervention, the key role of urban policies and spatial planning in addressing energy and environmental issues becomes of strategic importance for addressing climate change.

In this regard, two main policy implications are drawn from the results of both correlation and regression analysis; one at the building scale and one at the urban scale. (1) At the building scale, interventions should focus on buildings materials, especially for reducing the energy use of masonry buildings. (2) At the urban scale, planning strategies should encourage compact developments in order to reduce energy consumption and total CO_2 emissions. Furthermore, besides the lower energy footprint of compact cities, in Italy higher densities of housing units correspond to higher densities of jobs, which in turn are characterized by higher incomes, and therefore strategies for promoting urban compactness can also have positive economic effects.

This work confirms the complexity and multidimensionality of the relationship between cities and energy consumption and the importance of both building and urban interventions to increase energy saving and decrease CO_2 emissions on a city scale (Zanon and Verones 2013). Furthermore, the results of this research, which only partially support previous findings, suggest that important trade-offs exist between the different urban characteristics and the energy and carbon footprint of cities (Doherty et al. 2009; Lee and Lee 2014; Papa et al. 2014, 2016). Measuring all of the trade-offs is a very challenging task, and this research proposed a first step in this direction.

5.1 *Limitations and Future Developments*

This research has several limitations. The first limitation is data availability. Because of data limitations, indeed, (1) data on urban areas and energy consumption/CO_2 emissions refer to two different time periods, and moreover, (2) some urban sectors that significantly affect energy consumption and resulting CO_2 emissions, such as industry, could not been considered. Furthermore, if more data were available, a more numerous sample of cities would have allowed the construction of different regression models for each group of cities obtained with the cluster analysis, thus providing more detailed information about the energy behavior of different typologies of urban areas.

The second limitation concerns the statistical methods used to estimate the relationship between cities and energy consumption: the correlation analysis and the multiple regression analysis do not allow the identification of a causal link between the variables considered. In other words, a strong correlation between two variables does not imply a direct link between these variables but it could be the results of an indirect interaction that involves other variables. Therefore, future research should focus on using different statistical model to study the complex relationship between cities and energy consumption, such as, for example, a multilevel structural equation model, which simultaneously tests multiple causal relations (Lee and Lee 2014).

The third limitation concerns the research's sample: the results of the analyses previously described refer to a sample of Italian cities, therefore they may not apply to different geographical contexts. As previously highlighted, indeed, urban and energy features significantly differ from one country to another, and these differences may translate into different results.

An interesting future development of this work would be to apply the same methodology to different countries and compare the results in order to identify similarities and differences and better support policy makers in the definition of effective urban strategies for reducing energy consumption and CO_2 emissions on a city scale.

References

Banister D, Watson S, Wood C (1997) Sustainable cities: transport, energy, and urban form. Environ Plan 24(1):125–143

Battarra R, Gargiulo C, Pappalardo G, Boiano DA, Oliva JS (2016) Planning in the era of Information and communication technologies. Discussing the "label: Smart" in South European cities with environmental and socio-economic challenges. Cities J http://dx.doi.org/10.1016/j.cities.2016.05.007

Batty M (2008) Cities as complex systems: scaling, interactions, networks, dynamics and urban morphologies (UCL Working paper 131). Centre for Advanced Spatial Analysis, University College London

Baur AH, Thess M, Kleinschmit B, Creutzig F (2013) Urban climate change mitigation in Europe: looking at and beyond the role of population density. J Urban Plan Devel 140(1):04013003. https://doi.org/10.1061/(ASCE)UP.1943-5444.0000165

Bereitschaft B, Debbage K (2013) Urban form, air pollution, and CO_2 emissions in large US metropolitan areas. Prof Geogr 65(4):612–635. https://doi.org/10.1080/00330124.2013.799991

Brownstone D, Golob TF (2009) The impact of residential density on vehicle usage and energy consumption. J Urban Econ 65(1):91–98

Chen H, Jia B, Lau SSY (2008) Sustainable urban form for Chinese compact cities: Challenges of a rapid urbanized economy. Habitat Int 32(1):28–40. https://doi.org/10.1016/j.habitatint.2007.06.005

Clark TA (2013) Metropolitan density, energy efficiency and carbon emissions: multi-attribute tradeoffs and their policy implications. Energy Policy 53:413–428

Creutzig F, Baiocchi G, Bierkandt R, Pichler PP, Seto KC (2015) Global typology of urban energy use and potentials for an urbanization mitigation wedge. Proc Natl Acad Sci 112(20):6283–6288

Dimoudi A, Nikolopoulou M (2003) Vegetation in the urban environment: microclimatic analysis and benefits. Energy Build 35(1):69–76

Doherty M, Nakanishi H, Bai X, Meyers J (2009) Relationships between form, morphology, density and energy in urban environments. GEA Background Paper 28

Echenique MH, Hargreaves AJ, Mitchell G, Namdeo A (2012) Growing cities sustainably: does urban form really matter? J Am Plan Assoc 78(2):121–137. https://doi.org/10.1080/01944363.2012.666731

European Commission (2010) How to develop a sustainable energy action plan (SEAP)—Guidebook. Publications Office of the European Union, Luxembourg

Ewing R, Rong F (2008) The impact of urban form on US residential energy use. Hous Policy Debate 19(1):1–30. https://doi.org/10.1080/10511482.2008.9521624

Gargiulo C, Lombardi C (2016) Urban retrofit and resilience. The challenge of energy efficiency and vulnerability. TeMA J Land Use Mobility Environ [S.l.] 9(2):137–162. doi:http://dx.doi.org/10.6092/1970-9870/3922, ISSN 1970-9870

Gargiulo C, Papa R (1993) Caos e caos: la città come fenomeno complesso. Per il XXI Secolo: una enciclopedia e un progetto, pp 297–306. ISBN 978-88-97110-15-6

Gargiulo C, Russo L (2017) Cities and energy consumption: a critical review. TeMA J Land Use Mobility Environ 10(3):253–273. doi:http://dx.doi.org/10.6092/1970-9870/5182, ISSN 1970-9870

Gargiulo C, Tulisi A, Zucaro F (2016) Small green areas for energy saving: effects on different urban settlements. Arquitectura Ciudad y Entorno 11(32):81–94. https://doi.org/10.5821/ace.11.32.4659, ISSN 1886-4805

Gargiulo C, Tulisi A, Zucaro F (2017) Chapter 11: climate change-oriented urban green network design: a decision support tool. In: Gakis K, Pardalos P (eds) Network design and optimization for smart cities, world scientific, series on computers and operations research, vol 8. http://www.worldscientific.com/doi/abs/10.1142/9789813200012_0011, ISBN 978-981-3200-00-5 (hardcover), ISBN 978-981-3200-02-9 (ebook)

Habitat UN (2011) Cities and climate change: global report on human settlements 2011. Earthscan, London

Holden E, Norland IT (2005) Three challenges for the compact city as a sustainable urban form: household consumption of energy and transport in eight residential areas in the greater Oslo region. Urban Stud 42(12):2145–2166. https://doi.org/10.1080/00420980500332064

IEA—International Energy Agency (2016) Energy technology perspectives 2016. Towards Sustainable Urban Energy Systems

Kennedy C, Steinberger J, Gasson B, Hansen Y, Hillman T, Havranek M, Pataki D, Phdungsilp A, Ramaswami A, Mendez GV (2009) Greenhouse gas emissions from global cities. Environ Sci Technol 43(19):7297–7302. https://doi.org/10.1021/es900213p

Lee S, Lee B (2014) The influence of urban form on GHG emissions in the US household sector. Energy Policy 68:534–549. https://doi.org/10.1016/j.enpol.2014.01.024

Makido Y, Dhakal S, Yamagata Y (2012) Relationship between urban form and CO_2 emissions: evidence from fifty Japanese cities. Urban Climate 2:55–67. https://doi.org/10.1016/j.uclim.2012.10.006

Marrero GA (2010) Greenhouse gases emissions, growth and the energy mix in Europe. Energy Econ 32(6):1356–1363. http://dx.doi.org/10.1016/j.eneco.2010.09.007, ISSN 0140-9883
Marshall JD (2008) Energy-efficient urban form. Environ Sci Technol 42(9):3133–3137
Mindali O, Raveh A, Salomon I (2004) Urban density and energy consumption: a new look at old statistics. Transp Res Part A Policy Pract 38(2):143–162
Newman PW, Kenworthy JR (1989) Gasoline consumption and cities. J Am Plan Assoc 55(1):24–37. https://doi.org/10.1080/01944368908975398
Oliveira S, Andrade H, Vaz T (2011) The cooling effect of green spaces as a contribution to the mitigation of urban heat: a case study in Lisbon. Build Environ 46(11):2186–2194
Papa R, Fistola R (2016) Preface. In: Smart energy in the smart city. Springer International Publishing, Berlin, pp 151–175. https://doi.org/10.1007/978-3-319-31157-9, ISSN 1865-3529
Papa R, Battarra R, Fistola R, Gargiulo C (1995) La città come sistema complesso in crisi strutturale. In: Bertuglia CS, Fuccella R, Sartorio GL (eds) La città come sistema complesso in crisi strutturale - strumenti e tecniche per il governo metropolitano. Giuffré, Roma. ISBN 88-14-05263-8
Papa R, Gargiulo C, Zucaro F (2014) Climate change and energy sustainability. Which innovations in European strategies and plans. TeMA J Land Use Mobility Environ (Special Issue) 793–804. doi:http://dx.doi.org/, https://doi.org/10.6092/1970-9870/2554, ISSN 1970-9870
Papa R, Gargiulo C, Zucaro F (2016) Towards the definition of the Urban Saving Energy Model (UrbanSEM). In: Smart energy in the smart city, vol v–vi. Springer International Publishing, Berlin, pp 151–175. https://doi.org/10.1007/978-3-319-31157-9_9, ISSN 1865-3529
Rawlings JO, Pantula SG, Dickey DA (2001) Applied regression analysis. A research tool. Springer International Publishing, Berlin, p 213. ISBN 0-387-98454-2
Stead D, Marshall S (2001) The relationships between urban form and travel patterns. An international review and evaluation. Eur J Transp Infrastruct Res 1(2):113–141
von Bertalanffy L (1969) General system theory: foundations, development, applications (Revised edn)
Zanon B, Verones S (2013) Climate change, urban energy and planning practices: Italian experiences of innovation in land management tools. Land Use Policy 32:343–355
Zoulia I, Santamouris M, Dimoudi A (2009) Monitoring the effect of urban green areas on the heat island in Athens. Environ Monit Assess 156(1):275–292

Increasing Hydrological Resilience Employing Nature-Based Solutions: A Modelling Approach to Support Spatial Planning

Raffaele Pelorosso, Federica Gobattoni and Antonio Leone

Abstract Despite numerous studies on urban resilience, few practical applications of spatial explicit and quantitative resilience indicators in green infrastructure planning are present in literature. This paper presents a methodological framework to assess the hydrological resilience of an urban context employing modelling approach. The proposed resilience index is then used to support the definition of nature-based and engineered solutions aimed to increase resilience to floods as well as to enhance the green infrastructure multi-functionality in a densely populated district of Bari. The paper aims then to contribute to the introduction of resilience assessment and sustainable storm water management in practice urban planning in a context of climate adaptation plans.

1 Introduction

Nature-based solutions are a recent concept born to operationalise the ecosystem services approach within spatial planning policies and practices and fully integrate the ecological dimension within traditional planning concerns (Scott et al. 2016). Nature-based solutions (NBSs) are defined as engineered green/ecological systems inspired or supported by, or copied from, Nature (EU 2015). Multifunctional NBSs (such as green roofs or wetlands), where ecosystem service provisions are "designed-in", can therefore help the transition of social-ecological systems to more sustainable environments which are more resilient to changing future conditions (Lundy and Wade 2011; Pelorosso et al. 2013). Urban resilience is evaluated in terms of both climate and socio-economic changes (Gobattoni, Pelorosso, and Leone 2016; Pelorosso, Gobattoni, Lopez, et al. 2016a) and can be enhanced by the integration of NBS and ecosystem services into urban planning (Demuzere et al. 2014; EU

R. Pelorosso (✉) · F. Gobattoni · A. Leone
DAFNE, University of Tuscia, Viterbo, Italy
e-mail: pelorosso@unitus.it

R. Papa et al. (eds.), *Smart Planning: Sustainability and Mobility in the Age of Change*, Green Energy and Technology,
https://doi.org/10.1007/978-3-319-77682-8_5

2015; Gómez-Baggethun and Barton 2013; Leone, Gobattoni, and Pelorosso 2014; Pelorosso, Gobattoni, Geri, et al. 2016a, b, 2017a, b).

In particular sustainable drainage systems can increase the hydrological resilience of the urban system by a more soft, cost-effective and sustainable strategy than the conventional approaches based on the grey infrastructure development (e.g. pipes, conveyance systems and treatment plans) (Pappalardo et al. 2017; Pelorosso et al. 2013, 2016b). The objective of this paper is to contribute at the introduction of resilience assessment and sustainable storm water management in urban planning practice in the context of climate adaptation plans. Thus, a methodological framework is presented to assess hydrological resilience of an urban context employing a modelling approach.

A quantitative resilience index is then used to support the definition of nature-based and engineered solutions aimed to increase resilience to floods as well as to enhance the green infrastructure multi-functionality in a densely populated district of Bari.

2 Urban Green Infrastructure, Ecosystem Services and Resilience

Urban green infrastructure is "an interconnected network of natural systems and Nature-Based Solutions (NBSs), localized at landscape scale and fully integrated with the built environment, which provides a diversified array of Urban Ecosystem Services (UESs) to the urban socio-ecological system increasing its resilience" (Pelorosso et al. 2017a,b). NBSs bear a wide range of urban ecosystem services (UES). UESs are benefits that people derive directly or indirectly from natural and managed ecosystems (Gómez-Baggethun and Barton 2013; Haase et al. 2014) such as habitat provision, the cooling effect of green roofs and vegetated areas, or the recreational functions of wetland and detention basins.

UESs can provide metrics, numbers and monetary values about Nature in the city and show its functions for human activities and wellness (Pincetl 2015). UESs include (Hermann et al. 2011): (1) provisioning services such as food, wood and clean water; (2) regulating services such as flood, climate and CO_2 control; (3) cultural services such as social, recreational, educational and touristic benefits; and (4) supporting services, such as oxygen production and habitat provision, that maintain the conditions for a highly differentiated life on the Earth. Among the different categories of UES, the cultural services cover a relevant role in urban contexts. They are defined as non-material and/or socio-ecological benefits people obtain from the contact with ecosystems through spiritual enrichment, cognitive development, reflection, recreation, and aesthetic experiences (La Rosa et al. 2016; TEEB 2011).

A holistic assessment of the whole urban system by means of a framework analysis of the UESs can evaluate cultural and social aspects (which are) significant for life quality improvement and then support planners and designers in the definition of strategies and actions aimed at (i) increasing the complexity of urban systems, through the maximization of the exchange among natural and human components; (ii) reducing the ecological footprints and the ecological debts of cities. In other words, an effective UGI can reduce the negative effects of climate change by regulating storm water flows (to alleviate flooding and pollution runoff) and reducing air and surface temperature at the building, neighbourhood or city level (i.e. heat island effect). UGI functionalities can then improve the wealth state, the thermal comfort and energy use of the city. UGI can also support social capital and capacity to overcome periods of crisis by providing people with meeting opportunities or by allowing self-production of food. UGI also participates in the mitigation of climate change through carbon sequestration into biomass (Demuzere et al. 2014). This biomass can be also used for energy production, promoting energy self-sufficiency though local renewable energy and the limitation of dependence on fossil fuels (Leone et al. 2016). Moreover, proximity to urban green space has been shown to have a positive effect on the sale price of houses (Green et al. 2015). Finally, a UGI based on a UES framework could increase people's awareness of the way humans (and above all city dwellers) interact with the environment, of what a sustainable city might entail, and of the consequences of being a careless consumer of natural (and external) resources instead of a good UES steward. Thus, UES planning can help to redirect urbanization into becoming a driver of positive change for humanity and the life-supporting systems that we depend upon (Andersson et al. 2014).

3 Urban Resilience and Climate Change

In the last report, the European Environment Agency (EEA 2016), underlines how several European governments and organizations have already developed policies and laws at national and regional level to support cities in the implementation of appropriate measures for climate change adaptation, following the input by the European Strategy on Adaptation to Climate Change. The European Strategy represents the regulatory framework for Europe, it asks the Member States to undertake adaptation actions that are "cost-effective", and quick since these ones could be less expensive than repairing the damage caused by climate change (every Euro spent to protect us from flooding would save six Euros of damage). One of the main objectives that the European Strategy pursues is to promote the adoption of national, regional and local adaptation strategies and plans by Member States. This objective is actually being implemented with the recent development by European governments of National Strategies on Adaptation to climate change and the establishment in March 2014 of the mayors Adapt agreement (http://mayors-adapt.eu), the initiative of the Covenant of Mayors on adaptation, which involves municipalities on climate change and provides a support for the implementation of adaptation actions at the local level.

The Italian government completed in 2014 the development of a National Strategy (Castellari et al. 2014) which provides recommendations for its implementation at local level, in order to take into account the specificities of the national territory, adapting strategies to the sensing elements of every reality. According to the Strategy, some Italian municipalities elaborated a Climate Adaptation Plan while other municipalities have nevertheless undertake a path towards the development of an Adaptation Plan, which, for example in the City of Padua, has led to a Sustainable Energy Action Plan. It emerges, therefore, a clear need to rethink the settlement system as adaptive to changing climatic conditions, by implementing decision-making processes that go beyond the traditional objective of reducing levels of vulnerability of the exposed elements, and instead aimed at strengthening the resilience of the built environment in its entirety, in the interest of citizens and economic development. The main objective consists in enhancing the reactive capacities of the socio-ecological system, which, become more resilient, is able to adapt and change itself keeping its own characteristics and functions in the long-term period. In the last years, this approach has received a great attention conceptually but it still requires practical applications.

The concept of resilience includes capacities of persistence, recovery, transformability ad adaptation by socio-ecological systems and sub-systems (Biggs et al. 2012; Holling 2001; Walker et al. 2004). To achieve more resilient urban systems, the Italian National Strategy and the European Strategy suggest actions and technical solutions based on Nature, according to an ecosystem approach aimed at facilitating and encouraging innovative initiatives to increase public and private green areas able to provide ecosystem services (McPhearson et al. 2015). In particular, the Italian National Strategy underlines the importance of a "Climate Resilience Study" for every urban settlement in order to consciously identify adaptation strategies adjusted on the real local necessities. A "Climate Resilience Study" should include an evaluation of the vulnerability of urban systems to future climate change with the aim to define priority actions. Therefore, the investigation into the most important best practices for adaptation to climate change in urban areas can not be separated from the integration of the concepts of resilience, ecosystem services, vulnerability to the effects of climate change and social cohesion.

4 Study Case

The study area has an extension of about 330 ha and corresponds to the second municipality of the Bari Metropolitan Area (Fig. 1). The combined drainage network was built in the thirties-forties while in the last years the volumes generated by runoff increased due to the urban development and the consequent soil sealing. Thus, the drainage network is inadequate to manage storm water runoff efficiently. The combined sewer system, during heavy precipitation events, is not capable of managing all the volumes. Consequently, the overloads affect the urban system, flowing directly to the seafront, contaminating the adjacent public beaches with obvious consequences on health and fruition. The modernization of the network could be very expensive (about 120 million Euros, Barinedita 2014) and it represents a solution hard to realize and challenging on many aspects, for example the required

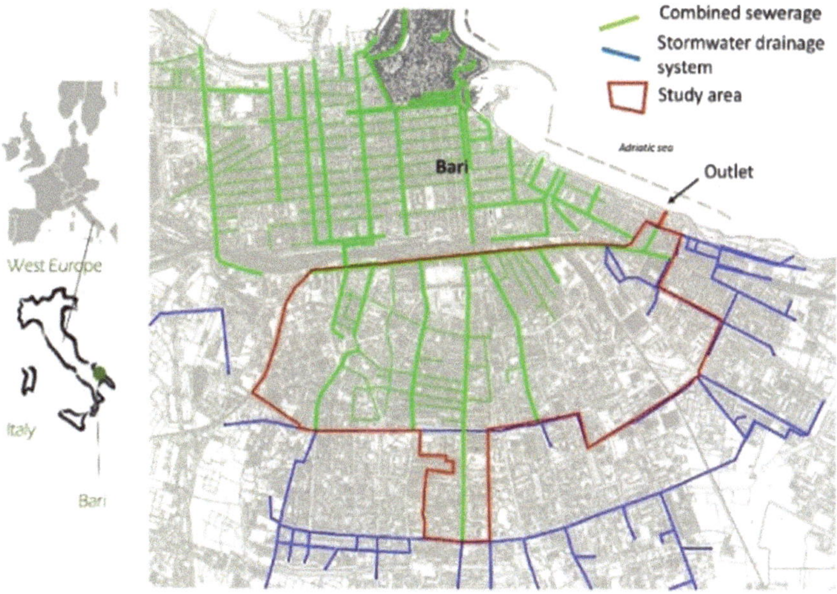

Fig. 1 Study area

funds are difficult to be found and this intervention could generate discomfort to citizens. Therefore, a holistic vision of the city is needed in order to go beyond the traditional approach of building infrastructures to merely manage storm water runoff taking into account the complex integration between natural and built environment while dealing with a changed and changing climate. Moreover, Bari Metropolitan Area shows a scarcity of green spaces (ISTAT 2016) and only 20% of people can reach green spaces bigger than 2 ha within the buffer of 500 m (Kabisch et al. 2016, Fig. 2). The study area represents therefore an exemplificative case of Mediterranean consolidated city where the necessity of green areas meets the risks related to an inadequate combined drainage system in a context of reduced economic resources and degraded sea ecosystems.

5 Methods

In this work, a preliminary resilience study was realised to support the definition of a NBS scenario aimed to increase the hydrological resilience of the study area (Sect. 5.1). The green scenario building (Sect. 5.2) has then taken into consideration the resilience study, the feasibility of the NBS and the possible multiple outcomes in terms of UES.

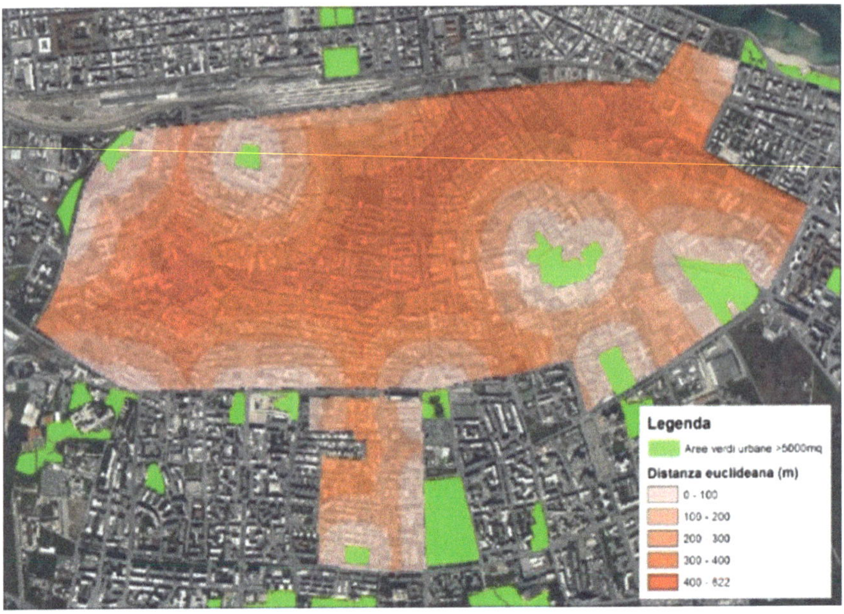

Fig. 2 Localization and analysis of accessibility to urban green areas bigger than 5000 mq (*Source* Urban Atlas)

5.1 *Resilience Assessment*

The concept of resilience has emerged in different discipline with different conceptions (Meerow et al. 2016). In the case of infrastructure systems, resilience implies "the ability of such infrastructure systems (including their interconnected ecosystems and social systems) to absorb disturbance and recover after a disturbance" (Joyce et al. 2017). Looking at drainage infrastructure systems the engineering resilience definition is usually used as "the degree to which the system minimises level of service failure magnitude and duration over its design life when subject to exceptional conditions" (Butler et al. 2014; Mugume et al. 2015; Wang et al. 2017). Only few studies investigated the resilience of coupled green and grey infrastructures (see Joyce et al. 2017). According to the above engineering resilience definition, the hydrological resilience of the study area was studied considering both green and grey infrastructure interaction (Pelorosso et al. 2016a). The methodological framework is based on two phases. In a first phase, the climatic factor relative to precipitation has been considered taking into account the rainfall trend from 1938 to 2015. According to the regional (Apulia region) law on urban drainage, a critical event of 3 h (equal to the time of concentration of the urban catchment) with a return period of five years has been identified (as requested by Italian technical legislation for urban drainage systems). To analyse the resilience of the urban system (phase 2), the selected rainfall event has been simulated using the US-EPA managerial model

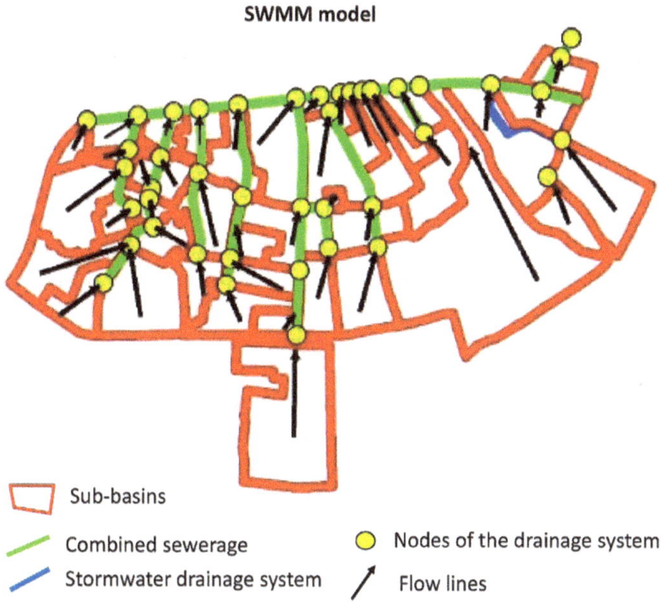

Fig. 3 SWMM model of study area

called Storm Water Management Model (SWMM, release 5.1.012), which is a well-recognized tool to deal with storm water management issue (Zhou 2014). The model works at catchment scale and allows to simulate a urban landscape structure as a dynamic system constituted by buildings, grey infrastructures (channels and pipes) and land-use (pervious and impervious surfaces, green infrastructure and NBS). In particular, SWMM is a dynamic rainfall-runoff simulation model specifically created for single event or long-term (continuous) simulation of runoff quantity and quality from primarily urban areas. The runoff component of SWMM operates on a collection of sub-catchment areas on which runoff is generated. The routing portion of SWMM transports this runoff through a conveyance system of pipes, channels, storage/treatment devices, pumps, and regulators. SWMM tracks the quantity and quality of runoff generated within each sub-catchment, and the flow rate, flow depth, the overflow and quality of water in each pipe and channel during a simulation period comprised of multiple time steps.

In the study area, each specific sub-basin outlet is defined with a node of the drainage network (Fig. 3). The critical rainfall event, the land use and the drainage system were simulated on a base of 43 sub-basins identified by high resolution data as LIDAR, aerial photos and technical maps (Pelorosso et al. 2016b), moreover, a base sewage load is calculated for each pipe on the basis of population density.

Hydrological resilience is calculated taking into account five factors of criticity related to the functionality of coupled green and grey infrastructure of the urban system. The factors derive from SWMM outputs for each sub-basin (runoff) and linked node (flooding). The five factors of criticity are: (1) node flooding volumes (i.e. overflow), (2) peak node flooding (3) sub-basin runoff volumes, (4) sub-basin peak runoff and (5) the total flood volume of the branches of the drainage system (Pelorosso et al. 2016a). These factors are proxy indicators of the water regulating service capacity of green and grey infrastructures considering different issues related to storm water management and hydrological functionality, as the water bomb effect (peak indicators), the dispersion of pollutants (runoff and overflow indicators) and the drainage system vulnerability (total overflow of linked pipes). Each factor was then standardized on scale 0–1 by dividing it for the maximum value registered in the whole study area. A hydrological resilience index was then derived as the sum of the five standardized factors of criticity. The value of the resilience index represents the intrinsic hydrological resilience of urban system consisting of grey and green infrastructure interactions. The resilience index varies between 0 and 5 where 0 represents the highest hydrological resilience, i.e. no runoff and overflow at nodes. Finally, the resilience map was then obtained through a reclassification of the hydrological resilience index of each subbasin in five classes.

5.2 NBS Scenario Definition

Low resilience sub-basins represent areas where water and water-transported pollutants are not fully managed by the drainage systems and the present green and permeable areas. The priorization of NBS in these critical areas will then increase both the resilience and the entropy of system. However, a feasibility study and a planning framework are necessary to avoid improper scenario building.

Taking into account the presence of unused and/or underused areas, parking lots, car and pedestrian/cycling circulation, surface water runoff, building typology and communities' needs, priority areas have been identified for the introduction of NBS and the increase of the hydrologic resilience of the system. On these areas, NBSs have been selected on the basis of costs, positive and negative externalities and efficiency. Initially, the presence of unused and/or underused spaces in the study area was analysed to design NBS falling in the less resilient sub-basins (see Fig. 4a). When unused and/or underused areas were not available in vulnerable sub-basins or not suitable to host effective NBS, the analysis was enlarged to the whole urban system, proposing the adoption of NBS compatible with the typology of buildings and the available space. For this purpose, surface runoff flow lines together with morphology, soil characteristics and suitability of the urban landscape to accommodate NBS were investigated according to well-recognized criteria in the literature (see for example Digman et al. 2012). Based on the efficiency database derived from the modelling simulations (Gobattoni et al. 2016; Pelorosso et al. 2016a), specific interventions were therefore proposed and dimensioned for the increase of hydrological resilience. Moreover, the project proposal took into account the multifunctional characteristics of the NBS and the previous use of the areas.

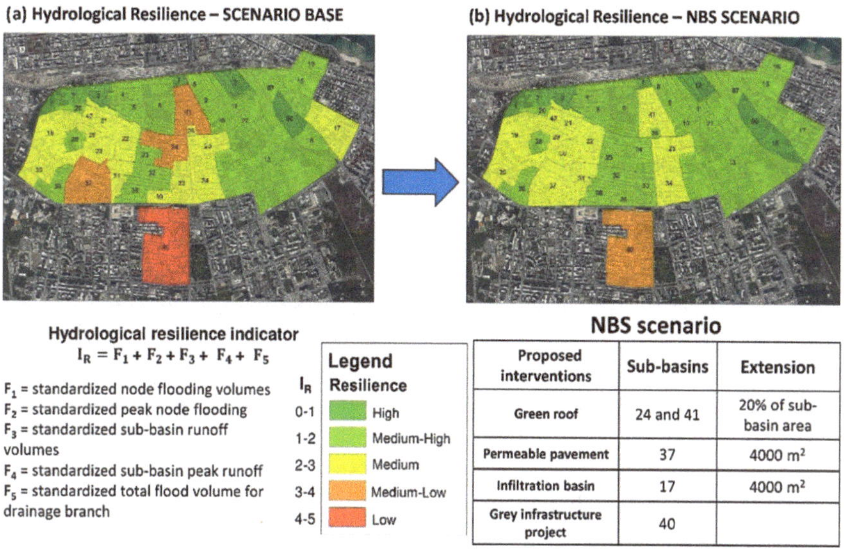

NBS scenario

Proposed interventions	Sub-basins	Extension
Green roof	24 and 41	20% of sub-basin area
Permeable pavement	37	4000 m^2
Infiltration basin	17	4000 m^2
Grey infrastructure project	40	

Fig. 4 Intrinsic hydrologic resilience of the urban system (grey and green infrastructure). **a** Base scenario base; **b** Nature-based solutions (NBS) scenario. The figure displays also the structure of the hydrological resilience indicator and the proposed interventions for the increase of this index

6 Results

Figure 4a shows the hydrological resilience map of the base scenario. In general, higher value of resilience are present in the northern part of the study area, but several sub-basins have high criticity due to inadequate grey infrastructure and high imperviousness of soils.

The NBS scenario is defined by four typologies of interventions (Fig. 4). A large greening of roofs is planned for the sub-basin 24 and 41 as they have not significant empty spaces and alternative NBS can not be realised easily in these high-density urban districts. In this case, the extensive green roof typology (15 cm of layer) is selected to minimize the load on the old building structures. In the sub-basin 37, a large parking lot at the service of a medical institution is redesigned using permeable pavements also to drain runoff coming from other impervious surfaces of the basin. An infiltration basin (bio-retention area) is planned at the east part of the district in order to storage the water in excess and create a further accessible green space. Finally, an engineered solution is hypothesized with the aim to reach a specific class of resilience. In the most critical sub-basin 40, not so adequate to significant interventions in terms of NBS, we propose to connect the discharge of storm water to the existing drainpipes thus reducing the load on the combined sewer system also in the upper branches and sub-basins. The proposed new UGI, integrated with the cited grey infrastructure intervention, allowed a new post-plan classification in terms of urban hydrologic resilience (Fig. 4b). It is worth noting the increasing in resilience class for several sub-basins, which reached a medium/high level except for the lower sub-basin for which the resilience raised from low to medium-low.

7 Discussion and Conclusions

The proposed methodological framework allows defining spatial localisation and general design characters of NBS for increasing the hydrological resilience considering a multifunctional reuse of urban spaces. Despite NBS offer a strategic scenario, it has to be verified and adapted in practice on the basis of local technical and socio-cultural characteristics which, in turn, determine the acceptance by communities. As an example, extensive green roofs represent the most appropriate solution in the considered high-density urban context but technical issues relative to loads on the roofs (to be verified in each case) and the citizen's willingness to realize them should be considered. Green roofs remain, anyway, a valid and interesting solution that urban planners should keep in mind as a concrete hypothesis of urban regeneration even offering incentives to private owners. The paper then demonstrates the feasibility of modelling approach to support spatial planning based on physical assessment of urban infrastructures. Further research will aim to better integrate the assessment methodology into practice UGI adaptation plan aimed to reduce climate vulnerability of cities.

References

Andersson E et al (2014) Reconnecting cities to the biosphere: stewardship of green infrastructure and urban ecosystem services. Ambio 43(4):445–453

Barinedita (2014) "Pane E Pomodoro: «Sì Agli Scarichi a Mare, Costa Troppo Rifare La Fogna» | Barinedita - Testata Giornalistica Online." Retrieved (https://www.barinedita.it/inchieste/n1437-pane-e-pomodoro--«si-agli-scarichi-a-mare-costa-troppo-rifare-la-fogna»)

Biggs R et al (2012) Toward principles for enhancing the resilience of ecosystem services. Ann Rev Environ Resour 37(1):421–448

Butler D et al (2014) A new approach to urban water management: safe and Sure. Procedia Eng 89:347–354

Castellari S et al (2014) Elementi per Una Strategia Nazionale Di Adattamento Ai Cambiamenti Climatici. Roma

Demuzere M et al (2014) Mitigating and adapting to climate change: multi-functional and multi-scale assessment of green urban infrastructure. J Environ Manage 146:107–115

EEA (2016) Urban Adaptation to climate change in Europe 2016. Transforming Cities in a Changing Climate

EU (2015) Towards an EU research and innovation policy agenda for nature-based solutions & re-naturing cities

Gobattoni F, Pelorosso R, and Leone A (2016) "Strategie Di Riqualificazione Urbana Clima-Adattiva: Nature-Based Solutions per Città Più Resilienti." Planum 101–10

Gómez-Baggethun E, Barton DN (2013) Classifying and valuing ecosystem services for urban planning. Ecol Econ 86:235–245

Green OO et al (2015) "Adaptive governance to promote ecosystem services in urban green spaces." Urban Ecosystems

Haase D et al (2014) A quantitative review of urban ecosystem service assessments: concepts, models, and implementation. Ambio 43(4):413–433

Hermann A, Schleifer S, Wrbka T (2011) The concept of ecosystem services regarding landscape research: a review. Living Rev Landscape Res 5(1):1–37

Holling CS (2001) Understanding the complexity of economic, ecological, and social systems. Ecosystems 4(5):390–405

ISTAT (2016) Verde Urbano

Joyce J, Chang NB, Harji R, Ruppert T, Imen S (2017) Developing a multi-scale modeling system for resilience assessment of green-grey drainage infrastructures under climate change and sea level rise impact. Environ Model Softw 90:1–26

Kabisch N, Strohbach M, Haase D, and Kronenberg J (2016) Urban green space availability in european cities. Ecological Indicators

La Rosa D, Spyra M, Inostroza L (2016) Indicators of cultural ecosystem services for urban planning: a review. Eco Ind 61:74–89

Leone A, Gobattoni F, and Pelorosso R (2014) Sustainability and Planning: Thinking and Acting according to Thermodinamics Laws. Journal of Land Use, Mobility and Environment (Special Issue Eighth International Conference INPUT)

Leone A, Gobattoni F, and Pelorosso R (2016) Energy supply, thermodynamics and territorial processes as a new paradigm of sustainability in planning science and practice. In: Papa R , Fistola R (eds) Smart energy in the smart city. Urban planning for a sustainable future, Berlin: Springer International Publishing, pp 83–101

Lundy L, Wade R (2011) Integrating sciences to sustain urban ecosystem services. Prog Phys Geogr 35(5):653–669

McPhearson T, Andersson E, Elmqvist T, Frantzeskaki N (2015) Resilience of and through urban ecosystem services. Ecosyst Serv 12:152–156

Meerow S, Newell JP, Stults M (2016) Defining urban resilience: a review. Landscape and Urban Planning 147:38–49

Mugume SN, Gomez DE, Fu G, Farmani R, Butler D (2015) A global analysis approach for investigating structural resilience in urban drainage systems. Water Res 81:15–26

Pappalardo V, La Rosa D, Campisano A, and La Greca P (2017) The potential of green infrastructure application in urban runoff control for land use planning: a preliminary evaluation from a southern Italy case study. Ecosyst Serv 1–10, (June 2016)

Pelorosso R, Gobattoni F, Lopez N, Leone A (2013) Verde Urbano E Processi Ambientali: Per Una Progettazione Di Paesaggio Multifunzionale. Journal of Land Use, Mobility and Environ 6(1):95–111

Pelorosso R, Gobattoni F, Geri F, Monaco R, Leone A (2016a) Evaluation of ecosystem services related to bio-energy landscape connectivity (BELC) for land use decision making across different planning scales. Ecol Ind 61:114–129

Pelorosso R, Gobattoni F, Lopez N, Leone A (2016b) Verde Urbano E Regolazione Delle Acque Meteoriche: L'approccio Modellistico Come Base per Nuovi Standard Urbanistici. Sentieri urbani 19:75–81

Pelorosso R, Gobattoni F, Lopez N, Leone A (2016c) Verde Urbano E Regolazione Delle Acque Meteoriche: L'approccio Modellistico Come Base per Nuovi Standard Urbanistici. Sentieri urbani 19:71–77

Pelorosso R, Gobattoni F, Geri F, Leone A (2017a) PANDORA 3. 0 Plugin : a new biodiversity ecosystem service assessment tool for urban green infrastructure connectivity planning. Ecosyst Serv 26:476–482

Pelorosso R, Gobattoni F, Leone A (2017b) Low-Entropy cities: a thermodynamic approach to reconnect urban systems with nature. Landscape and Urban Planning 168:22–30

Pincetl S (2015) Cities as novel biomes: recognizing urban ecosystem services as anthropogenic. Front Ecol Evol 3(December):1–5

Scott M et al (2016) Nature-based solutions for the contemporary city/Re-Naturing the city/Reflections on urban landscapes, ecosystems services and Nature-based solutions in cities/Multifunctional green infrastructure and climate change adaptation: brownfield greening as an adaptation strategy for vulnerable communities?/Delivering green infrastructure through planning: insights from practice in Fingal, Ireland/Planning for biophilic cities: from theory to practice. Plann Theor Pract 17(2):267–300

TEEB (2011) TEEB manual for cities: ecosystem services in urban management

Walker B, Holling CS, Carpenter S, Kinzig A (2004) Resilience, adaptability and transformability in social—ecological systems. Ecology Soc 9(2):5

Wang M, Sweetapple C, Fu G, Farmani R, Butler D (2017) A framework to support decision making in the selection of sustainable drainage system design alternatives. J Environ Manage 201:145–152

Indicators and Actions for the Smart and Sustainable City: A Study on Italian Metropolitan Cities

Rosaria Battarra, Fulvia Pinto and Maria Rosa Tremiterra

Abstract In recent years, there has been a wide diffusion of two concepts, the Sustainable City and the Smart City. Considering the several definitions of Smart City, it arises that the two concepts can have many commonalities. For instance, the most obvious refers to the role of ICT—main characteristic of the Smart City—for reaching the goal of sustainability. Hence, the purpose of this paper is to compare the current status of environmental sustainability in some Italian cities with the strategies regarding the "Environment" dimension, which are adopted in accordance with the Smart City concept. In the first part, the paper proposes a brief summary of the cities surveyed from an environmental perspective. In the second part, the methodology implemented during research is illustrated. The third part highlights the results obtained in relation both to the definition of the cities' status in terms of environmental sustainability and of assessing their propensity towards the implementation of a "smart" approach in the Environmental dimension, thanks also to a detailed analysis of projects underway in the various cities. Finally, the last part is aimed at highlighting the specific experiments underway and challenges identified during the research.

1 Sustainable City and Smart City: Two Approaches for Dealing with Urban Challenges

In studying urban phenomena two concepts took more and more root over the last thirty years: that of the sustainability of urban development and that of the use of

R. Battarra (✉)
ISSM, National Research Council, Naples, Italy
e-mail: battarra@unina.it; battarra@issm.cnr.it

F. Pinto
DASTU, Politecnico di Milano, Milan, Italy

M. R. Tremiterra
DICEA, University of Naples Federico II, Naples, Italy

R. Papa et al. (eds.), *Smart Planning: Sustainability and Mobility in the Age of Change*, Green Energy and Technology,
https://doi.org/10.1007/978-3-319-77682-8_6

technological innovation for dealing with the increasingly complex challenges posed by cities. According to the most recent theories these concepts have been identified as that of the Sustainable City and the Smart City, two labels that are as appealing as they are generic.

The concept of sustainability in the field of urban studies has roots that go far back: nowadays its acceptance as a key to interpretation for guiding transformations of urban systems is undisputed (Wheeler 2000; Berke and Conroy 2000; Gohar 2016) and this contribution does not include a close examination of the evolution of this approach.

What we want to stress is that the topical interest of the themes related to sustainability is confirmed by the more recent establishment of new approaches such as that of the Resilient City (Pickett et al. 2004; Jabareen 2013; Papa et al. 2015a, b) and of the Sustainable Energy City (Webb et al. 2016; Song et al. 2016) which are associated with the need for defining new action strategies for limiting the impact of emerging and pervasive phenomena such as environmental pollution, reduction of resources, and climate change.

Likewise, since the end of the nineties the Smart City has been the subject of many deep studies in the scientific field (Mahizhnan 1999; Caves and Walshok 1999; Graham and Marvin 2001; Komninos 2002) with a view to demarcating this approach procedure and defining its typical features (inter alia: Caragliu et al. 2011; Batty et al. 2012; Papa et al. 2013; Anthopoulos 2015; Papa et al. 2015a, b; Ramaprasad et al. 2017; Vanolo 2017). Many studies assigned information and communication technologies (ICTs) the central role for improving the quality of life and making cities more sustainable (Chourabi et al. 2012), even though there are also approaches that radically criticise this point of view (Hollands 2008, 2015) and others that contribute other aspects to this "technological" outlook, such as those related to social capital, environmental sustainability, public services, etc. (Caragliu et al. 2011; Morelli et al. 2013; Albino et al. 2015).

The increasing number of academic and other contributions on the subject of the smart city shows how pervasive this approach to urban development is, which might be attributed to its ability to summarise by means of a label the various subjects that have been permeating urban policies during the last fifty years (environmental challenges and sustainability, new technologies and cities, climate change and resilience, etc.). Nevertheless, even without a single definition it can be confirmed that in the Smart City the technological element is only one of the tools that can be used for pursuing the aims of urban development, among which sustainability (environmental, social, economic) plays a leading role.

Moreover, an analysis of the considerable literature reveals how there are many connections and overlaps between the two concepts. The most obvious one refers to the role of the ICTs—which even though not the sole and determining factor that connotes a Smart City, are certainly a characteristic feature of it—in dealing effectively with some environmental challenges: eliminating pollution, reducing the consumption of non-renewable resources, improving energy efficiency (Bibri and Krogstie 2017a; Morelli et al. 2013). Having resource to the ICTs is deemed to be an indispensable and necessary element for guaranteeing the sustainability of the city regarded at the same time as the main cause of environmental problems but also as a place where they can be dealt with more effectively (European Commission 2011). In this sense the field of action of the Smart City overlaps so much with that of the sustainable city that it can be said that the minimum common denominator of intelligent cities is environmental sustainability broken down into its various components (The European House-Ambrosetti 2012).

Recently this subject was dealt with by Mosannenzadeh et al. (2017), who state that the Smart City, the specific characteristic of which is promoting actions based on the use of the ICTs, can be regarded as a "subset" of the sustainable city. The difference between a city classified as smart and a city that is "truly" smart should be attributed to whether it is able to pursue sustainable development.

Ahvenniemi et al. (2017), on the other hand, in stating that «Ultimately, a city that is not sustainable is not really "smart"» (p. 242) and that the main aim of the smart city is the pursuit of sustainability by using ICTs, propose the adoption of the term "Smart Sustainable Cities"[1].

The connections between the two approaches also emerge from an analysis of the characteristic used for describing smart cities in various studies. For example, in the one put forward by Giffinger et al. (2007), one of the most widespread and used models, the Smart City is broken down into six dimensions (Smart Economy, Smart People, Smart Governance, Smart Mobility, Smart Environment and Smart Living), each of which is associated with one or more key aims to be pursued using the ICTs.

In short, through the concept of the Smart City, people try to promote innovative approaches to the management of urban areas that are compatible with the demand for sustainable development of the urban environment and with the needs of the communities that are settled there.

From this point of view, an analysis of the various Smart City models proposed by the literature shows that, even with different classifications, the dimension of environmental sustainability, mainly known as Environment, almost always exists (Table 1), which emphasise the increasing attention paid by cities to dealing with the main questions related to environmental matters at the urban level (Barrionuevo et al. 2012).

In particular, the Environment denomination mainly includes elements related to the efficient management of urban utilities, such as water and waste; to the improvement of air quality; to the construction of green and energy-efficient buildings, as well as to the promotion of alternative energy sources to fossil fuels and to a more sustainable use of non-renewable resources such as soil.

Dirks and Keeling (2009) propose a Smart City model that pays particular importance to the management of the main urban utilities, where two dimensions related to environmental sustainability (Water and Energy) are identified; meanwhile, environmental sustainability aspects proposed by Lombardi et al. (2012), Baron (2012), Barrionuevo et al. (2012), Zygiaris (2013) and the Japanese Institute for Urban Strate-

[1]On the development and propagation of the term (already adopted by Kramers et al. 2014), cfr. Bibri and Krogstie 2017b.

Table 1 Denomination and categories of Smart Environment in the Smart City literature

Denomination	Categories	References
Water energy	Water cycle Water supply and sanitation Power generation and transmission infrastructure Waste disposal	Dirks and Keeling (2009)
Smart environment	Attractivity of natural conditions Pollution Environmental protection Sustainable resource management	Giffinger et al. (2007)
Smart environment	Reduction of CO_2 Energy efficiency and use of renewable resources Water management Land Air pollution Waste management Green buildings	Lombardi et al. (2012)
Energy and environment	Energy efficiency Reduction of carbon footprint Re-use of water Achievement of energy performance Renewable resources Waste management	Baron (2012)
Environmental sustainability	Air pollution and noise pollution Water management Green buildings and alternative energy Waste management Landscape	Barrionuevo et al. (2012)
Environment	Ecology Air quality Natural environment	Japanese Institute for Urban Strategies (2010)
Green city layer	Alternative Energy master planning Water conservation Green transport practices Green building policies CO_2 reduction master planning	Zygiaris (2013)
Smart environment	Sustainability-certified buildings Smart homes Energy Carbon footprint Air quality Waste generation Water consumption Climate resilience planning Density Green space per capita	Cohen (2012)

(continued)

Table 1 (continued)

Denomination	Categories	References
Gebäude & Siedlungsstrukturen	Innen- und. Außenentwicklung Veränderung der Bevölkerungsdichte Kompaktheit der genutzten Wohngebiete Energieeffiziente Gebäude	Smart City Profiles (2013)
Structure	Settlement Biodiversity Air Soil Water Communication Water cycle Energy cycle Matter cycle Mobility network Nature Dwellings Building & blocks Neighborhood & district City/metropolis Public space Land use	City Protocol Society (2015)
Planet	Energy and mitigation Material, water and land Climate resilience Pollution & waste Ecosystem	Bosch et al. (2016)

gies (2010) are more akin to the Smart Environment model proposed by Giffinger et al. (2007).

In recent years, as highlighted by other Smart City models, including Cohen (2012), Smart City Profiles (2013) and City Protocol (2015), the Environmental dimension has been further expanded upon. In fact, it no longer refers to the natural environment and its resources, but also to the physical characteristics of the city system, including, among the aspects of this dimension, not only buildings' green features, but even features of the settled urban environment, including urban density, urban compactness, etc., which can have a positive impact on a more efficient and sustainable management of the city (Pinto et al. 2013; Galderisi et al. 2016).

One aspect that has recently been taking on an increasingly important role, especially considering the global challenge posed by climate change and the local one of its negative effects on the urban environment, concerns Environmental Protection (e.g. Bosch et al. 2016). Along these lines, the use of ICTs for the protection of the environment against natural hazards and those related to climate change can contribute to further define the Environmental dimension of a Smart City. The central role of sustainability and of environmental issues in the Smart City approach is further confirmed by this dimension's increased focus on a number of projects and activities in cities labelled "smart". In the most widespread interpretation, the Smart Environment dimension mainly includes experiments aimed at improving energy efficiency at the construction and city planning scale, but also interventions involving the use of ICTs to build smart grids for energy transport and, more broadly, to improve city services, such as innovative public lighting systems, or to reduce air pollution resulting from urban transportation (European Parliament 2014).

Thus, the main objective of the work presented is to compare the current status of some Italian cities in terms of environmental sustainability with the strategies implemented to apply the Smart City paradigm under its "Environment" dimension.

Based on these premises, below is a brief summary of the cities surveyed from an environmental perspective (paragraph 2), of the methodology implemented during research (paragraph 3) and of results obtained (paragraph 4) in relation both to the definition of the cities' status in terms of environmental sustainability (through the use of 3 aggregate indexes and a composite index) and of assessing their propensity towards the implementation of a "smart" approach in the Environmental dimension, thanks also to a detailed analysis of projects underway in the various cities. Conclusions (paragraph 4) are aimed at highlighting the specific experiments underway and challenges identified during research.

2 Smart Environment: Indicators and Activities in a Few Italian Cities

Within this theoretical-scientific framework, this contribution illustrates the results of research conducted on the administrative centres of 14 cities: Milan, Turin, Genoa, Venice, Bologna, Florence, Rome, Naples, Bari, Reggio Calabria, Cagliari, Palermo, Catania and Messina; established in Italy with Law no. 56/2014 (Papa et al. 2016). The cities subject of this research were selected due to their size and demographics, and are cities where we find all the main environmental problems (air pollution, efficient management of urban utilities, etc.). On the other hand, it is in these cities that the economic and social capital is concentrated, and which becomes a catalyst for the development of innovative solutions aimed at solving the main urban issues (European Commission 2011).

Thus, the research was divided into the following phases:

1. overview of the cities and the main characteristics of the built urban environment;
2. definition of macro-categories for the analysis of the Environmental dimension;
3. selection of a set of indicators and development of a compound index to define the cities in terms of the Environmental dimension;
4. selection of the main Smart Environment initiatives implemented in each city, structured as per the previously defined macro-categories;

5. comparative analysis of the cities regarding the selected initiatives and the aggregate indexes for each macro-category.

In the first phase, the main characteristics of the built environment were identified, referring in particular to the level of urbanization of the cities, so as to take into account the heterogeneity of the urban frameworks analysed from the perspective of urban development and so as to allow for their comparison.

Then, based on the most popular Smart City models used in the scientific literature, and in order to better manage the results and analyses for the study of the 14 cities, three summary categories for the Environmental dimension were identified:

- Energy (EN): improve the energy efficiency of buildings and networks and promote the use of alternative energy sources to fossil ones;
- Resource Management (RM): efficiently manage the main urban utilities linked to natural and environmental resources and contain/reduce potential pollution;
- Hazard Mitigation (HM): protect the urban environment from the impacts of natural events, including those related to climate change.

Once the categories for the Environmental dimension were defined through the analysis of the literature, which uses qualitative-quantitative variables to define the smartness of cities (see Table 1), 18 indicators were selected (Table 2).

These indicators, useful for measuring the Environmental dimension of Italian cities, were chosen for: (i) their capacity to effectively describe aspects that contribute to define the four macro-categories comprised in the Environmental dimension; (ii) the frequency with which these indicators are used in both national and international studies; (iii) the availability of reliable data at the municipal level. They refer to the years between 2011 and 2017 and are provided by official sources (Table 2).

Once the set of indicators for each of the 14 cities was defined, an aggregate index was constructed for each aspect comprising the Environmental dimension and, finally, a total compound index was developed. Although the development of such indexes involves the loss of part of the information collected, they are an effective way to compare the status of the Environmental dimension of the 14 cities. The indexes were developed in three sub-phases: data collection, standardization and aggregation. During the first phase, the data required for each city was collected, resulting in a 14×18 data matrix.

Then the data collected were standardized using the following formula:

$$\mathrm{r_{ij}} = \frac{\mathrm{x_{ij}} - \min \mathrm{x_j}}{\max \mathrm{x_j} - \min \mathrm{x_j}} \times 10$$

where:

r_{ij} represents the standardized value
x_{ij} the indicator's value;
$\min \mathrm{x_j}$ the minimum value of the j-nth indicator;
$\max \mathrm{x_j}$ the maximum value of the j-nth indicator.

Table 2 Indicators selected for the environmental dimension

Macro-category	ID	Parameter	Unit	Source	Year
Energy	EN1	Energy consumption per capita	kWh/inhab	ISTAT	2011
	EN2	Number of electric vehicle charging stations	n° per 10 m^2	ISTAT	2014
	EN3	Energy derived from photovoltaic source	kW/1000 inhab	ISTAT	2014
	EN4	Inhabitants receiving Teleheating	%	ISTAT	2011
Resource management	RM1	Water consumption for household uses	cu.m/inhab	ISTAT	2011
	RM2	Real network losses	%	ISTAT	2015
	RM3	Pop. served by waste water treatment plants	%	ISTAT	2011
	RM4	Collected load conveyed into the drainage system	%	ARPA	2012
	RM5	PM10—exceedance days	n° of days	ISTAT	2012
	RM6	Ozone—exceedance days	n° of days	Legambiente	2013
	RM7	Air monitoring stations	N° per 100,000 inhab	ISTAT	2011
	RM8	Noise pollution: exceedance of limit values	%	ISTAT	2011
	RM9	City's solid waste that is recycled	%	ISPRA	2011
	RM10	Total collected municipal solid waste city	t/inhab	ISPRA	2011
Hazard mitigation	HM1	Climate change strategy or plan	n°	Covenant of Mayors	2017
	HM2	Landslides with high risk of landslides	%	#ItaliaSicura	2017
	HM3	Landslides with high risk of floods	%	#ItaliaSicura	2017
	HM4	Investments for environmental protection	Euro/sq km	#ItaliaSicura	2017

Table 3 Types of Smart Environment actions and corresponding categories

Types of Smart Environment actions	Categories
A. Energy efficiency (buildings and networks) and consumption reduction B. Promoting the use of renewable energy	Energy
C. Pollution control and monitoring D. Efficient use of water resources E. Improving urban waste management	Resource management
F. Territorial control and monitoring G. Reducing the impact of climate change	Hazard mitigation

Finally, the I_{ij} aggregate indexes referring to the three macro-categories of the Environment were developed, as the arithmetic average of the standardized values of the indicators linked to each macro-category, using the following formula:

$$I_{Aj} = \frac{1}{N}\sum_{A=1}^{N} r_{ij}$$

where:

r_{ij} represents the standardized value of the j-nth indicator for the i-nth city;
N represents the number of indicators in the macro-category.

After calculating the three aggregate indices, the total I_{En}, compound index was developed, obtained as the geometric mean[2] of the three aggregate indexes for each city:

$$I_{En} = \prod_{j=1}^{4} (I_{Aj})^{\frac{1}{4}}$$

where:

I_{SE} represents the value of the total compound index for the j-nth city

At the same time, to analyse how large Italian cities are applying ICTs to the Environmental dimension in line with their development as smart cities and taking into account the definitions of the six dimensions of smartness provided by the literature (Manville et al. 2014; Giffinger,…), a database was developed for the initiatives and projects implemented in recent years. The criteria used for their selection were the level of technological innovation[3], their replicability in other territorial frame-

[2]The method for aggregating the composite index was selected for its ability to immediately use and interpret its results (Massoli et al. 2014).

[3]Most selected initiatives have a high technological content, even if in some cases, activities that do not directly involve ICT devices and products were chosen, because they can still be traced back to the Smart Environment dimension, such as sustainable energy and climate action plans, which are aimed at making cities more sustainable even through the use of ICTs.

works and the current state of implementation. Such initiatives have been selected by the examination of several indirect sources (i.e. instruments for urban and territorial government, web sites, publications).

The initiatives were then classified into eight types of activities, structured into the three aforementioned summary categories (Energy, Resource Management and Hazard mitigation), so as to be able to proceed, in the last phase, to the comparative analysis of the cities with regards to both the initiatives and the aggregate indexes (Table 3).

3 Overview on Urban and Environmental Characteristics of the Italian Metropolitan Cities

The cities analysed in this study reveal heterogeneous urban characteristics not only as far as demographic, economic and social aspects are concerned, but also in relation to urban development and environmental parameters.

As far as territorial surface area is concerned, Rome is the city with the largest extension (1287 km^2), while the largest metropolitan area in Italy is that of Turin (6827 km^2). Looking at the demographics of the 14 capitals, Rome, Milan and Naples have the highest number of inhabitants both at the metropolitan and municipal level, while the least populated metropolitan cities (and capitals) are Cagliari, Reggio Calabria and Messina. As far as population density is concerned, however, both at the municipal and metropolitan level Naples and Milan have the highest values, much higher than the average of the other capitals; while the lowest values apply to Venice, Reggio Calabria and Messina. The latter two have low population density values even at the metropolitan level, together with the Metropolitan City of Cagliari (Table 4).

The territorial surface and density of metropolitan cities is closely linked to their degree of urbanization, although this has also been impacted by an urban development that is strongly linked to the physical structure of each city and to the history of its communities. In particular, the Impervious Surface analysis reveals that Naples is the city with the highest percentage of urbanized area, together with Turin and Milan. This is even more evident when compared to the percentage of green space found in the three capitals, below the average of the 14 metropolitan cities analysed (26.74%). In fact, among the metropolitan cities with the lowest population density, Reggio Calabria is the one with the highest green space value and, at the same time, which has a low Impervious surface value (Table 5).

As far as the building age of the cities is concerned, Rome has the highest percentage value for the last fifty years, followed by the metropolitan cities of the South and the Islands, like Reggio Calabria, Cagliari, Messina, Palermo, Bari, Naples and Catania. The Cities with the least construction expansion in the last fifty years are Florence, Genoa and Bologna, while cities that are large as far as surface and pop-

Table 4 Population, territorial surface and density of the Metropolitan Cities

City	Population (2017)		City (km^2)	Metropolitan area (km^2)	City density ($inhab./km^2$)	Metropolitan city density ($inhab./km^2$)
	City	Metropolitan area				
Rome	2873,494	4353,738	1287.36	5363	2232	812
Milan	1351,562	3218,201	181.67	1576	7440	2042
Naples	970,185	3107,006	119.02	1179	8151	2635
Turin	886,837	2,277,857	130.01	6827	6821	334
Palermo	673,735	1,268,217	160.59	5009	4195	253
Bari	324,198	1,260,142	117.39	3863	2762	326
Catania	313,396	1,113,303	182.9	3574	1713	312
Florence	382,258	1,014,423	102.32	3514	3736	289
Bologna	388,367	1,009,210	140.86	3702	2757	273
Venice	261,905	854,275	415.90	2473	630	345
Genoa	583,601	850,071	240.29	1834	2429	464
Cagliari	154,083	560,373	85.01	4570	1812	123
Reggio Calabria	182,551	553,861	239.04	3210	764	173
Messina	236,962	636,653	213.75	3266	1109	195
Average	684,510	1,576,952	258.29	3569	3325	612
Total	9,583,134	22,077,373	3616	49,660	2650	442

Source ISTAT

ulation, such as Turin and Milan, have a lower percentage compared to the average of metropolitan cities.

To better understand the Environment dimension of sample cities, it is worth assessing the tendency of their populations towards the adoption of "environmentally friendly" behaviours. Table 6 reveals that in northern cities there is a greater tendency to use sustainable modes of transport, favoured also by a greater and improved range of transport options aimed at developing sustainable mobility.

Whereas, in Southern cities like Palermo and Reggio Calabria we find the highest number of volunteers for environmental protection: 99.2 volunteers out of every 1000 citizens for Palermo and 37.2 every 1000 for Reggio Calabria.

Table 5 Impervious surface, green space share and building age of the metropolitan capitals

City	Impervious surface (%)	Green space share (%)	Building age (%)
Naples	88.93	12.30	48.72
Turin	87.61	24.50	36.83
Milan	81.46	17.60	36.47
Florence	60.58	19.90	16.53
Bologna	55.37	30.50	28.89
Palermo	55.15	9.90	52.58
Bari	50.85	7.90	51.16
Rome	39.55	16.70	68.39
Catania	34.79	16.90	45.12
Genoa	33.11	6.30	17.56
Cagliari	32.80	56.80	56.53
Reggio Calabria	24.80	105.00	65.92
Messina	20.84	13.00	54.04
Venice	18.46	37.10	42.27
Average	48.88	26.74	44.36

Source Ispra (2017), ISTAT (2011)

Table 6 Working population traveling to work with sustainable modes and volunteers for the environmental protection of the metropolitan capitals

City	Working population traveling to work with sustainable modes (%)	Volunteers for the environmental protection (n° for 1000 volunteers)
Milan	20.8	7.0
Venice	17.3	18.8
Genoa	15.3	17.2
Bologna	14.7	25.2
Turin	14.4	7.2
Florence	13.1	11.7
Rome	11.3	22.7
Naples	9.8	14.6
Bari	8.2	9.3
Palermo	6.0	99.2
Cagliari	5.7	16.5
Messina	5.4	11.6
Reggio Calabria	5.0	37.2
Catania	4.8	8.2
Average	10.8	21.9

Source ISTAT (2011)

4 The Environmental Dimension in Metropolitan Cities: A Field Research

The results of the research are divided into two sections. In the first, through 18 indicators related to the environmental and descriptive characteristics of the Environmental dimension, as interpreted by the main Smart City models, a cognitive framework for the 14 cities is provided, structured into the three macro-categories: Energy, Resource Management and Hazard Mitigation. The aim is not to build a ranking of the most "sustainable" cities, but rather is to define them from an environmental perspective, highlighting their strengths and weaknesses, while also conscious of the great heterogeneity (demographic, social, economic, etc.) that distinguishes them. In the second section, the cities' tendency towards a Smart Environment approach is described, inferred from the study of activities involving ICTs.

4.1 The Analysis of the Environment Dimension's Status of Italian Cities

With regards to the cities' environmental dimension, by aggregating the 18 indicators, 3 aggregated indexes were developed that, thanks to the method used for their development (data normalization and average of the values obtained), allow us to obtain a sort of "basis for comparison" for the Environmental dimension of Italian cities.

Furthermore, beyond the inevitable biases deriving from the selection of indicators sets, it can be stated that, the analysis of the performance of the summary indexes for each macro-category reveals that the cities with the best performances are Milan, Bologna, Florence and Bari, although with some differences.

As far as "Energy" is concerned, taking into account the individual indicators (energy consumption per capita, energy from photovoltaic source, etc.) and taking into account that the aggregation of individual indicators was made by adding the standardized values of each one and adding the "minus" sign to those that recorded characteristics that negatively impacted environmental sustainability (such as energy consumption, the number of days exceeding threshold values of pollution, etc.), painted a very diverse picture for the 14 cities. This is mainly due to energy consumption, which varies greatly: Milan, with the lowest value (around 1000 kwh/inhabitant) and Cagliari, with the highest (approximately 1600 kwh/inhabitant); and due to the production of photovoltaic energy (Bari and Cagliari are the Italian cities with the highest production of energy from this source). Moreover, Milan and Florence stand out due to the number of electric top-up stations for cars, far superior to that of all other cities; this reflects the fact that these cities are investing in a more sustainable development of the urban environment and of urban mobility, focused on electric vehicles and not just car-pooling.

Fig. 1 Energy indicators

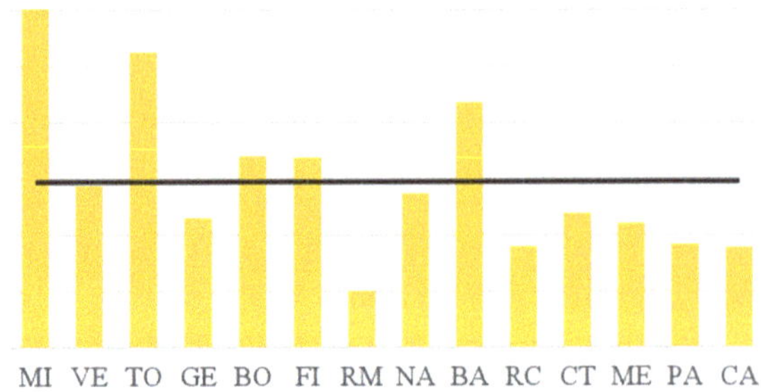

Fig. 2 Resource Management indicators

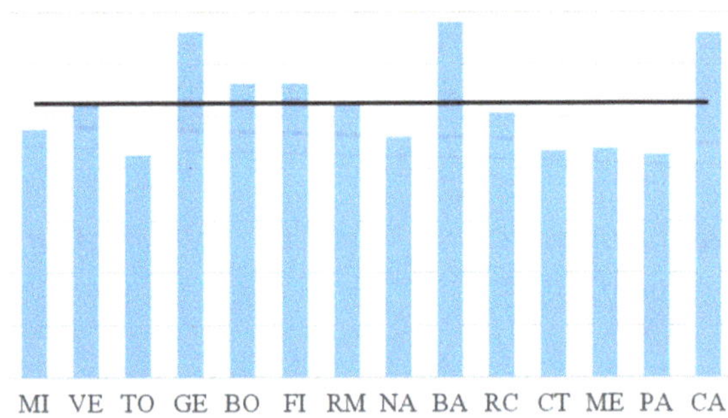

In general, most of the Southern cities reveal heterogeneous trends in all parameters and, for the most part, below the sample's average. The calculation of the aggregate index for this macro-category revealed that the city with the worst performance among those selected is Rome, with values well below the average in all parameters, along with all the other cities in Southern Italy, except for Bari. The latter, in fact, together with Milan and Turin, has above average values. Bologna and Florence also perform well, while Genoa is the only northern city with lower than average values (Fig. 1).

With regards to the Resource Management macro-category (Fig. 2), which collects indicators linked to water, air and urban waste management, the aggregate index's performance seems less varied. The best performances are recorded in Genoa for the North and in 2 Southern cities (Bari and Cagliari).

But here too, an analysis of individual indicators reveals that the cities have considerable differences. In what concerns the use and management of water resources, northern cities, which have higher per-capita consumptions, show an efficient management of the grids with very low loss values[4] and high purification capacities, unlike southern cities. Catania is particularly noteworthy, with very high water-consumption and grid-losses and the lowest percentages of population serviced by waste water treatment plants (24%) and of load conveyed into the sewage network (41%).

[4]Istat measures the total losses in municipal drinking water grids by subtracting the volumes dispensed to the volumes fed into the network.

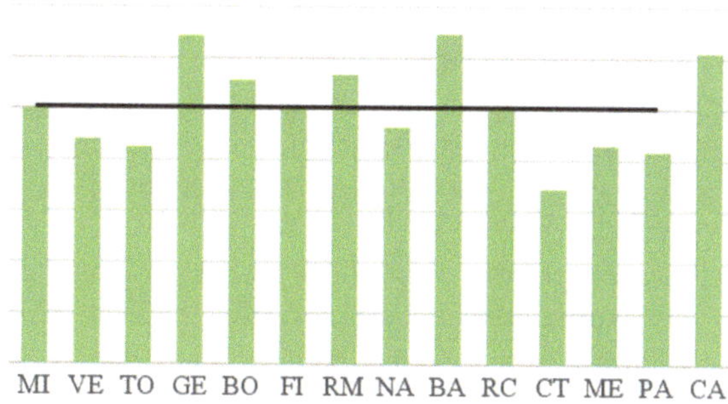

Fig. 3 Hazard Mitigation indicators

Air pollution is a different story: here Milan, Turin and Bologna present the worst environmental conditions, while amongst southern cities, Naples is the only one to exceed the number of days above threshold for PM10.

Italian cities still have low rates of separated collection, although with significant fluctuations (from 6% in Messina to 43% in Turin). Once again, northern cities and Cagliari perform better, while southern ones vary between 18% in Naples and 7% in Messina. Finally, we have Hazard Mitigation (Fig. 3), defined, on the one hand, by the exposure of cities to the impacts of climate change in relation to the geomorphological characteristics of the territory and, on the other, by the attention paid to these phenomena, measured by looking at the presence (or lack thereof) of tools for the development of intervention strategies and through investments in the territorial protection sector.

The performance of individual indicators reveals that because of the greater exposure to risks of landslides and floods recorded in Catania, Cagliari, Genoa, Bologna, Naples and Palermo, these cities are implementing a climate plan, and that Genoa, Milan and Naples invested more resources in environmental protection in 2017.

4.2 Limitations and Weaknesses of the Experiments in the Smart Environment Dimension

To look at how the main Italian cities are moving towards the implementation of a smart city approach using ICTs to improve environmental sustainability, information on the experiments in progress was gathered. The classification system implemented entailed assigning the initiatives to a particular dimension of smartness, according to who implemented them and their type.

In detail, the initiative promoters were classified into the following 4 categories; universities and research entities, local authorities and institutions, companies and associations. The initiative typologies can be articulated as follows:

- Research, carried out by universities, research centres, etc. located in the city;

- Interventions that involve the "physical" transformation of the object of application, with an advanced stage of implementation or already completed;
- Projects, both intangible and non-initiated interventions; technologies, products and innovations tested or developed in the studied city and which include control and monitoring tools, open sensor networks, open data portals, wi-fi networks, apps, etc.;
- Plans and programs specifically aimed at adopting the Smart City paradigm; initiatives promoting and disseminating the Smart City approach to a wide audience (info-points, exhibitions, conferences, university and training courses, etc.).

The initiatives under study refer to the period 2012–2017.

Thus, over 1000 initiatives (1075) were collected in the database; highly heterogeneous (as far as their level of implementation, innovativeness and objectives pursued) in nature.

Despite the limitations of using information derived mainly from indirect sources and the impossibility of conducting a comprehensive survey of the wide range of initiatives underway, the large survey does shed some light on some of the trends in the 14 subject cities, also supported by quantitative data.

Milan, Bologna and Florence, along with Bari, are the cities with the best performances regarding indicators of the Environmental dimension, and this is where approximately 30% of initiatives are concentrated, while in the other cities the number fluctuates between 33 in Messina and 115 in Turin.

Out of the 6 smartness dimensions (Fig. 4), the one with the largest number of initiatives (approximately 30% of the total) is the Environment, followed by Smart Mobility (22% of the total) and Smart Governance (18% of the total). It is worth noting that citizens and administrators on environmental issues are paying greater attention to the investments of companies for the development of new ICT products to be applied to this field, but also because Italian cities are drawing from the vast European Community funding in this sector. And it is precisely within the Environment dimension that the Energy issue has gained a lot of importance, interpreted as both energy savings and efficiency as well as the use of renewable sources (Battarra et al. 2016a, b).

The city with the highest recorded number of initiatives in the Environment dimension is Bologna (with 38 initiatives), followed by Turin and Genoa. Comparing the number of Smart Environment initiatives to the total number of initiatives for each city (Fig. 5), Reggio Calabria and Palermo seem to have focused their actions firmly on this sector.

When we look at the distribution of initiatives in relation to their "type" (Fig. 6), we can see that the "projects" categories encompass approximately 30% of initiatives and about 20% of those in the "research" category, confirming that the European Community's financing of researches and networks of partnerships between cities for the exchange of good practices has played an important role as incentive. The number of initiatives is rather low (approximately 17% of total initiatives) and while this may be attributed to the difficulty of finding reliable and updated information from indirect sources on the effective implementation of certain initiatives, it is worth

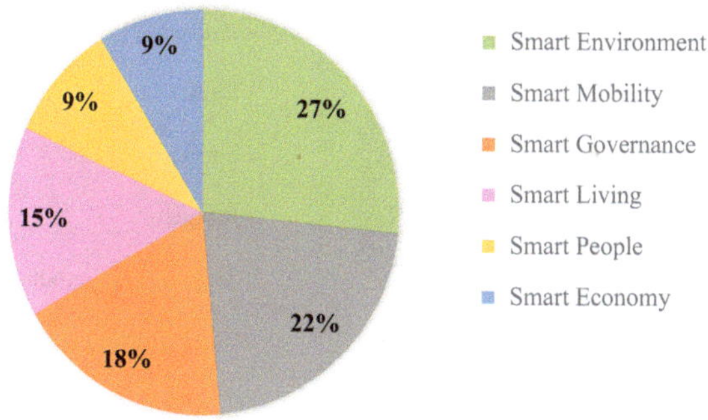

Fig. 4 Classification of initiatives into the 6 dimensions of smartness

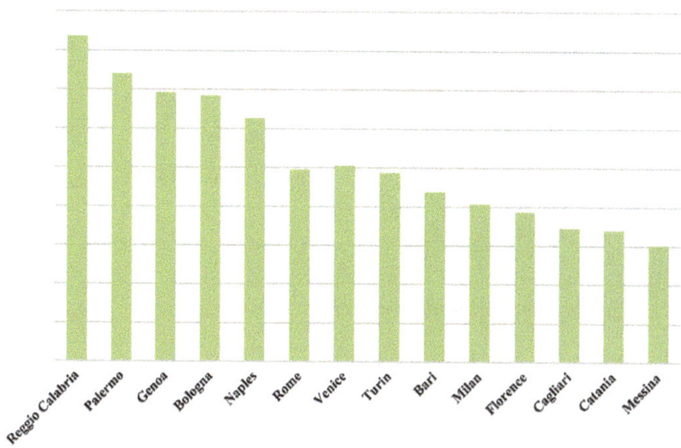

Fig. 5 Smart Environment initiatives (% of total initiatives for each city)

noting that many initiatives have yet to reach an advanced stage of implementation. In fact, often the experiments are quite prototypical in nature, and are aimed at being "pilot initiatives" focused on very small territorial areas. The cities with the highest number of initiatives were Bologna, Turin and Milan, which are also the first cities that started the process of implementing high-tech content initiatives.

In what concerns the "Technological products" type (13% of the total), often they are applications and sensors applied to energy grids, whose data converge in platforms that translate data in real time (e.g. project Eden in Turin and Smart ring in Milan) or other devices for monitoring pollution or the waste sector (e.g. electronic key for garbage bins trialled in 3 districts of Genoa).

Local authorities are the main "promoters" of Smart Environment initiatives (Fig. 7): over 60% of initiatives were in fact implemented by Municipalities, although this could partly be influenced by the fact that the main source of information was official websites.

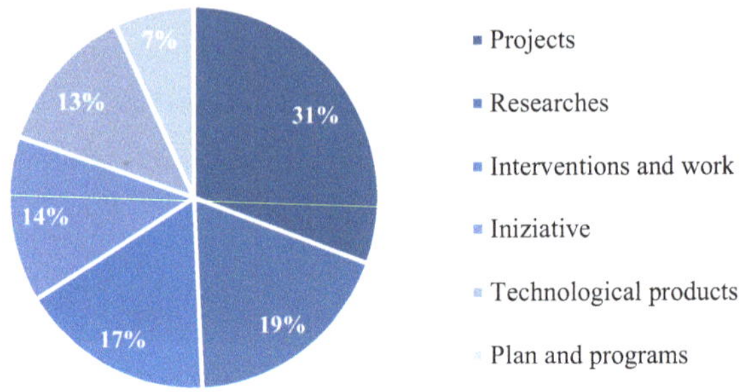

Fig. 6 Smart Environment initiatives by type

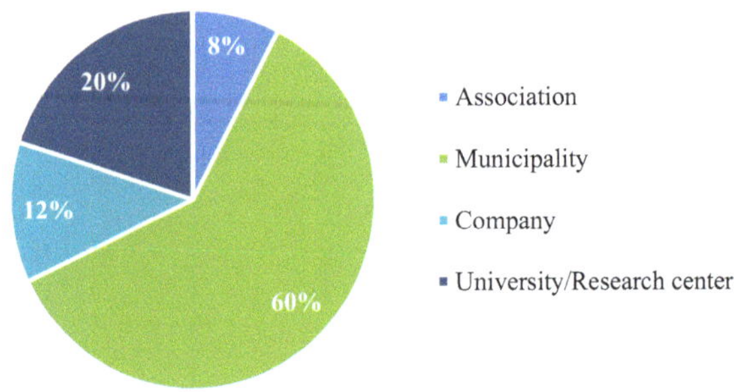

Fig. 7 Smart Environment initiatives by promoter

The analysis of the initiatives in the three categories of the Smart Environment—Energy, Resource Management and Hazard Mitigation—reveals that those implementing actions aimed at energy savings and efficiency and at using renewable energy sources are by far the most widespread. In fact, over 65% of initiatives in the Environmental dimension are aimed at the energy efficiency of buildings and grids and at promoting the use of renewable energy sources, photovoltaic being the most popular. Bologna is the city with the most initiatives in the energy sector (40, or 18% of all energy initiatives), backed by its framing within a carefully structured energy plan, both at the municipal and metropolitan scale (Fig. 8).

Other cities, such as Genoa, focused on smartness by participating in bids from the European Commission. Such is the case with the project R2 Cities and Energy Efficiency in Low Income Housing in the Mediterranean (ELI_Med), currently being implemented in Genoa, or European cities serving as Green Urban Gate towards Leadership in Sustainable Energy (Eu-Gugle) in Milan.

Another field of experimentation is the one related to ICTs in energy grid management networks, and for streamlining the use of renewable energy sources. This category spans many initiatives aimed at reducing consumption from public lighting systems and public buildings (schools and municipal offices).

As far as the Resource management category is concerned, initiatives focused on optimizing the supply of public utilities can undoubtedly benefit by using ICTs in the

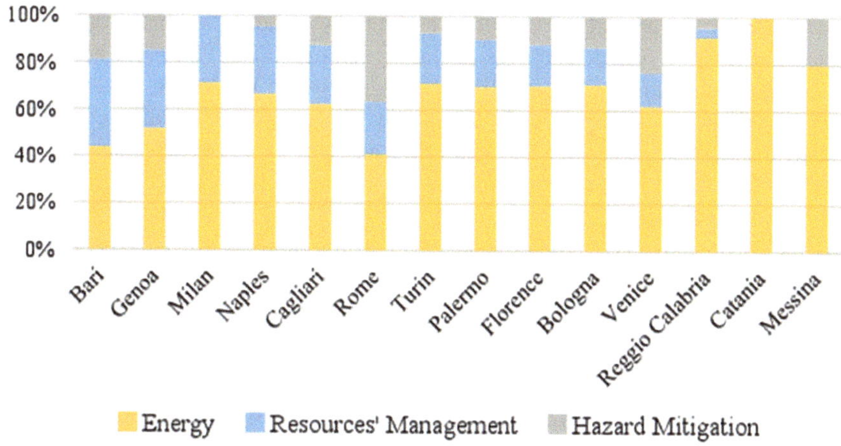

Fig. 8 Percentage distribution of initiatives related to Energy, Resource Management and Hazard mitigation

different stages of their processes: from supply assessment to real-time control and monitoring of grids and the related need to manage large amounts of data. Therefore, cities that have experimented with products and technologies aimed at a smart and more efficient management of this important aspect of urban life that impacts not only the Environment dimension across the board (but also more broadly a better quality urban environment, and therefore also impacting the Living dimension) is surely of interest.

Bari, Genoa and Milan are the cities with the highest number of initiatives in this sector. Specifically, initiatives aimed at managing the waste cycle include research projects funded by the National Operating Program PON for Research & Competitiveness (such as SIGLOD-Intelligent system to support the management and localisation of landfills and waste management plants in Palermo) and a few pilot projects trialled in Genoa as part of partnerships financed by the European Community (e.g. WASMAN-Waste Management as policy tools for corporate governance, MED-3R-Platform stratégique euro-méditerranéenne pour une gestion des déchets, SCOW—Selective Collection of the Organic Waste in tourist areas and valorisation in farm). With approximately 10 million euro in funding from the Lombardy Region in the framework of the ROP ERDF 2007–2013 on Smart Cities and Communities, two projects implemented in Milan by universities, research centres and companies and are focused on the management of RSU (CITY-WISE-NET—energy optimization of industrial sites via waste treatment; E-WASTE strengthening the WEEE recycling chain). The Osserva-WEEE Palermo project (once again part of the Ministry of Education Smart Cities and Communities bid) also focuses on the recycling of WEEE, and is aimed at creating a platform for the computerized management of the collection and recovery of waste from electrical and electronic equipment.

There are many projects, also within the Ministry of Education bids, focused on the management of water resources, including, among others, SWaRM-net-Smart Water Resource Management in Milan and Aquasystem in Naples, which aim at streamlining the use of water resources also by means of experimenting with technological devices such as sensors, smart meters, "electronic noses", smart kits, as well as iPhone and tablet apps.

Finally, the Hazard Mitigation category includes initiatives focused on natural hazards and the impacts of climate change on the city. Bologna, Venice and Florence are the ones that are more focused on this issue, developing specific plans, such as is the case of Bologna and Venice, and by promoting research projects on natural hazards such as BLUEAP (Bologna Local Urban Environment Adaptation Plan for a Resilient City) and the THESEUS (Innovative technologies for safer European coasts in a changing climate) project in Venice.

5 Conclusions

The application of indicators to cities surveyed and the review of initiatives, highlighted several weaknesses regarding how Italian metropolitan cities are interpreting the Smart Environment dimension.

First, it is worth noting the lack of alignment between the current condition of cities from an environmental perspective (as described via the indicators) and the initiatives currently being implemented. While keeping in mind the limitations of applying a system of quantitative indicators, which are inherently biased and reductive, and the shortcomings of the system for classifying and surveying initiatives[5], this comparison reveals some empirical evidence that can support qualitative considerations.

Despite indicators related to air pollution, land use, water management and waste disposal, reveal critical situations in most cities, activities in the Smart Environment dimension are mainly focused on improving efficiency and energy savings. Even accounting for the fact that this is the result of European policies that in recent years have given priority to energy issues and have made significant funding available, it should be noted that ICTs are still scarcely applied to the management and supply of network services, to pollution control, but above all to territorial protection. Although the reduction in energy consumption is the main long-term measure of mitigating the effects of climate change, Italian cities are struggling to define integrated and differentiated strategies to face the various environmental challenges.

The cities that are investing more in this direction are Venice, Bologna and Genoa. The latter, in particular, as part of the "National Operational Program for Metropolitan

[5]Referring to the classification criteria for initiatives used during the research, which foresees that the Smart Environment dimension include only initiatives directly related to environmental aspects, thus leaving out interventions that, although have positive impact on the environment, are classified under other dimensions of smartness. Such as the case of sustainable mobility initiatives aimed at reducing pollution, which were not accounted for because they are classified under the Smart Mobility dimension

Cities 2014–2020" (PON Metro) obtained funding to develop a digital system for risk management (meteorological, hydrogeological and hydraulic risks) that uses sophisticated technology systems. Bologna and Venice, in addition to supporting a number of researches on climate change carried out by universities and research institutions, have started drafting a Climate Plan.

In general terms, the "smarter" cities—Milan, Turin, Bologna and Florence—do seem to align the highest number of initiatives with the areas in which they face critical environmental circumstances (like air pollution in Milan). This would suggest that these cities defined their priorities in relation with their most critical sectors, defining initiatives that, thanks also to ICTs, can impact these aspects.

Continuing the comparison between indicators and initiatives reveals a non-homogeneous picture in the country's various geographical areas. While the smart city debate has highlighted the need to take into account the specificities and identities of different urban contexts, both in the application as in the recognition of the smart model (which, like all models, cannot be generalized and replicated as a one-size-fits-all for every territorial reality), still it is possible to acknowledge, without needing to resort to political-administrative propaganda-like statements, that sectors considered strategic in defining a smart model are more developed in some urban realities than in others. This is the case, for example, of northern cities, which have reached a more advanced stage of ICT implementation in the reduction of energy consumption or to improve the management of urban utilities. While in some cases initiatives may seem to be driven by the interests of large business groups (Milan) and in others they are led by the public administration and trials were developed within the framework of European projects (Genoa), northern cities are undoubtedly witness to a more active set of "concrete" interventions. Moreover, some of these cities are investing on initiatives that, at least in terms of intentions, would seem to incorporate experimenting with sets of activities that, across-the-board and in an integrated manner, are aimed at making the city more sustainable. Two projects, one in Milan (Sharing Cities) and another in Florence (Replicate), both funded by the Horizon 2020 Program, are particularly representative of this: these programs are considered best practices for urban-scale intervention, and operate in a complex and integrated manner aimed at improving the energy efficiency of buildings, to encourage the use of sustainable transport modes and to promote virtuous behaviour among users.

On the other hand, in southern cities, it is not so much a smaller number of initiatives launched, but rather the scarcity of initiatives integrated as part of a clear innovation strategy for the urban system, that make it so that even initiatives that could give a boost to the Environmental dimension, lose their impact.

This can partly be attributed to how initiatives are implemented: many of the initiatives surveyed were developed within European funding programs and are "pilot experiments", applied only to contexts of limited size. The prototypal nature of many initiatives, not only in Italy, articulated in all dimensions of the smart city model, has been questioned by many (including van Winden and van den Buuse 2017) and in this sense, it should be explored under which conditions projects could be "scaled up" for their application to a larger territory, also because of the huge public investments that they often are a product of.

Within this landscape it is possible to identify experiences that bear witness to how even southern cities are implementing policies that, by also using ICTs, are aligned with the smart city paradigm (Battarra et al. 2016a, b). For example, Cagliari and Bari, which already show a positive trend in many environmental indicators compared to other southern cities (both in the Energy and Resources Management categories), several initiatives in the Energy category have been promoted, characterized not only by clear vision and objectives, but above all by the involvement of important stakeholders, which are important factors for their success and credibility. In other cases (Naples, Catania and Palermo) it is mainly in the field of research that a number of projects have been developed; if they pass the experimentation phase, they could have important impacts on the city, especially with regards to the Resource management category.

This reveals that in light of the risks highlighted by some of the literature, regarding the increased differences between more advanced contexts, where ICTs find fertile ground for their diffusion, and less advanced cities, which end up being excluded from this "new urban renaissance" (The European House-Ambrosetti 2012), the key (especially for southern cities) to addressing the environmental issues that affect them should not be the implementation of an "ideal model" of smart city focused on the use of sophisticated technological devices that cost more than their often ephemeral benefits; rather, there is a need to define a clear vision of the city's smart evolution, which starts with the specificities of local contexts and, equipped with the appropriate planning tools that, for large cities, will have to focus on a metropolitan scale, involving all social components in the innovation process of its urban systems (Pettit et al.2017).

Acknowledgements The authors have made a joint contribution to the paper's conception and design, background and methodological approach to the problem. Rosaria Battarra designed the results on experiments underway in the Smart Environment dimension and concluding remarks (§§ 4.2 and 5). Fulvia Pinto developed the section on urban and environmental characteristics of the Italian Metropolitan Cities (§ 3) and Maria Rosa Tremiterra designed the knowledge framework and the Environment dimension's status of Italian cities (§§ 1 and 4.1).

References

Ahvenniemi H, Huovila A, Pinto-Seppä I, Airaksinen M (2017) What are the differences between sustainable and smart cities? Cities 60:234–245, Part A. https://doi.org/10.1016/j.cities.2016.09.009

Albino V, Berardi U, Dangelico RM (2015) Smart cities: definitions, dimensions, performance, and initiatives. J Urban Technol 22(1). https://doi.org/10.1080/10630732.2014.942092

Anthopoulos LG (2015) Understanding the smart city domain: a literature review. In: Rodríguez-Bolívar MP (ed) Transforming city governments for successful smart cities, public administration and information technology 8. Springer International Publishing, Switzerland. https://doi.org/10.1007/978-3-319-03167-5_2

Baron M (2012) Do we need smart cities for resilience? J Econ Manage 10:32–46. https://s3.amazonaws.com/academia.edu.documents/32597350/Journal_010.pdf?AWSAccessKeyId=AKIAIWOWYYGZ2Y53UL3A&Expires=1514986788&Signature=x%2FUl%2B7dg5TkOfl%2BqGIEbiBt0kSk%3D&response-content-disposition=inline%3B%20filename%3DPublic_Space_Development_in_the_Context.pdf#page=31. (Last Access 03/01/2018)

Barrionuevo JM, Berrone P, Ricart JE (2012) Smart cities, sustainable progress. IESE Insight 14:50–57. http://www.iese.edu/en/files/Smart%20cities%20sustainable%20progress_tcm4-87325.pdf. (Last Access 03/01/2018)

Battarra R, Fistola R, La Rocca RA (2016) City Smartness: the energy dimension of the urban system. In Papa R, Fistola R (eds) Smart energy in the smart city. Urban planning for a sustainable future. Springer International Publishing, Switzerland. ISBN: 978-3-319-31155-5, ISBN: 978-3-319-31157-9 (eBook). https://doi.org/10.1007/978-3-319-31157-9

Battarra R, Gargiulo C, Pappalardo G, Oliva JS, Boiano DA (2016) Planning in the era of information and communication technologies. Discussing the "label: Smart" in South-European cities with environmental and socio-economic challenges. Cities 59:1–7. https://doi.org/10.1016/j.cities.2016.05.007

Batty M, Axhausen KW, Giannotti F, Pozdnoukhov A, Bazzani A, Wachowicz M, Ouzounis G, Portugali Y (2012) Smart cities of the future. Eur Phys J Spec Topics 214:481–518. https://doi.org/10.1140/epjst/e2012-01703-3

Berke PR, Conroy MM (2000) Are we planning for sustainable development? An evaluation of 30 comprehensive plans. J Am Plan Assoc 66(1):21–33. https://doi.org/10.1080/01944360008976081

Bibri SE, Krogstie J (2017a) ICT of the new wave of computing for sustainable urban forms: their big data and context-aware augmented typologies and de-sign concepts. Sustain Cities Soc 32:449–474. https://doi.org/10.1016/j.scs.2017.04.012

Bibri SE, Krogstie J (2017b) Smart sustainable cities of the future: an extensive interdisciplinary literature review. Sustain Cities Soc 31:183–212. https://doi.org/10.1016/j.scs.2017.02.016

Bosch P, Jongeneel S, Rovers V, Neumann H-M, Airaksinen M, Huovila A (2016) Deliverable 1.4 smart city KPIs and related methodology—final. CITYkeys project. http://nws.eurocities.eu/MediaShell/media/D1.4-CITYkeys_D14_Smart_City_KPIs_Final_20160201.pdf. (Last Access 03/01/2018)

Caragliu A, Del Bo C, Nijkamp P (2011) Smart cities in Europe. J Urban Technol 18(2):65–82. https://doi.org/10.1080/10630732.2011.601117

Caves RW, Walshok MG (1999) Adopting innovations in information technology: the California municipal experience. Cities 16(1):3–12. https://doi.org/10.1016/s0264-2751(98)00049-3

City Protocol Society (2015). City anatomy indicators. City protocol agreement CPA-I_ 001-v2 of the city protocol society. http://www.cptf.cityprotocol.org/CPAPR/CPAPR_002_Anatomy_Indicators.pdf. (Last Access 03/01/2018)

Cohen B (2012) What exactly is a smart city? https://www.fastcodesign.com/1680538/what-exactly-is-a-smart-city. (Last Access 03/01/2018)

Chourabi H, Nam T, Walker S, Gil-Garcia JR, Mellouli S, Nahon K, Pardo TA, Scholl HJ (2012) Understanding smart cities: An integrative framework. In: 2012 45th Hawaii international conference on system science (HICSS), pp 2289–2297. IEEE. https://doi.org/10.1109/HICSS.2012.615

Dirks S, Keeling M (2009) A vision of smarter cities: how cities can lead the way into a prosperous and sustainable future. IBM Inst Bus Value. https://www-03.ibm.com/press/attachments/IBV_Smarter_Cities_-_Final.pdf (Last Access 03/01/2018)

European Commission (2011) Cities of tomorrow. Challenges, visions, ways forward. Publications Office of the European Union, Luxembourg. https://doi.org/10.2776/41803

Galderisi A, Mazzeo G, Pinto F (2016) Cities dealing with energy issues and climate-related impacts: approaches, strategies and tools for a sustainable urban development. In Papa R, Fistola R (eds) Smart energy in the smart city. Urban planning for a sustainable future. Springer International Pub-

lishing, Switzerland. ISSN: 1865–3529. https://doi.org/10.1007/978-3-319-31157-9_11. ISBN 978-3-319-31155-5
Giffinger R, Fertner C, Kramar H, Kalasek R, Pichler-Milanovic N, Meijers E (2007) Smart cities: ranking of European medium-sized cities. Centre of Regional Science (SRF), Vienna University of Technology, Vienna, AU. http://www.smart-cities.eu/download/smart_cities_final_report.pdf. (Last Access 03/01/2018)
Gohar A (2016) Urbanization & sustainable development: evolution and contemporary challenges. J Civil Eng Archit Res 3(12).http://www.ethanpublishing.com/index.php?m=content&c=index&a=show&catid=271&id=800
Graham S, Marvin S (2001) Splintering urbanism: networked infrastructures, technological Mobilities and the urban conditions. Routledge, London
Hollands RG (2008) Will the real smart city please stand up? City 12(3):303/320. https://doi.org/10.1080/13604810802479126
Hollands RG (2015) Critical interventions into the corporate smart city. Camb J Reg, Econ Soc 8:61–77. https://doi.org/10.1093/cjres/rsu011
ISPRA (2017) Consumo di suolo, dinamiche territoriali e servizi ecosistemici. http://www.isprambiente.gov.it/public_files/RapportoConsumoSuolo2017_rev.pdf. (Last Access 03/01/2018)
Jabareen Y (2013) Planning the resilient city: concepts and strategies for coping with climate change and environmental risk. Cities 31:220–229. https://doi.org/10.1016/j.cities.2012.05.004
Japanese Institute for Urban Strategies (2010) Global power city index. http://mori-m-foundation.or.jp/wordpress/english/publications2. (Last Access 03/01/2018)
Komninos N (2002) Intelligent cities. Innovation, knowledge systems and digital spaces. Spon Press, London
Kramers A, Höjer M, Lövehagen N, Wangel J (2014) Smart sustainable cities—exploring ICT solutions for reduced energy use in cities. Environ Model Softw 56:52–62. https://doi.org/10.1016/j.envsoft.2013.12.019
Lombardi P, Giordano S, Farouh H, Yousef W (2012) Modelling the smart city performance. Innov: Eur J Social Sci Res 25(2):137–149. https://doi.org/10.1080/13511610.2012.660325
Mahizhnan A (1999) Smart cities: the Singapore case. Cities 16(1):13–18. https://doi.org/10.1016/S0264-2751(98)00050-X
Manville C, Cochrane G, Cave J, Millard J, Pederson JK, Thaarup RK, Liebe A, Wissner M, Massink R, Kotterink B (2014) Mapping smart cities in the EU. European Parliament, Policy Department A: Economic and Scientific Policy, Bruxelles. doi:http://dx.doi.org/10.2861/3408
Massoli P, Mazziotta M, Pareto A, Rinaldelli C (2014) Indici compositi per il BES. http://www.istat.it/it/files/2014/10/Paper_Sessione-IV_Massoli_Mazziotta_Pareto_Rinaldelli.pdf. (Last Access 03/01/2018)
Morelli VG, Weijnen M, Van Bueren E, Wenzler I, De Reuver M, Salvati L (2013) Towards intelligently-sustainable cities. TeMA J Land Use, Mobility Environ 6(1):73–86. doi:http://dx.doi.org/10.6092/1970-9870/1496
Mosannenzadeh F, Bisello A, Vaccaro R, D'Alonzo V, Hunter GW, Vettorato D (2017) Smart energy city development: a story told by urban planners. Cities 64:54–65. https://doi.org/10.1016/j.cities.2017.02.001
Papa R, Gargiulo C, Galderisi A (2013) Towards an urban planners, perspective on smart city. TeMA. J Land Use, Mobility Environ 6(l):517. https://doi.org/10.6092/1970-9870/1536
Papa R, Galderisi A, Vigo Majello M, Saretta E (2015) Smart and resilient cities. A systemic approach for developing cross-sectoral strategies in the face of climate change. TeMA. J Land Use, Mobility Environ 8(1):19–49. doi:http://dx.doi.org/10.6092/1970-9870/2883
Papa R, Gargiulo C, Cristiano M, Di Francesco I, Tulisi A (2015) Less smart more city. TeMA. J Land Use, Mobility Environ 8(2):159–182. https://doi.org/10.6092/1970-9870/3012
Papa R, Gargiulo C, Battarra R (a cura di) (2016) *Città metropolitane e Smart Govemance*. FedOA-Press, Napoli. https://doi.org/10.6093/978-88-6887-005-8

Pettit C, Bakelmun A, Lieske SN, Glackin S, Hargroves KC, Thomson G, Shearer H, Dia H, Newman P (2017) Planning support systems for smart cities. City, Cult Soc. doi:https://doi.org/10.1016/j.ccs.2017.10.002

Pickett STA, Cadenasso ML, Grove JM (2004) Resilient cities: meaning, models, and metaphor for integrating the ecological, socio-economic, and planning realms. Landscape Urban Plan 69:369–384. https://doi.org/10.1016/j.landurbplan.2003.10.035

Pinto F, Maltese I, Mariotti I (2013) Sustainable planning and housing market prices. In Oktay U, Pizzi E, Croce S (eds) Changing needs, adaptive buildings, smart cities. Proceedings of the XXXIX IAHS world congress, 17–20 Sept 2013, Politecnico di Milano, Italy. ISBN: 9788864930138

Ramaprasad A., Sánchez-Ortiz e Syn T. (2017). A unified definition of a smart city. In Janssen M et al (eds.) EGOV 2017 LNCS 10428, pp 13–24. Springer International Publishing, Switzerland. https://doi.org/10.1007/978-3-319-64677-0_2

Smart City Profiles (2013) Ergebnisse. 7.6.2013. http://www.smartcities.at/assets/03-Begleitmassnahmen/SmartCity-PDF-INTRO.pdf. (Last Access 03/01/2018)

Song Q, Li J, Duan H, Yu D, Wang Z (2016) Towards to sustainable energy-efficient city: a case study of Macau. Renew Sustain Energy Rev. https://doi.org/10.1016/j.rser.2016.11.018

The European House-Ambrosetti (2012) Smart Cities in Italia: un'opportunità nello spirito del Rinascimento per una nuova qualità della vita. https://www.ambrosetti.eu

Vanolo A (2017) Politica e cittadinanza nella smart city: alcune riflessioni sugli immaginari della città intelligente. Rivista Geografica Italiana 124(1):1–16

van Winden W, van den Buuse D (2017) Smart city pilot projects: exploring the dimensions and conditions of scaling up. J Urban Technol 24(4):51–72. https://doi.org/10.1080/10630732.2017.1348884

Webb J, Hawkey D, Tingey M (2016) Governing cities for sustainable energy: the UK case. Cities 54:28–35. https://doi.org/10.1016/j.cities.2015.10.014

Wheeler SM (2000) Planning for metropolitan sustainability. J Plann Educ Res 20(2):133–145. https://doi.org/10.1177/0739456X0002000201

Zygiaris S (2013) Smart city reference model: assisting planners to conceptualize the building of smart city innovation ecosystems. J Knowl Econ 4(2):217–231. https://doi.org/10.1007/s13132-012-0089-4

Effect of Urban Greenspaces on Residential Buildings' Energy Consumption: Case Study in a Mediterranean Climate

Carmela Gargiulo, Ahmed Ayad, Andrea Tulisi and Floriana Zucaro

Abstract The paper is part of the scientific research sector concerning the government of urban transformations in order to promote efficiency and reduction of energy consumption in urban areas. In this study, urban greenspaces (green areas) are proposed as a strategy for cities to achieve both urban sustainability and resilience while addressing the issues of energy reduction and climate change adaptation. The study investigated the microclimate impact of greenspaces on the cooling energy needs of residential buildings in Naples, Italy, given different urban fabric characteristics by coupling the microclimate model ENVI-met with the building energy model EnergyPlus. The charts resulted from the study could represent an useful decision support tool for urban planners and policy-makers to locate and size greenspaces based on their effectiveness in terms of energy consumption reduction. The study found that—in general—a medium-size green area (4900 m^2) would reduce the cooling energy consumption by 9.20% which is more than double the effect of a large green area (32,400 m^2).

1 Introduction

Since the 1987 Brundtland Report defined sustainable development and pointed out that cities should be pivotal to this development as the majority of the world's population will live in urban areas (Brundtland 1987), the concepts of sustainable cities and urban sustainability have gained much attention with an increasing interest

C. Gargiulo (✉) · A. Tulisi · F. Zucaro
Department of Civil, Architectural and Environmental Engineering, University of Naples Federico II, Naples, Italy
e-mail: gargiulo@unina.it

A. Ayad
Construction & Building Engineering Department, College of Engineering & Technology, Arab Academy for Science, Technology & Maritime Transport, Alexandria, Egypt
e-mail: ayad@aast.edu

R. Papa et al. (eds.), *Smart Planning: Sustainability and Mobility in the Age of Change*, Green Energy and Technology,
https://doi.org/10.1007/978-3-319-77682-8_7

in the role of cities in addressing global environmental challenges such as climate change (Bulkeley and Betsill 2005).

This role was further emphasized in the 1992 United Nations Conference on Sustainable Development where local authorities worldwide were encouraged to develop plans for applying sustainable development (Bulkeley and Betsill 2005).

In order to achieve urban sustainability, a set of goals and strategies matching the urban context of the city has been developed in collaboration with the practitioners responsible for applying them (National Academies 2016).

Nevertheless, as cities continue to grow, challenges like: climate change, economic fluctuations, etc. started to be increasingly pressing, causing cities to struggle to provide basic urban services to its residents and making achieving urban sustainability alone not enough (Harter et al. 2010; Da Silva and Moench 2014). In this context, the concept of urban resilience has emerged to work hand-in-hand with urban sustainability plans to achieve rational urban development that can help cities adapt and transform in the face of the challenges (Papa et al. 2014a; Zhang and Li 2017).

Therefore, building urban resilience requires a holistic view of the city and its systems to have a good understanding of the interacting relationships between them and both expected and unexpected shocks they might face, in order to prepare cities to be more flexible and have the ability to bounce back in case of acute shocks (Papa et al. 2014b; Gargiulo and Lombardi 2016; 100 Resilient Cities 2017).

This challenge is even more complex if we consider that cities are not just the victims of the consequences of climate change but one of the main causes. In fact, urban environments have a significant impact on climate change since 35% of global greenhouse gas emissions are produced in urban areas (excluding urban transport); one of the main consequences is the global warming (van der Heijden 2014) which emphasizes the negative impact of the UHI phenomenon by increasing the number of heat waves affecting most of the cities worldwide, determining an increased death rates due to heat peaks in urban areas compared to rural areas (Tan et al. 2010; Echevarria Icaza et al. 2016).

Furthermore, urban environments also affect global energy consumption which is expected to increase by 48% in year 2040 compared to 2012 (U.S. EIA 2016). Between 60 and 80% of this consumption happens within cities (Grubler et al. 2013) and is mainly consumed by buildings which is considered the largest energy consuming sector in the world (IEA 2013).

The issues of energy consumption and climate change are strongly related since increased energy consumption in urban areas and the associated CO_2 emissions intensify the greenhouse effect leading to increased warming of the climate which can be considered one of the main reasons for climate change. On the other hand, increased global warming due to climate change will cause a bigger energy demand mainly due to an increased cooling needs (Sharifi and Yamagata 2016).

In light of the previous discussion on the challenges facing the world and the recognition of the need for a more sustainable city that reduces its impact on the environment while being more resilient to natural and man-made stresses, there is an urgent need to facilitate the transition of cities towards less dependence on energy and other resources and producing fewer greenhouse gases (van der Heijden 2014).

This transition can only be achieved through, clear goals and specific—applicable—strategies (Papa et al. 2015). In this context, urban greenspaces (green areas) such as parks, street trees, etc. can be proposed as a strategy for cities to achieve both urban sustainability and urban resilience while addressing the issues of energy and climate change.

In a previous study done by the authors (Gargiulo et al. 2016, 2017), the urban greenspace cooling effect and its influence on the urban microclimate was investigated as a proposed mitigation measure to UHI and climate change. In that study, the main goal was to establish greenspace dimensions threshold values which influence the urban microclimate by lowering the temperature given the challenging space constraints provided by the need for a more compact city.

In the current study, this work was further extended by adding the dimension of building energy consumption with the aim to investigate the impact of the cooling effect of urban greenspaces on the energy consumption of the surrounding buildings, given different building density and green area size contexts. This study—along with the previously mentioned study—can provide a more comprehensive evaluation of the effectiveness of urban greenspaces as a mitigation measure to both climate change and global energy consumption increase which can help to derive planning and design guidelines for urban greenspaces and verify its efficiency in achieving urban sustainability and resilience. The study focuses on the cooling energy needs for the residential buildings in the Municipality of Naples (referred to in the rest of the study as Naples), the capital of the Metropolitan City of Naples located in Campania region in the south of Italy which is characterized by a Mediterranean hot climate. The choice to focus only on residential buildings was due to the fact that they consume about three-quarters of the total global energy used in the buildings sector (the largest energy consuming sector worldwide) (IEA 2013). Also, while heating needs dominate the residential energy use worldwide, cooling needs was chosen as the focus of this study since it is projected to more than double worldwide by 2050 (IEA 2013) and also due to the hot nature of Naples climate which requires more cooling than heating (Palombo et al. 2012).

2 State of the Art

Various approaches have been used to link the microclimate effect of green areas and the energy consumption of buildings. These approaches include: descriptive case studies, mathematical modelling, analytical modelling, empirical modelling, and remote sensing (Skelhorn 2013; Papa et al. 2014c). Since mathematical modelling was chosen as the methodology for this study (see Sect. 3), only these studies were reviewed with a focus on the ones that utilized the coupling of a microclimate model with a building energy model.

By reviewing literature, it appeared that very few studies have used modeling-based methodologies supported by the coupling of microclimate and building energy models to investigate the impacts of the cooling effect of urban greenspaces on building energy consumption. Among these studies, (Akbari and Konopacki 2005) produced summary tables to estimate the potential of heat island reduction (HIR) strategies (i.e., solar-reflective roofs, shade trees, reflective pavements, and urban vegetation) to reduce cooling energy use in buildings in all U.S. cities based on a meteorological simulation model developed by Taha et al. (2000) and the DOE-2.1E building energy simulation model. Although this study was comprehensive as it addressed all U.S. cities and presented the results in tabular formats for easy interpolation, one drawback was that the effect of urban vegetation alone was not identified because the study assumed that all HIR measures have been fully implemented. The same methodology of including vegetation among a set of other HIR measures was followed in (Bouyer et al. 2011) where a dynamic simulation platform was developed and tested on a real urban context to study the effect of vegetation (shading and cooling effect), nature of the soil and building coatings on the energy consumption of a theoretical office building with no particular focus on the effect of vegetation alone. The platform developed relied on the coupling of: (1) Fluent, a commercial CFD software and (2) Solene, a simulation tool to compute a comprehensive thermoradiative balance of the urban surfaces with a building thermal sub-model developed by the authors.

As for the studies that investigated the effect of vegetation—alone—on building energy consumption, (Gros et al. 2016) evaluated the effect of two cooling strategies: vegetation (trees, green walls and roofs) and high albedo values (cool roofs and façades) on the cooling energy demand of two urban blocks located in Part-Dieu, in Lyon (France). Two indices, Energy Performance Index (EPI) and Ambient Temperature Mitigation Index (ATMI), were defined to evaluate the efficiency of each strategy. The evaluation was done through the coupling of microclimate and building energy models (EnviBatE, SOLENE & QUIC). Although the study was unique in its approach of defining indices to evaluate the efficiency of vegetation, the study used sophisticated software and models that cannot be easily used by urban planners. Another study that investigated the effect of vegetation in isolation was (Skelhorn et al. 2016) in which the impact of shading, air temperature, and wind effects of trees on commercial buildings' cooling energy consumption in Manchester, UK was studied. The study modeled microclimate changes due to different types of greenspace and then used the results in combination with measured UHI intensity results to develop customized weather files for building energy modeling. The used models were ENVI-met for microclimate and IES-VE for building energy simulation. One issue of this study was the quality of data inputs for the model since numerous datasets were used that were not developed by the authors and hence complete accuracy cannot be guaranteed.

It was interesting to see that only two studies coupled the ENVI-met microclimate model with the EnergyPlus building energy simulation model despite each of these models is considered the best in its field (Crawley et al. 2008; Roset and Vidmar 2013). One of these studies (Pastore et al. 2017) explored the benefits of vegetation for microclimate mitigation and thermal comfort improvement in a case study of neighborhood renewal design in the city of Palermo (Italy). The study focused only on the effect of vegetation on residential building's indoor thermal comfort without considering its effects on the building's energy consumption. The other study (Fahmy et al. 2017) investigated how the application of green cover (trees, green roofs and green walls) can affect the residential buildings' energy consumption in present and future by conducting two case studies located in different climatic zones in Egypt (New Borg El-Arab and 6th of October cities). One innovative aspect of the study was that the weather data used to simulate case studies' site conditions were developed not only for the present but also for the future (2020, 2050 and 2080) using the Climate Change World Weather Generator tool in order to verify the efficiency of green cover as an adaptation measure of Egyptian urban communities for climate change. Also, pedestrian thermal comfort (PMV) and air temperature (Ta) maps were developed for both case studies. While the study showed innovative approach for investigating the relationship between vegetation and building energy consumption by including the dimension of climate change, the focus of the study on real case studies that represent specific urban contexts make it difficult to generalize the results and derive guidelines from it.

In the context of the reviewed literature, the research presented in this paper contributes to the knowledge of the impacts of the cooling effect of urban greenspaces on residential buildings' energy consumption by providing simple design guidelines (in the form of charts and general rules of thumb) that can empower urban planners and policy makers to make decisions related to the appropriate green area size and location for a certain urban context with taking into account the effects of that decision on the energy consumption of the surrounding residential buildings. To the authors' knowledge, this study is the first to address the impacts of vegetation on building's energy consumption in a Mediterranean climate and the only one that investigated the combined effects of building density and green area size on building's energy consumption.

3 Methodology

Empirical studies and modelling are frequently used in investigating the effect of urban greenspaces on building energy consumption (Skelhorn 2013). Given the aim of this study, modelling was more appropriate as it allowed the flexibility to change the size of the green area and the building density surrounding it. This wouldn't be possible in an empirical study because an empirical study is restricted to the characteristics of the case study selected for measurement. Also, choosing modeling facilitated the connection to the previous research work done by the authors (Gargiulo et al. 2016) upon which this study was based. A four-step modeling-based methodology was developed by coupling the microclimate model ENVI-met with the building energy model EnergyPlus since there is no software package that is capable of modeling both the microclimate effect of vegetation and the impact of that effect on building's energy consumption (Skelhorn 2013).

3.1 Step 1: Identifying a Typical Neapolitan Residential Building

It was necessary to start the study by investigating the residential building stock in Naples in order to come up with a conceptual building representative of that stock to be used in building energy simulations. Based on the classification of Italian residential buildings into three main categories: Single-Family Houses, Terraced Houses, Multi-Family Houses and Apartment Blocks proposed by (Corrado et al. 2012), It was assumed that the conceptual building used in this study was an apartment block. Then, the general characteristics of the building in terms of the building type (reinforced concrete or load-bearing wall) and the construction period were investigated.

Building Type Although 51.13% of the residential buildings in Naples are constructed using the load-bearing wall system (ISTAT 2011), the conceptual building was assumed to be built using reinforced concrete (RC). This assumption was based on that, while the total number of load-bearing wall residential buildings is higher than the RC buildings, since 1946 the number of RC buildings built each year is higher than that of the load-bearing wall buildings (Fig. 1) which reflects the tendency of the Neapolitan building stock to abandon the load-bearing wall buildings and shift towards the RC buildings.

Construction Period Around half of the RC residential buildings in Naples (49.8%) were built in the period from 1961 to 1980 (ISTAT 2011) and hence it was assumed to be the construction period of the conceptual building.

After the general characteristics of the conceptual building were determined, it was then necessary to determine the characteristics of each apartment in the building. The most important characteristic was the area of the apartment. More than half of the apartments in Naples (51.1%) have an area from 60 m^2 up to 99 m^2 (ISTAT 2011) and so the apartments in the conceptual building were assumed to have an average area of 80 m^2. For the number of rooms, the highest percentage of apartments in Naples have 4 rooms (ISTAT 2011) and hence it was assumed that each apartment in the conceptual building has 4 rooms that serve the basic needs of the occupants (i.e. bedroom, living room, bathroom and kitchen).

3.2 Step 2: Creating Building Energy Model

In this step, a building energy model was created for the conceptual building identified in Sect. 3.1 using DesignBuilder software. DesignBuilder is a user-friendly modelling

Fig. 1 The construction trend of for the load-bearing and RC residential buildings in Naples (ISTAT 2011)

environment that provides various environmental performance data such as: energy consumption, carbon emissions, comfort conditions … etc. (DesignBuilder 2017). DesignBuilder is based on the building energy simulation engine EnergyPlus which was developed by the Lawrence Berkeley National Laboratory and others for the U.S. Department of Energy and is considered the most accurate and bug-free energy simulation engine available due to the various tests and industry standards it passed (EnergyPlus 2017). In this study DesignBuilder version 5.0.3.007 was used which was equipped with EnergyPlus version 8.5.

Building Geometry The first part of the model is to define the building geometry. Because this study is based on the microclimate model conducted in (Gargiulo et al. 2016), the dimensions of the building energy model had to be the same as the ones used in the microclimate model. The dimension of the building in the microclimate model was 20 m × 20 m. In this study, the 20 m × 20 m area was assumed to contain two similar adjacent buildings to reduce in the running time of the model as only one building will be simulated and its results will be multiplied by two. As for the height of the building, it was dependent on the building density with the low building density being represented by a nine-meter-high building (three floors) while the medium and high building densities were represented by a 24-meter-high building (eight floors) according to (Gargiulo et al. 2016). Figure 2 shows the two building energy models used to represent different building densities with each model containing two adjacent buildings, while Fig. 3 shows a floor plan for one of the buildings showing the different rooms (zones).

Building Orientation To determine the building orientation, multiple energy simulation trials were performed on the two building models shown in Fig. 3 to test the different alternatives available (north-facing building, south-facing building,

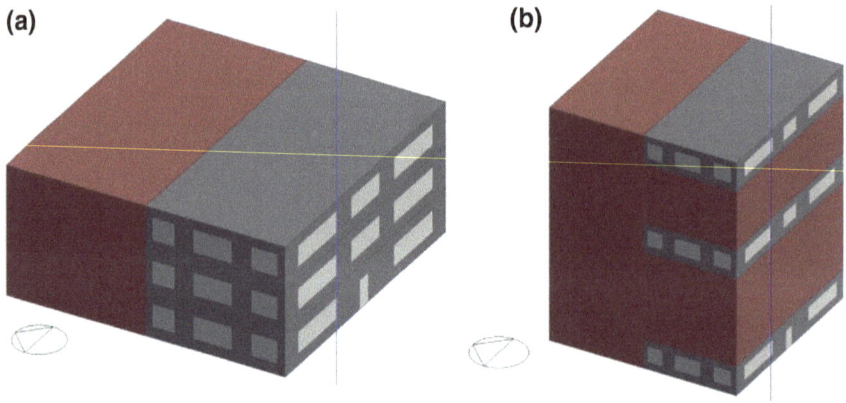

Fig. 2 **a** Nine-meter-high building (three floors) to represent low building density, **b** 24-meter-high building (eight floors) to represent medium and high building densities

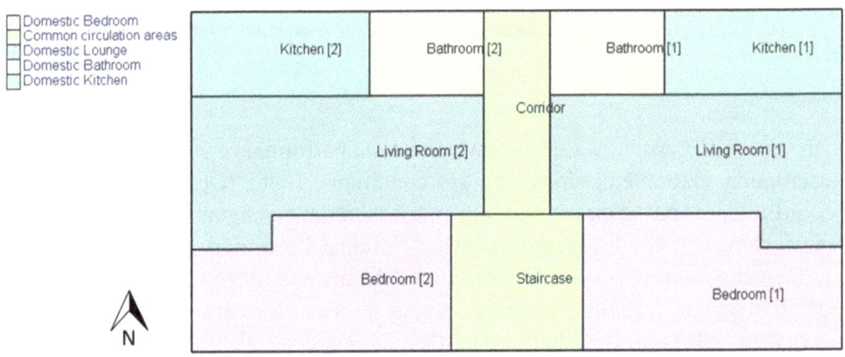

Fig. 3 Floor plan of the building used in energy simulation

east-facing building and west-facing building). These trials have indicated that the highest energy consumption in summer (the period considered for building energy simulations in this study) is for the south-facing building and hence the south orientation was chosen as the worst-case scenario upon which the influence of the cooling effect of green areas on the building's energy consumption will be much apparent.

Building's Location Relative to The Green Area Each green area has a maximum cooling distance—which is dependent upon the size of the green area—beyond which the effect of the green areas to reduce the air temperature is not sensible (Gargiulo et al. 2016). So, it was necessary to locate the building at a distance within the maximum cooling distances of the different green area sizes included in this study (see Sect. 3.3 for the classification of the green area sizes used in the study). In this study, it was assumed that the modeled building was located 50 m away from the green area. This distance is less than the maximum cooling distance of all the green area sizes considered in the study and hence the building was within the cooling effect of the green area.

Model Parameters The data warehouse of the 2011 population and housing census produced by the Italian National Institute of Statistics (ISTAT), Naples Sustainable Energy Action Plan (SEAP)—Annex B, and the National Scientific Report on the TABULA activities in Italy (ISTAT 2011; Corrado et al. 2012; Palombo et al. 2012) were used to define building occupancy, the elements of the building envelop, and the HVAC system used while the Italian standards (UNI/TS 11300-1:2014) and (UNI EN 15251:2008) were used to define the indoor environmental conditions (temperatures, ventilation rates … etc.).

Model Calibration To keep the model as simple as possible while still maintaining sufficient accuracy and since the objective of this study was to conduct a comparative energy modeling to report the percentage of savings in building energy use due to the cooling effect of urban greenspaces with no interest in the actual energy use of the building in absolute terms (predictive energy modeling), a forward engineering model was created using DesignBuilder software. The forward engineering model eliminates the need for calibration as its results are validated through the verification status of the simulation engine used (EnergyPlus) and is more suitable for comparative energy modeling (IBPSA-USA 2012; Fumo 2014).

3.3 Step 3: Utilizing TeMALab Urban Microclimate Model for Naples

The microclimate model discussed in (Gargiulo et al. 2016) was utilized in this study. This model was conducted for Naples by the authors as a part of *Smart Energy Master for the energy governance of the territory (SEM)* research project undertaken by the Laboratory of Territory, Mobility and Environment (TeMALab) in University of Naples Federico II, Italy (Papa et al. 2016). In (Gargiulo et al. 2016), the study area was divided into three main urban fabric typologies representing low building density (2.084 m^3/m^2), medium building density (8.482 m^3/m^2) and high building density (13.280 m^3/m^2) areas. Also, the green areas located in the study area were divided into small green areas (900 m^2), medium green areas (4900 m^2) and large green areas (32,400 m^2). This classification resulted in nine different scenarios that were simulated in ENVI-met software. In this study, the results of (Gargiulo et al. 2016) were used to calculate the reduction in air temperature due to the cooling effect of green areas (*ΔT for Building Energy Simulation*). Then—for each modeling scenario—the relevant temperature difference (*ΔT for Building Energy Simulation*) was subtracted from each hourly value in the original weather file for Naples and a set of nine weather files were created using the "Weather Data Analysis Tool" available in DesignBuilder software. It should be noted that this process was based on the assumption that *ΔT for Building Energy Simulation* was the same for each

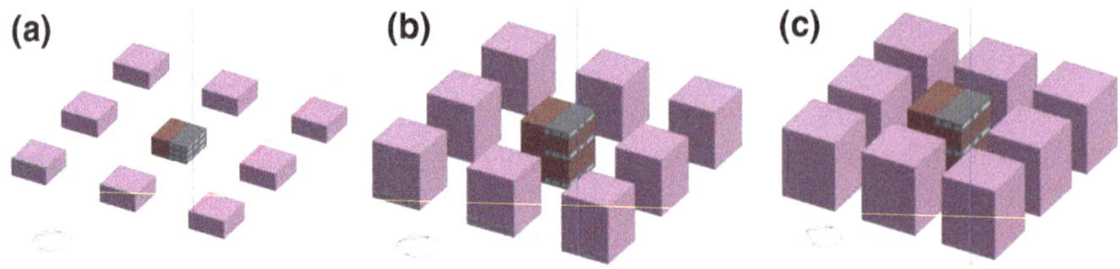

Fig. 4 Building energy models created in DesignBuilder: **a** low building density, **b** medium building density, **c** high building density

hour in the day for each day in the year which is not true as the it should vary from hour to hour but this assumption was made for simplification.

3.4 Step 4: Building Energy Simulation Scenarios

In this step, the coupling of the microclimate model ENVI-met with the building energy simulation model EnergyPlus was achieved. The coupling procedure was performed in the following two steps:

1. Three building energy models that represent different building densities were created in DesignBuilder software as shown in Fig. 4.
2. Nine building energy simulation scenarios—that reflect the different combinations between building density and green area size—were run using DesignBuilder software. Each of these scenarios utilized one of the building energy models to represent the building density (in terms of the shading effect) and one of the weather files created in Step 3 (see Sect. 3.3) to represent the effect of the green area size and the building density on the outdoor air temperature. In addition to these nine scenarios, three other scenarios that reflect the different building densities were run using the original weather file for Naples to calculate the energy consumption of each scenario in the absence of the green areas.

In Naples, the HVAC system is normally switched on in the period between the 1st of June and the 15th of September (Palombo et al. 2012) and hence this was chosen as the period for running the building energy simulation scenarios.

4 Results & Discussion

Since a *forward engineering model* was chosen for the building energy simulations (as discussed in Sect. 3.2), it was necessary to present the results in a comparative form that report the percentage of savings in cooling energy use due to the cooling effect of the green areas as shown in Table 1. In order to calculate these saving percentages,

Table 1 Reduction in cooling energy consumption due to the cooling effect of urban greenspaces

Modelling scenario		Cooling energy consumption reduction (%)
L.B.D.[a]	S.G.A.[b]	2.05
	M.G.A.	3.72
	L.G.A.	6.50
M.B.D.	S.G.A.	2.18
	M.G.A.	3.68
	L.G.A.	5.96
H.B.D.	S.G.A.	0.38
	M.G.A.	16.72
	L.G.A.	3.66

[a]L.B.D.: Low building density/M.B.D.: Medium building density/H.B.D.: High building density
[b]S.G.A.: Small green area/M.G.A.: Medium green Area/L.G.A.: Large green area

the cooling energy consumption of each modeling scenario was compared to the base case scenario (with no green area).

We can see from Table 1 that in both low and medium building density scenarios, the larger the green area, the more the effect it has on reducing the energy consumption. However—in high building density scenario—the medium green area has significantly higher energy reduction effect compared to the small and large green areas. These interesting findings indicate that there is a relationship between the effect of the green area on reducing the cooling energy consumption of the buildings and the urban morphology in which the green area is located. In order to understand this relationship, it was necessary to include the distance from the green area as an important factor in the analysis.

By referring to the reductions in cooling energy consumption reported in Table 1, it is important to note that these reductions were based on the assumption that the modeled building was located at a fixed distance of 50 m away from the green area (see Sect. 3.2). In order to see the effect of varying this distance, two possible methods were proposed: the first method was to rerun the nine building energy simulation scenarios where each scenario would be run multiple times at different distances from the green area (a different weather file—reflecting the distance—will be used in each time). The second method was to predict the reduction in cooling energy consumption when the distance from the green area is varied using statistical analysis. In order to vary the distance from the green area at a minimum of 5 m steps to provide sufficient accuracy, the simulations would be run more than 200 times and thus the first method was considered impracticable due to the huge amount of time, effort and computing power needed to rerun the simulations and the second method was used.

In the second method, the relationship between the reductions in cooling energy consumption reported in Table 1 and the reductions in outside air temperature (ΔT *for Building Energy Simulation*) calculated in Sect. 3.3 was analyzed using regression analysis for each building density. In all building densities, the relationship was linear with an R-squared value of 0.99 for low building density and 1.00 for medium and high building densities. The distance from the green area was varied at 5 m intervals and the corresponding *ΔT for Building Energy Simulation* was calculated using the microclimate model results from (Gargiulo et al. 2016). Then, the reductions in cooling energy consumption expected to result from each *ΔT for Building Energy Simulation* were calculated based on the linear regression equations. Figures 5, 6 and 7 show the relationship between the reduction in cooling energy consumption and the distance from the green area in each building density.

From the figures, we can see the strong influence the distance from the green area has on reducing the cooling energy consumption. In low building density (Fig. 5), the results show that up to 32.7 m distance, the medium green area has higher effect in reducing the cooling energy consumption than the large green area. This is reversed for distances more than 32.7 m. Regardless of the distance, the small green area always has the lowest effect compared to the other sizes.

In medium building density (Fig. 6), it is obvious that the large green area has a higher effect than the medium green area regardless of the distance. This effect remains up to a distance of 160 m away from the green area—after which the large green area has no effect. The small green area was neglected from the analysis because it has nearly no impact on lowering the air temperature as reported in (Gargiulo et al. 2016) and hence it would have no effect on reducing the cooling energy consumption. In high building density (Fig. 7), the medium green area appears to has significantly higher cooling energy consumption reduction effect compared to both large and small green areas. This remains true up to a distance of 115 m away from the green area after which the large green area has a higher effect but this effect remains only to a distance of 190 m.

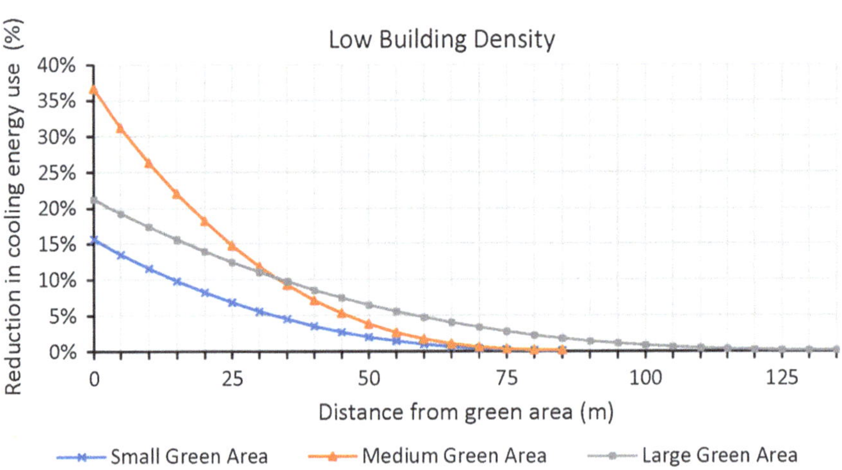

Fig. 5 Relationship between the reduction in cooling energy use and the distance from the green area for different green area sizes located in a low building density

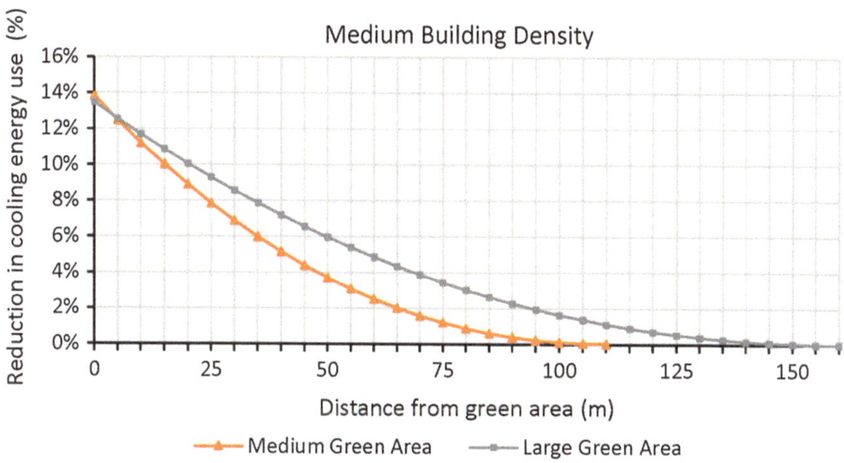

Fig. 6 Relationship between the reduction in cooling energy use and the distance from the green area for different green area sizes located in a medium building density

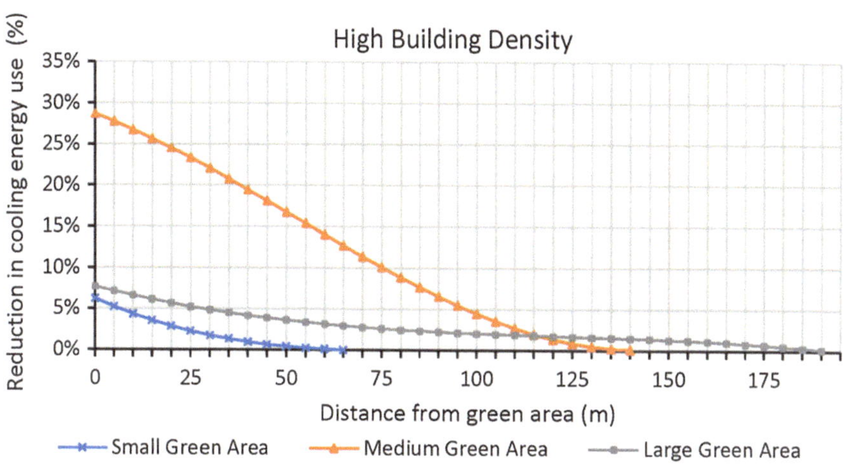

Fig. 7 Relationship between the reduction in cooling energy use and the distance from the green area for different green area sizes located in a high building density

By averaging the results for all the building densities and all the possible distances from the green area, the medium green area appears to has the highest effect in reducing the cooling energy consumption (9.20%) followed by the large green area (4.43%) and the small green area (2.98%) as shown in Fig. 8.

The results of this study conforms with the results of (Gargiulo et al. 2016) in terms of the influence range that each green area size has and in that—in general—the medium green area has the highest effect followed by the large and small green areas respectively. But in terms of which green area size has the highest effect and which has the lowest when the distance is varied, it appears at first glance that

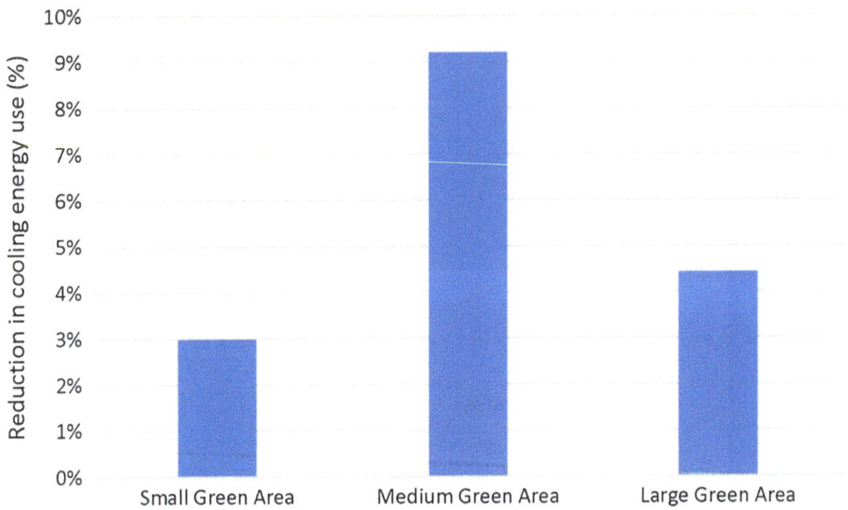

Fig. 8 Percentage reduction in cooling energy use due to the cooling effect of different green area sizes

there is a discrepancy between the results of the two studies (for example, the results of (Gargiulo et al. 2016) showed a higher effect for the medium green area located in a medium building density when compared to a large green area located in the same density which is reversed in the results of the current study). This discrepancy can be explained by referring to that each study looks at the topic from a different perspective, while (Gargiulo et al. 2016) investigated the cooling effect of the green area in terms of the difference between the air temperature and the green area temperature to determine the maximum influence range of each green area size, this study investigated the same effect but in terms of the difference in air temperature in the presence and absence of the green area and the effect of that on the building's cooling energy consumption. So, there is no conflict between the results of the two studies but rather each study integrates with the other to give a better understanding of the complex relationship between green area size, building energy consumption, and urban morphology.

5 Conclusion

This study was aimed at investigating the effect of urban green areas on the energy consumption of the surrounding residential buildings. One important aspect of the study was the inclusion of the urban morphology defined by building density and building characteristics in Naples as an important factor in the investigation. Also, the study proposed an innovative technique for the investigation by using a modeling-based methodology which facilitated the analysis which would be difficult to achieve using empirical studies.

The charts introduced in this study (Figs. 5, 6 and 7)—along with the charts from (Gargiulo et al. 2016)—can help in defining standards and guidelines for Naples to locate and size new green areas taking into consideration the effects on the urban microclimate and on the residential buildings' energy consumption as these charts can be easily used to select the optimal green area size given its distance from the built environment and the building density in which it would be located. Also, the study introduced a general rule of thumb for the optimal green area size in Naples regardless of the building density or the distance from the green area by confirming that using a medium-size green area (4900 m^2) would reduce the cooling energy consumption of the surrounding residential buildings by 9.20% which is more than twice as much as the effect of a large green area (32,400 m^2). This general rule would be useful in the initial urban planning stages where the urban fabric is still not clearly defined.

The complex nature of the interactions between the processes of evapotranspiration and the numerous physical characteristics of urban areas, suggests further development for this research work, by investigating how changes in the climate type and the urban morphology would affect the relationship between the green areas and the energy consumption of the residential buildings; for this reason, it is planned to extend this study in the future by conducting new case studies in different cities with different climate conditions and different urban fabric than Naples.

Acknowledgements The work was carried out under the scientific and methodological guidance of C. Gargiulo. Sharing objectives, approaches and method of work, Ayad wrote § 3, Gargiulo § 4, Tulisi § 2 and Zucaro §§1 and 5.

References

Resilient Cities (2017) Urban resilience. http://www.100resilientcities.org/resources. Accessed 23 Sep 2017

Akbari H, Konopacki S (2005) Calculating energy-saving potentials of heat-island reduction strategies. Energy Policy. https://doi.org/10.1016/j.enpol.2003.10.001

Bouyer J, Inard C, Musy M (2011) Microclimatic coupling as a solution to improve building energy simulation in an urban context. Energy Build. https://doi.org/10.1016/j.enbuild.2011.02.010

Brundtland GH (1987) Our common future: report of the world commission on environment and development. United Nations Comm 4:300. https://doi.org/10.1080/07488008808408783

Bulkeley H, Betsill MM (2005) Rethinking sustainable cities: multilevel governance and the "urban" politics of climate change. Env Polit 14:42–63. https://doi.org/10.1080/0964401042000310178

Corrado V, Ballarini I, Corgnati SP, Dipartimento T (2012) Typology approach for building stock national scientific report on the TABULA activities in Italy

Crawley DB, Hand JW, Kummert M, Griffith BT (2008) Contrasting the capabilities of building energy performance simulation programs. Build Environ 43:661–673. https://doi.org/10.1016/j.buildenv.2006.10.027

Da Silva J, Moench M (2014) City resilience framework. Arup 2014. http://www.seachangecop.org/files/documents/URF_Bo

DesignBuilder (2017) DesignBuilder Help v5.0. https://www.designbuilder.co.uk/helpv5.0/. Accessed 24 Sep 2017
Echevarria Icaza L, Van Der Hoeven F, Van Den Dobbelsteen A (2016) Surface thermal analysis of North Brabant cities and neighbourhoods during heat waves. Tema J L Use, Mobil Environ 9:63–87. https://doi.org/10.6092/1970-9870/3741
EnergyPlus (2017) Testing and validation. https://energyplus.net/testing. Accessed 19 Sep 2017
Fahmy M, El-Hady H, Mahdy M, Abdelalim MF (2017) On the green adaptation of urban developments in Egypt; predicting community future energy efficiency using coupled outdoor-indoor simulations. Energy Build 153:241–261. https://doi.org/10.1016/j.enbuild.2017.08.008
Fumo N (2014) A review on the basics of building energy estimation. Renew Sustain Energy Rev 31:53–60. https://doi.org/10.1016/j.rser.2013.11.040
Gargiulo C, Lombardi C (2016) Urban retrofit and resilience: the challenge of energy efficiency and vulnerability. Tema J L Use, Mobil Environ 9:137–162. https://doi.org/10.6092/1970-9870/3922
Gargiulo C, Tulisi A, Zucaro F (2016) Small green areas for energy saving: effects on different urban settlements. ACE Archit City Environ. https://doi.org/10.5821/ace.11.32.4659
Gargiulo C, Tulisi A, Zucaro F (2017) Chapter 11: climate change-oriented urban green network design: a decision support tool. In: Gakis K, Pardalos P (eds) Network design and optimization for smart cities, world scientific, series on computers and operations research: volume 8, ISBN: 978-981-3200-00-5 (hardcover), ISBN: 978-981-3200-02-9 (ebook), http://www.worldscientific.com/doi/abs/10.1142/9789813200012_0011
Gros A, Bozonnet E, Inard C, Musy M (2016) A new performance indicator to assess building and district cooling strategies. In: Procedia engineering
Grubler A, Bai X, Buettner T, et al (2013) Urban energy systems. Glob Energy Assess Towar a Sustain Futur 1307–1400. https://doi.org/10.4324/9780203066782
Harter G, Sinha J, Sharma A, Dave S (2010) Sustainable urbanization: the role of ICT in city development. Booz Co Inc 1–13
van der Heijden J (2014) Governance for urban sustainability and resilience: responding to climate change and the relevance of the built environment. Edward Elgar, UK
IBPSA-USA (2012) BEMBook wiki: energy modeling methodology (Comparison vs. Prediction). http://bembook.ibpsa.us/index.php/Comparison_vs._Prediction. Accessed 19 Sep 2017
IEA (2013) Transition to sustainable buildings—strategies and opportunities to 2050
ISTAT (2011) Data warehouse of the 2011 population and housing census. http://dati-censimentopopolazione.istat.it
Academies National (2016) Pathways to urban sustainability: challenges and opportunities for the United States. The National Academies Press, Washington, DC
Palombo A, Buonomano A, De Luca G (2012) Relazione specialistica sull'efficienza energetica negli edifici (climatizzazione e acqua calda sanitaria)
Papa R, Gargiulo C, Cristiano M et al (2015) Less Smart More City. Tema J L Use, Mobil Environ 8:159–182. https://doi.org/10.6092/1970-9870/3012
Papa R, Gargiulo C, Zucaro F (2014a) Climate change and energy sustainability. Which innovations in European strategies and plans. Tema J L Use, Mobil Environ. https://doi.org/10.6092/1970-9870/2554
Papa R, Gargiulo C, Zucaro F (2014b) Urban systems and energy consumptions: a critical approach. Tema J L Use, Mobil Environ. https://doi.org/10.6092/1970-9870/2552
Papa R, Gargiulo C, Carpentieri G (2014c) Integrated urban system and energy consumption model: residential buildings. Tema J L Use, Mobil Environ. https://doi.org/10.6092/1970-9870/2473
Papa R, Gargiulo C, Zucaro F (2016) Towards the definition of the urban saving energy model (UrbanSEM). Springer, Berlin, pp 151–175
Pastore L, Corrao R, Heiselberg PK (2017) The effects of vegetation on indoor thermal comfort: the application of a multi-scale simulation methodology on a residential neighborhood renovation case study. Energy Build. https://doi.org/10.1016/j.enbuild.2017.04.022

Roset J, Vidmar J (2013) Evaluation of simulation tools for assessment of urban form based on physical performance. Environ Phys Archit 18

Sharifi A, Yamagata Y (2016) Principles and criteria for assessing urban energy resilience: a literature review. Renew Sustain Energy Rev 60:1654–1677. https://doi.org/10.1016/j.rser.2016.03.028

Skelhorn C (2013) A fine scale assessment of urban greenspace impacts on microclimate and building energy in Manchester. The University of Manchester

Skelhorn CP, Levermore G, Lindley SJ (2016) Impacts on cooling energy consumption due to the UHI and vegetation changes in Manchester, UK. Energy Build. https://doi.org/10.1016/j.enbuild.2016.01.035

Taha H, Chieh Chang S, Akbari H (2000) Meteological and air quality impacts of heat island mitigation measures in three U.S. Cities, pp 1–61

Tan J, Zheng Y, Tang X et al (2010) The urban heat island and its impact on heat waves and human health in Shanghai. Int J Biometeorol 54:75–84. https://doi.org/10.1007/s00484-009-0256-x

U.S. EIA (2016) International energy outlook 2016

UNI/TS 11300-1:2014 Energy performance of buildings—part 1: evaluation of energy need for space heating and cooling

UNI EN 15251:2008 Indoor environmental input parameters for design and assessment of energy performance of buildings addressing indoor air quality, thermal environment, lighting and acoustics

Zhang X, Li H (2017) Urban resilience and urban sustainability: what we know and what do not know? Cities 72:141–148. https://doi.org/10.1016/j.cities.2017.08.009

Carmela Gargiulo Full professor of Urban Planning Techniques at the University of Naples Federico II. Since 1987 she has been involved in studies on the management of urban and territorial transformations. Since 2004, she has been Member of the Researcher Doctorate in Hydraulic, Transport and Territorial Systems Engineering of the University of Naples "Federico II". She has been Member of the Committee of the Civil, Architectural and Environmental Engineering Department of the University of Naples "Federico II". Her research interests focus on the processes of urban requalification, on relationships between urban transformations and mobility, on the estate exploitation produced by urban transformations, and on the complex connections between land use, energy consumption and climate change. On these subjects she has co-ordinated research teams within National Project such as Progetto Finalizzato Edilizia—Sottoprogetto "Processi e procedure" (Targeted Project on Building—Subproject "Processes and procedures"), from 1992 to 1994; Progetto Strategico Aree Metropolitane e Ambiente, (Strategic Project Metropolitan Areas and Environment) from 1994 to 1995; PRIN project on the "Impacts of mobility policies on urban transformability, environment and property market" from 2011 to 2013; Project Smart Energy Master for the energy management of territory financed by PON 04A2_00120 R&C Axis II, from 2012 to 2015; Project "Mobilage" 2018-2021, financed by Cariplo Foundation. She has been editor of the Scientific Journal TeMA—Land Use, Mobility and Environment since 2007. She is author of more than 120 publications.

Integrating Green Infrastructure and Ecological Corridors: A Study Concerning the Metropolitan Area of Cagliari (Italy)

Ignazio Cannas, Sabrina Lai, Federica Leone and Corrado Zoppi

Abstract Green Infrastructure (GI) is defined as a network of natural and semi-natural areas that need to be planned in a strategic way to deliver ecosystem services (ESs). This definition highlights two important concepts: multifunctionality and connectivity. Multifunctionality concerns the capacity of a single area to deliver several benefits and to perform multiple functions. The concept of connectivity is often linked to ecological corridors (ECs), defined as spatial elements that connect habitats, allowing species migration and genetic exchange. Consequently, the spatial identification of GIs and ECs in spatial planning is a noteworthy topic. In this study, we propose a three-step methodological approach to identify a multifunctional GI and ECs connecting Natura 2000 sites (N2Ss) in the case study of the Metropolitan City of Cagliari, Italy.

1 Introduction

The concept of Green Infrastructure (GI) has become increasingly important within the international debate (Garmendia et al. 2016). Several disciplines analyze and interpret the concept of GI providing different definitions (Benedict and McMahon 2002; Madureira et al. 2011). The Communication of the European Commission (2013) "Green Infrastructure: Enhancing Europe's Natural Capital" represents an important conceptual reference. In this document, the European Commission proposes an innovative vision according to which GI is a tool to provide social, economic and ecological benefits and, in contrast with grey infrastructures, it promotes multifunctional nature-based solutions that are regarded as comparatively more sustainable in both economic and social terms.

I. Cannas · S. Lai (✉) · F. Leone · C. Zoppi
Dipartimento di Ingegneria Civile, Ambientale e Architettura, Università di Cagliari, Cagliari, Italy
e-mail: sabrinalai@unica.it

R. Papa et al. (eds.), *Smart Planning: Sustainability and Mobility in the Age of Change*, Green Energy and Technology,
https://doi.org/10.1007/978-3-319-77682-8_8

In particular, GI is defined as "... a strategically planned network of natural and semi-natural areas with other environmental features designed and managed to deliver a wide range of ecosystem services" (European Commission 2013, p. 3). This definition is grounded on three important concepts: planning of a network of natural and semi-natural areas, environmental management of this network, and ecosystem service (ES) provision. Therefore, GIs integrate ecological connectivity, conservation of biodiversity and multifunctionality of ecosystems (Liquete et al. 2015).

The concept of GI is closely connected with two principles: multifunctionality and connectivity (Liquete et al. 2015; Hansen and Pauleit 2014).

The European Commission (2012) defines the concept of multifunctionality in relation to GIs as the capacity of a single area to deliver several benefits and to perform multiple functions such as protection of biodiversity (environmental function), provision of green spaces (social function) or supply of new jobs (economic function). Therefore, the concept of multifunctionality in relation to GIs concerns the spatial integration of various functions (environmental, economic, cultural and esthetic) (Madureira and Andresen 2014) in order to manage space in a more effective way (Ahern 2011).

The concept of connectivity in relation to GIs is often linked to ecological corridors (ECs) (Hansen and Pauleit 2014; Chang et al. 2012), defined as spatial elements that connect habitats, allowing species migration and, consequently, their genetic exchange (Jordán 2000). The identification and appropriate management of ECs is likely to generate positive impacts on habitat fragmentation, by minimizing human-induced pressures related to decreased agricultural production, water pollution and introduction of exotic species (D'Ambrogi and Nazzini 2013).

Moreover, Baudry and Merriam (1988) analyze theories on connections among landscape elements, and define and discuss the concepts of connectedness and connectivity. According to Battisti (2004), connectedness regards physical connections between ecosystems and/or species, whereas connectivity depends on two components: (i) a structural component, concerning physical and territorial elements, such as the spatial distribution of ecosystems and typology and size of natural and artificial elements; and, (ii) a functional component, related to the intrinsic ecological and behavioral characteristics of species. Therefore, although structural variables can potentially influence movements of some species, wild species mobility is affected by functional and behavioral aspects as well (D'Ambrogi and Nazzini 2013; D'Ambrogi et al. 2015).

Hence, the spatial identification of GIs and ECs can be considered a key-issue in spatial and landscape planning as regards the definition and implementation of policies related to protection of nature and natural resources.

In this study, we propose a methodological approach to identify a multifunctional GI and ECs connecting Natura 2000 sites (N2Ss) and we implement this approach with reference to the Metropolitan City of Cagliari, Italy, through a three-step process. First, we define the spatial configuration of a multifunctional GI based on four aspects: conservation value, natural value, recreational value and anthropic heritage, building on a previous study by Arcidiacono et al. (2016). The second step concerns the identification of potential ECs in relation to the conceptual category of habitat suitability and ecological integrity, integrated in a resistance map. Finally, we assess the degree of inclusion of the ECs in a metropolitan GI by merging the spatial taxonomy of the first step and the ECs identified in the second step.

This study is structured as follows. The second section describes the main spatial characteristics of the study area. The following three sections discuss the three methodological steps defined above and their implementation as regards the Metropolitan City of Cagliari. In the conclusion, we highlight planning policy implications of our results and propose directions for future research.

2 Case Study

The Metropolitan City of Cagliari, located in Southern part of the island of Sardinia (Italy), includes seventeen local municipalities (1247 km^2). It was officially established under the provisions of Sardinian Regional Law no. 2/2016. The Metropolitan City of Cagliari represents a strategic node due to the presence of a port and of an airport of national importance. The metropolitan area is characterized by highly urbanized zones surrounded by wetlands and water bodies, mainly located in its central part. Rural and semi-natural areas are mainly located in the peripheral zones.

In order to accomplish with Law enacted by Decree no. 42/2004 (the Italian Code of cultural heritage and landscape), in 2006 the Sardinian regional administration approved a Regional Landscape Plan (RLP). In particular, the Sardinian RLP identifies 27 coastal landscape units, out of which three partially include the Metropolitan City of Cagliari. Although the Sardinian RLP does not state explicit rules for identifying GIs or ECs, articles nos. 23 and 26 severely restrict activities in sensitive contexts and article no. 34 promotes the inclusion of N2Ss in a network consistent with the provisions of the Habitats[1] and Birds[2] Directives, and, by doing so, it indirectly prepares the ground for the establishment of ECs.

Moreover, Sardinia has an extensive Natura 2000 network that covers nearly 19% of the regional land.

Our study area (Fig. 1) amounts to approximately 1800 km^2 and includes:

1. the area of the Metropolitan City of Cagliari;
2. the three coastal landscape units that overlap the Metropolitan City of Cagliari;
3. the twenty N2Ss (seven special areas of conservation under the Birds Directive and thirteen either sites of community importance or special areas of conservation under the Habitats Directive) that overlap the Metropolitan City of Cagliari and the three coastal landscape units.

---

[1]Council Directive 92/43/EEC of 21 May 1992 on the conservation of natural habitats and of wild fauna and flora.

[2]Council Directive 79/409/EEC of 2 April 1979 on the conservation of wild birds, substantially amended several times up to the current codified version, i.e. Directive 2009/147/EC of the European Parliament and of the Council of 30 November 2009 on the conservation of wild birds.

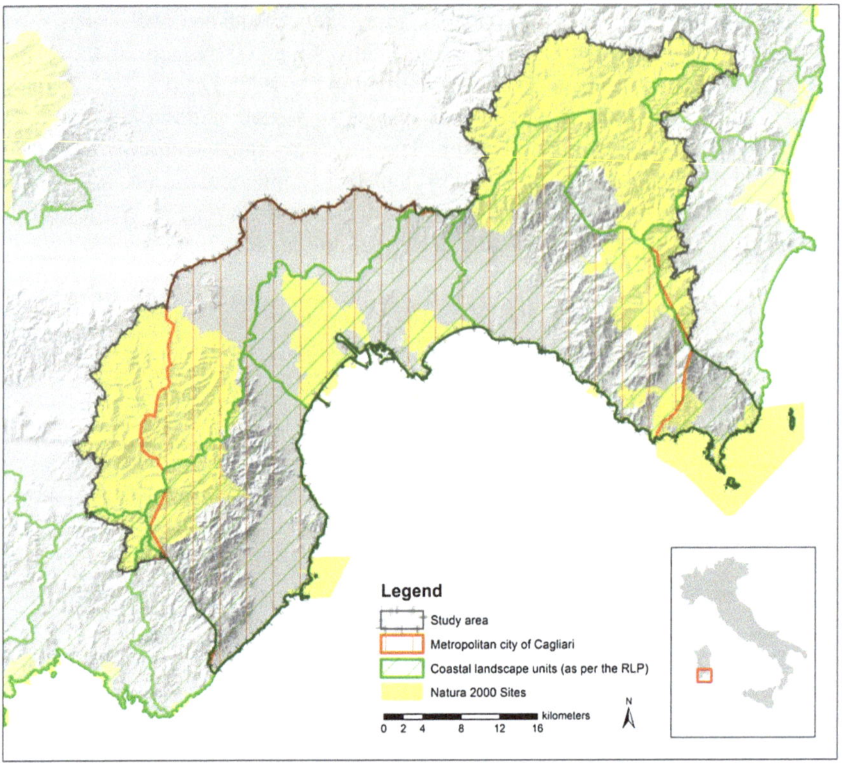

Fig. 1 The study area

3 Multifunctional Green Infrastructures as a Means to Integrate Biodiversity Protection into Spatial Planning

Although the multifunctionality concept has increasingly become popular in relation to GIs, some authors (Madureira and Andresen 2014) raise doubts about their intrinsic ability to exert important functions such as the provision of ESs (European Commission 2012; Horwood 2011). Indeed, conflicting functions are likely to negatively impact on biodiversity protection (Garmendia et al. 2016).

Within this conceptual framework, in this section we propose a methodology to identify a metropolitan GI based on four values: conservation value, natural value, recreational value and anthropic heritage. Each value accounts for a specific aspect that characterizes landscape, as summarized in Table 1.

In the following subsections, we describe the methodology, and present and discuss the results of its implementation in our study area.

Table 1 Description of specific aspects covered by the four values (conservation value, natural value, recreational value and anthropic heritage)

Value	Specific aspect
Conservation value	Presence of natural habitat types of Community interest, identified in the Habitats Directive (Annex I)
Natural value	Quality of biodiversity in relation to ecological integrity, ecosystem functions, provision of ESs and level of degradation
Recreational value	Interactions between landscape and recreational activities
Anthropic heritage	Interactions between natural capital and human beings

3.1 Methodology

We assume the spatial configuration of a GI as characterized by four values that are spatially and quantitatively implemented within the software ArcGis® ESRI.

Conservation value (CONSERV) is assessed building upon a method proposed in a report by CRITERIA and TEMI (2014a).

For areas that do not include habitats of Community interest, CONSERV is zero, otherwise CONSERV is computed as follows:

$$CONSERV = P * (R + T + K) \tag{1}$$

In Eq. (1), each letter accounts for an aspect that contributes to increase the relevance of a habitat of Community interest. In particular, P denotes priority of the habitat, with reference to its classification as a priority habitat in the Habitats Directive. R accounts for the rarity of the habitat, assessed with regard to the frequency of its presence in N2Ss: a higher value of rarity is associated to a lower number of presences. T concerns the vulnerability of the habitat, in relation to the number of threats recorded in N2Ss: the higher the number of threats, the higher T. K, which accounts for the level of knowledge, is qualitatively classified by the expert judgments presented in a report by CRITERIA and TEMI (2014b). As for the scores, P equals 1.5 in case of priority habitats, and 1 otherwise; R and T vary in the [1–5] interval, while K ranges in the [1–4] interval. By applying the above equation, the total score would therefore range between 1 and 21; the score is next normalized in the [0–1] interval.

Natural value (NATURAL) is calculated through the model "Habitat quality" of the software "InVEST".[3] This model integrates data concerning land covers and threats to biodiversity in order to define first a spatial representation of habitat degradation, related to a given spatial unit and a given time period, and next a habitat quality map, with the background assumption that biodiversity is better supported by high levels of habitat quality. NATURAL takes values in the [0–1] interval.

Recreational value (RECREAT) is calculated through the InVEST model "Visitation: Recreation and Tourism", which evaluates the attractiveness of the natural capital for recreation activities through data from Flickr, a social media that supplies geotagged photographs uploaded by its users. For each cell or polygon located in a given spatial unit, the model quantifies the total photo-user-days (PUD) in relation to three parameters, namely location, username and date. One PUD represents the number of photographers who took at least one picture on a specific day in a given location. RECREAT hence equals PUD normalized in the [0–1] interval.

Anthropic heritage (ANTHROP) is assessed in relation to the restrictions imposed by the implementation code of the RLP and by other national or regional planning codes: a score ranging from 0 to 1 is assigned to each protection level identified in the RLP.

Finally, the total value is calculated using a GIS geoprocessing tool, as the sum of the above described four values; therefore, it takes values in the [0–4] interval.

3.2 Results and Discussion

The spatial distributions of the four values are mapped in Fig. 2.

Conservation value and natural value have a similar behavior. Areas with higher values are mostly concentrated in two peripheral zones, one located in the eastern part and the other in the western part of the study area, and in two small zones located in the central study area. The two small zones overlap two wetlands that surround the core built-up area of Cagliari. However, unlike the distribution of conservation value, characterized by non-zero values within and in proximity of N2Ss and by zero values in the rest of the study area, natural value does not show zero values. This entails that, despite the absence of habitats of Community interest, areas characterized by a medium quality level perform an important function since they help connecting the spatial units characterized by higher quality levels.

Recreational value shows zero values in a 96% of the study area, while the highest values are concentrated within the city of Cagliari and along the coastline.

Anthropic heritage takes either zero or one values in the majority of the cells. Maximum values correspond mainly to areas located in the coastal strip and in rivers, creeks and their 150-m buffers, and, to a lesser extent, in areas classified as

---

[3]InVEST (Integrated Valuation of Ecosystem Services and Tradeoffs) is a free software program. Further information is available online at http://data.naturalcapitalproject.org/nightly-build/invest-users-guide/html/index.html. Accessed 05 Oct 2017.

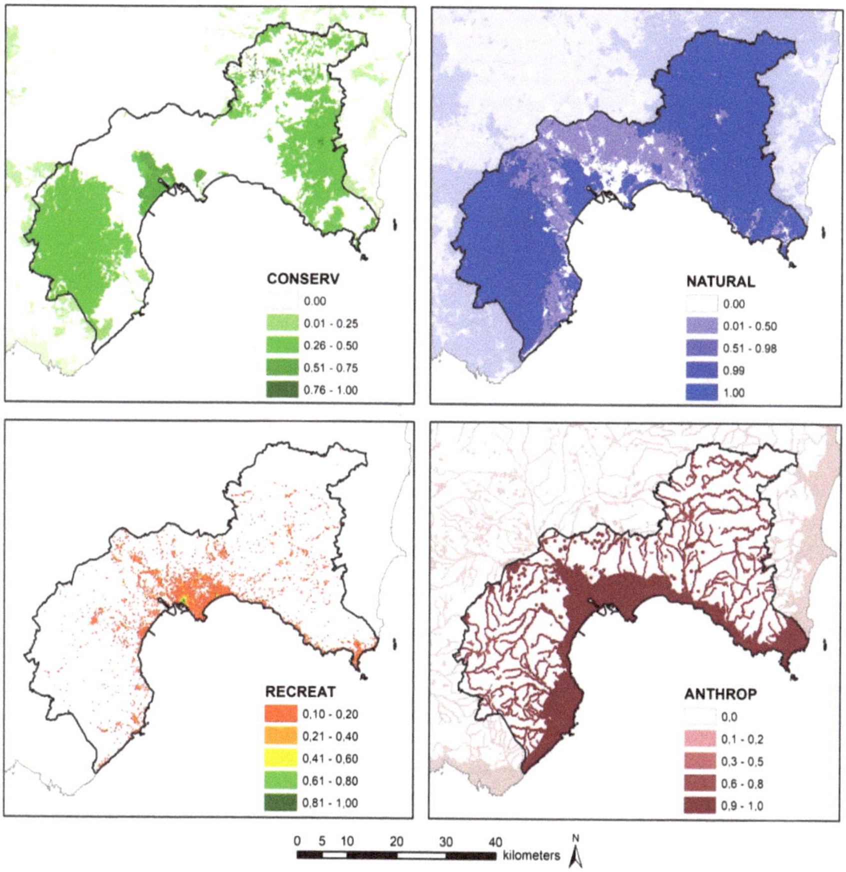

Fig. 2 Spatial distribution of conservation value (upper right), natural value (upper left), recreational value (bottom right) and anthropic heritage (bottom left)

archaeological heritage or characterized by prehistoric, historic, cultural remnants. Under the provisions of the current planning codes, these zones are featured by severe limitations that make them unsuitable for any spatial transformation.

Figure 3 shows the spatial distribution of the total value, whose highest values correspond to rivers, creeks and their 150-m buffers, wetlands, forests and, to a lesser extent, to the coastal strip.

The total value does not achieve the maximum score (4) in any part of the study area, since no single point shows simultaneously the highest values. This outcome is consistent with theories according to which multifunctionality is an "elusive" objective (Meerow and Newell 2017). Planning policies related to GIs are complex political processes, and, that being so, they need several discussion rounds and negotiating tables, especially in order to make decisions on the spatial definition of GIs, which entail the inclusion/exclusion of land parcels in/from a GI. The spatial representation

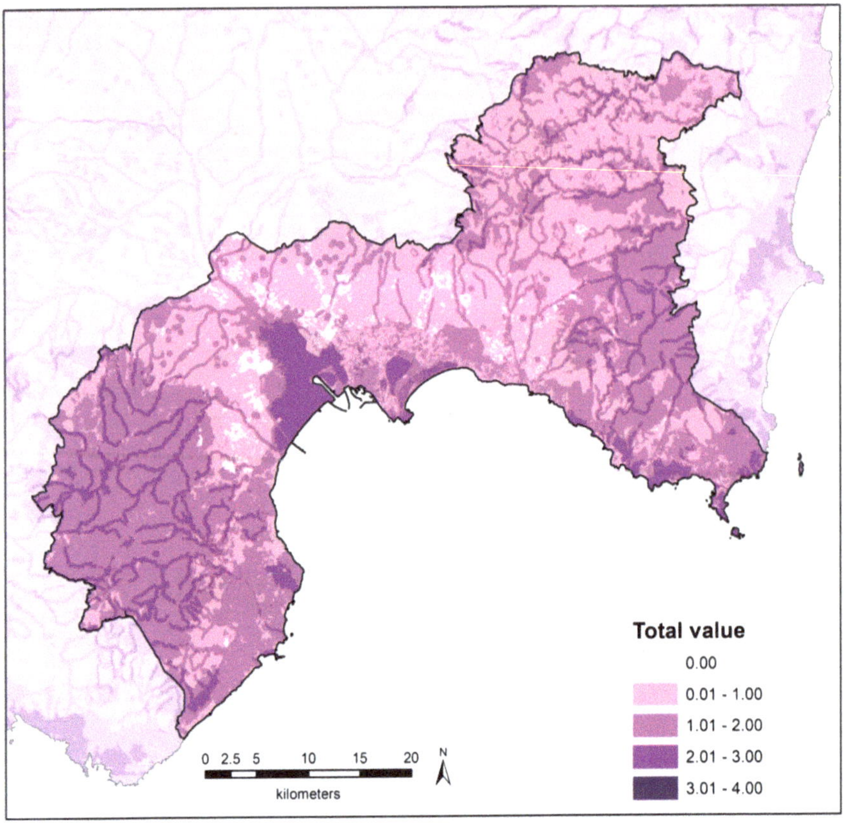

Fig. 3 Spatial representation of the total value

of the total value can be considered an effective visual tool to show the configuration of a GI.

4 The Identification of ECs in the Metropolitan City of Cagliari

ECs are spatial elements that connect habitats, support biodiversity protection and, by doing so, implement ESs supply. The spatial identification of ECs is based on two conditions: (i) protection of biodiversity functions; and, (ii) long-term conservation of biodiversity functions (Snäll et al. 2016). The prioritization of spatial elements as ECs aims at maximizing the availability of ESs and, at the same time, at ensuring and supporting species movement. To this end, we select ECs by using two input layers (potential habitat suitability and ecological integrity), and combine them into a resistance map.

In the following subsections, we describe the methodology, and present and discuss the results of its implementation as regards the study area.

4.1 Methodology

In order to identify ECs in the metropolitan context, we implement a model based on the "Least-cost path" algorithms (LCPs). LCPs are used to identify potential ECs and aim at prioritizing patches among N2Ss (Adriaensen et al. 2003; Beier et al. 2009; European Environment Agency 2014; Lechner et al. 2017; Liquete et al. 2015; Sawyer et al. 2011; Zeller et al. 2012). An LCP requires two raster layers as inputs: a source layer and a friction/resistance layer (Adriaensen et al. 2003). The former represents the single or multiple patches upon which the model calculates connectivity. The latter offers two types of information for each cell of a grid, namely a resistance value and values concerning its spatial position and orientation. The resistance value defines the cost of movement in relation to the land cover attributes of a cell. The LCP identifies paths featured by the least effort or the lowest cost, in terms of species movement. The movement is considered a cost, that is the resistance to species' movements. The movement is characterized by three elements: energy needed, mortality risk, and negative powerful effect on future reproductive potential.

In this study, the identification of potential ECs is structured in four phases. The first phase evaluates which patches, located outside the borders of the N2Ss, can be regarded as habitat; the output of this phase is a habitat suitability map. Habitat suitability represents the probability that a given habitat is used by a given species (Boitani et al. 2002; Wang et al. 2008). Although, in the literature, habitat suitability indexes are generally defined through expert opinions (Graves et al. 2014; Zeller et al. 2012), in our study we build upon a study commissioned by the Sardinian regional administration to monitor the environmental status of the N2Ss (AGRISTUDIO et al. 2011). We assign a global value of potential habitat suitability to each land cover class (classed according to the CORINE Land Cover taxonomy,[4] CLC) of the Sardinian region, within and outside N2Ss. This value is calculated as the weighted mean of the values of habitat suitability associated to each CLC class in relation to the species of the Sardinian N2Ss.

In the second phase, we map the ecological integrity values, following Burkhard et al. (2009, 2012), where the concept of ecological integrity concerns the maintenance of structures and processes essential for ecosystems in terms of self-organization. We assume that patches that show high values of ecological integrity are suitable to species' movements.

In the third phase, we map resistance values; these represent the resistance to species' movement due to physical characteristics of the environmental context

[4]The CORINE Land Cover taxonomy is described in European Environment Agency (1995).

(European Environment Agency 2014; Forman 1995; Graves et al. 2014; Lechner et al. 2017). We generate two resistance maps by inverting habitat suitability and ecological integrity values, scaled in the 0–100 interval, where 100 represents the highest resistance (European Environment Agency 2014). Next, we combine the two maps into a single map, rescaled in the 0–100 interval. The resulting map provides information on resistance to species' movement in relation to a combined effect of landscape morphological aspects (from ecological integrity) and species' use of the environmental context in terms of land cover types (from habitat suitability).

The last phase concerns the identification of potential ECs through the GIS tool *Linkage Mapper*.[5] We start from a resistance map and a map of the core areas (the N2Ss), then we define least-cost corridors between core areas. The definition is based on targeted adjacent core areas and on the identification of a network where least-cost corridors are identified through cost-weighted distance (CWD) analyses and LCPs.

4.2 Results and Discussion

Figure 4 shows the ECs identification through Linkage Mapper within the study area. Our analysis generates a composite raster map by overlapping all of the normalized corridors. Each cell of the composite raster map represents the minimum value of all CWDs. CWDs can take values between 0 and 260,746 km. The second output is a shape file containing twenty-four normalized least-cost corridors.

According to McRae and Kavanagh (2011), linear corridors can be analyzed and compared through a variable (the cost-weighted ratio) computed by means of the ratio of CWDs to the Euclidean distances. For instance, the presence of built-up areas, which are related to high values of resistance to movement in the initial resistance map, justifies a very high value of cost-weighted ratio (32.3) of the linear corridor that connects core area 7 to core area 9 (Fig. 4). Moreover, core areas 1 and 9 are separated by zones characterized by medium-low resistance values and, for this reason, their connecting corridor takes a 5.4 cost-weighted ratio, which is a medium value in the variable range, despite its short Euclidean distance.

The least-cost corridors are defined in one dimension only. Therefore, we re-classify the normalized composite raster CWD map in ten deciles, in order to identify a two-dimensional areal dimension of ECs. All of the cells whose values are included within the first decile are conceived as a part of ECs. The first decile includes normalized least-cost corridors having a length of 0 to 2.4 km. Around 14% of the study area belongs to the first decile. The identified ECs are mainly agricultural areas (36.24%) and forest and semi-natural areas (61.07%).

---

[5]Linkage Mapper is a GIS tool that analyzes habitat connectivity. It is available online at http://www.circuitscape.org/linkagemapper. Accessed 05 Oct 2017.

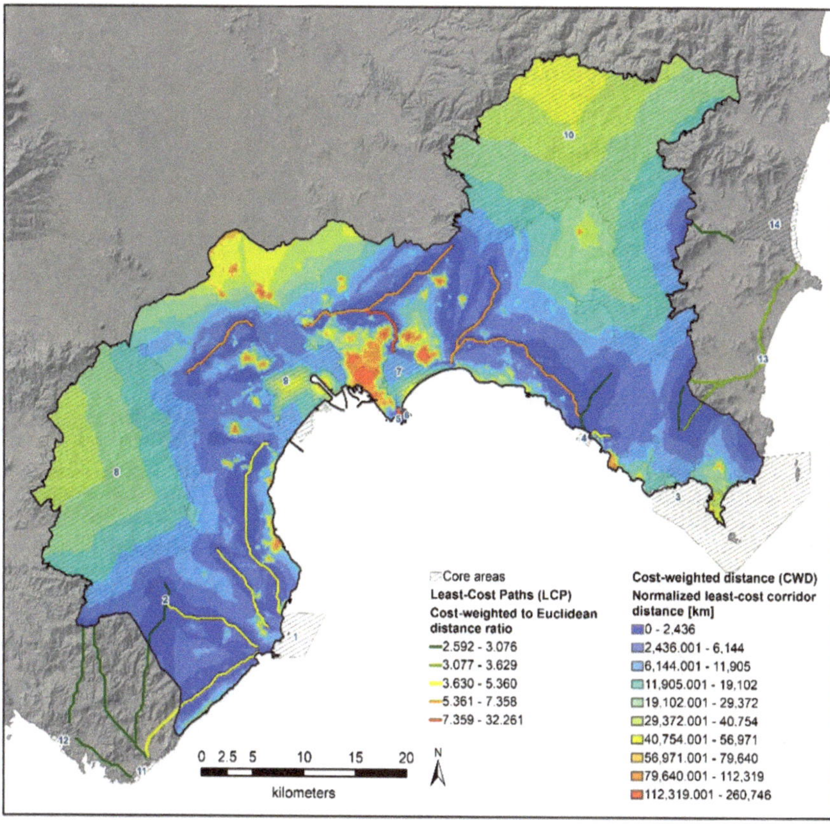

Fig. 4 The ECs identified through Linkage Mapper

5 Merging Metropolitan GI and ECs

In this section we assess if and to what extent ECs, defined in the fourth section, can be included in the metropolitan GI, as defined by the taxonomy proposed in the third section.

In Fig. 5 the land patches, identified as a part of ECs, are classed in relation to the taxonomy proposed in the third section, by using a GIS geoprocessing tool that intersects the ECs' map (i.e., the areas belong to the first decile) with the spatial distribution of the total value.

For this purpose, we use a specific category of discrete-choice-model (DCM), named Logit DCM model (LM). The use of this type of model in the context of spatial analysis and planning concerning regional and local scales represents an innovative element in the implementation of DCMs.

In our study, a dichotomous DCM represents the best choice because patches that belongs to the ECs are categorized into two groups depending on whether they

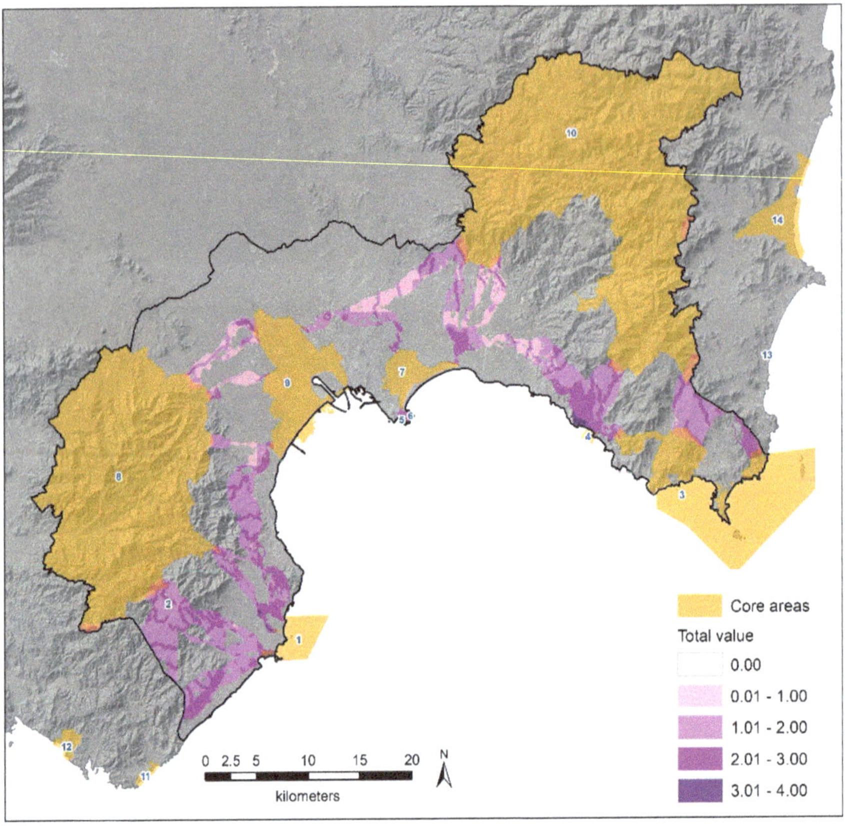

Fig. 5 Intersection between the ECs' map and the spatial distribution of the total value

are included in the metropolitan GI (value 1 if included and total value ≥ 2, and 0 otherwise).

In the following subsections, we explain the methodology, and present and discuss the results of its implementation.

5.1 Methodology

DCMs describe and interpret phenomena, characterized by discrete variables, concerning a set of mutually exclusive alternatives. McFadden's works (1978; 1980) represent a theoretical foundation for agent choice modeling. McFadden (1978, 1980, 2000) generalizes Williams' study (1977) by implementing behavioral models derived from classic microeconomic theory and by including features that are

Table 2 Variables and descriptive statistics

Variable	Mean	St.dev.
GIEC	0.401	0.490
CONSERV	0.0637	0.146
NATURAL	0.624	0.314
RECREAT	0.361	0.299

heterogeneous across agents and that might either be known by the modeler or not; were they not known, they would be random characteristics.

DCMs are usually based on textbook references (Ben-Akiva and Lerman 1985; Ortúzar and Willumsen 2001; Train 2009) which entail that the modeler is aware that agents are endowed with incomplete information and behaviorally characterized by imperfect rationality (Tversky 1972).

Moreover, we assume that the random element of the utility function of an agent is not correlated to the other agents' random elements. This restriction is related to the definition of the dependent and independent variables of the DCM, which we independently and deterministically identify. Therefore, the random terms should be featured as follows: (i) $E(\varepsilon|x) = 0$ (i.e., the random terms have a 0 conditional mean), (ii) $Var(\varepsilon) = \sigma^2$ (i.e.: the random terms always show the same variance), and, (iii) $E[\varepsilon_i \varepsilon_j | X] = 0$ (i.e.: the random terms are not correlated)[6] (Cannas and Zoppi 2017; Cherchi 2009).

In this study, we implement a Logit dichotomous choice model in order to assess the inclusion of a land patch, which belongs to the ECs identified through the methodological approach defined above, into a metropolitan GI. In order to operationalize the LM, we work on Greene's (1993, pp. 666–672), Nerlove and Press' (1973), and Zoppi and Lai's (2013) studies.

The inputs included in the model are:

a dichotomous variable (GIEC), related to land parcels, which equals 1 if a land parcel, included into an EC, belongs to the metropolitan GI identified in the third section, or 0 otherwise;

three explanatory variables (CONSERV, NATURAL, RECREAT) related to conservation value, natural value and recreational value, that is the characteristics of a land parcel which settle the decision over its inclusion into the metropolitan GI. Variables and descriptive statistics are reported in Table 2.

[6]Where: x is the set of independent variables; X is the matrix of observations concerning the independent variables.

Table 3 Marginal impacts on the probabilities of Y = 1 of the variables representing conservation, natural and recreational values

Variable	Marginal effect	z-statistic	Hypothesis test: marginal effect = 0
Marginal impact on Y = 1 probability, dProb (Y = 1)/dx, Prob (Y = 1) = 0.399			
CONSERV	1.148	15.331	0.0000
NATURAL	1.630	40.002	0.0000
RECREAT	1.337	33.700	0.0000

Log-likelihood goodness-of-fit test
Log-likelihood ratio = 4441.012 – Prob. > chi-square = 0.00000 (3 degrees of freedom)
Hosmer and Lemeshow (1989) goodness-of-fit test
HL = 493.67531 – Prob. > chi-square = 0.00000 (8 degrees of freedom)

5.2 *Results and Discussion*

The overlay mapping of the ECs and of the metropolitan GI returns 6233 land parcels. The model estimates the probability of each land parcel to be included into the metropolitan GI, that is, the probability that Y equals 1. Table 3 shows the marginal impacts of each independent variable on the probability of Y = 1 implied by the implementation of the LM.

In relation to the goodness of fit, according to the results of log-likelihood ratio and Hosmer and Lemeshow's (1989) chi square-based tests, the observed values of the probability that the event Y = 1 occurs do not differ significantly from the values estimated by the LM. Moreover, observed and estimated values are consistent in terms of both size and sign.

With regard to conservation value (CONSERV in Tables 2 and 3), two factors influence the probability that a land parcel is included into the metropolitan GI: (i) the presence of habitats of Community interest; and, (ii) the rarity, threats and level of knowledge on the habitats status. In relation to the first factor, if a ten percent increase of CONSERV does occur at the mean value, the probability that a land parcel is included into the metropolitan GI grows by 11.4%. Moreover, if a priority habitat overlaps a land parcel, this implies an increase in the probability of inclusion a 50% higher than in the opposite case (see Table 3). In relation to the second factor, a higher level of rarity and threats and poorer documentation and studies are correlated to a higher probability of inclusion.

With regard to natural value (NATURAL in Tables 2 and 3), supply of final ESs generated by biodiversity positively affects the probability of inclusion. This result is not surprising because ECs contain land parcels comparatively more suitable to host species protected by the Habitats Directive and to provide ESs related to biodiversity. For example, our estimates show that a 10% increase in natural value is correlated to a 16.3% increase in the probability that a land parcel is included into the metropolitan GI.

Recreational value (RECREAT in Tables 2 and 3) depends on the interest shown by the users, mainly tourists and local residents, as regards natural resources and historic heritage located in the Metropolitan City of Cagliari.

Our estimates show that the impact of RECREAT on the probability of inclusion is higher than that of both CONSERV and NATURAL. Moreover, NATURAL entails a higher effect than CONSERV. On the other hand, although RECREAT can hardly represent an issue which planning policies can easily deal with, our outcomes put in evidence that environmental attractiveness should be carefully considered in decision-making processes concerning environmental planning and protection.

6 Conclusions

In this study, we propose a methodological approach to identify a multifunctional GI and ECs connecting N2Ss implemented in three steps with reference to the Metropolitan City of Cagliari. In the first step, we build the spatial configuration of the GI in relation to four functions related to natural resources, environmental structure and historic and cultural heritage. In the second step, we identify a set of ECs through a conceptual framework based on connectivity and ecological integrity. Finally, we assess the probability of including land parcels that belong to ECs into the GI.

Our results put in evidence (see Table 3) that although the European Commission (2012) identifies ECs as part of GIs, however, in the case of the Metropolitan City of Cagliari, the share of ECs included into the metropolitan GI is quite low (only 40%). This is a critical issue that needs to be addressed.

Natura 2000 network includes sites of community importance, special areas of conservation and special protection areas, which correspond to the nodes of the network. The ECs we have identified in our study are related to priority habitats and, more generally, to habitats of Community interest located outside the borders of the N2Ss. Without the restrictive protection measures established under the provisions of the Habitats and Birds Directives, these habitats are negatively affected by anthropic developments, such as new residential and productive settlements. Land uptaken by urbanization needs major attention in terms of appropriate planning actions and protection measures, based on sound scientific knowledge and technical expertise. In fact, the protection regime defined by the Habitats and Birds Directives should be extended to habitats located outside the N2Ss boundaries.

Secondly, our study highlights the importance of biodiversity in supplying ESs. This issue is crucial to promoting the inclusion of ECs into the metropolitan GI. National, regional and metropolitan administrations should enhance and promote protection of areas characterized by a high productive potential of ESs, by means of: (i) sound scientific knowledge framework related to the multifaceted disciplinary fields concerning ESs and their relationships with land cover types; (ii) prevention- and mitigation-oriented policies related to potential and ongoing land-taking processes; and, (iii) planning actions aimed at maintaining and increasing the ESs supply.

Increasing or maintaining the productive capacity of ESs may possibly generate negative impacts on other ESs, by undermining the effectiveness of conservation measures established to grant their protection. For instance, improving and promoting recreational services (classed by Millenium Ecosystem Assessment 2003, under the cultural services category) or increasing agricultural production (classed by Millenium Ecosystem Assessment 2003, under the category of provisioning services) may generate negative impacts on habitats and species within and outside the N2Ss, and by doing so, they may cause a decrease in their productive capacity in terms of supporting services (another category of the classification stated by Millennium Ecosystem Assessment 2003). Therefore, a future development of this research work can be identified with the analysis and assessment of the trade-offs between the increase/decrease in the productive capacity of different types of ESs as a consequence of the implementation of conservation measures related to N2Ss. Several authors analyze the trade-off question: see, among many, Kovács et al. (2015), who assess ESs trade-offs in non-monetary terms as regards three Hungarian protected areas.

Moreover, our study shows a significant impact of recreational value on the probability of land parcels being included into the metropolitan GI. Indeed, its marginal effect is 40% higher than conservation value. On the other hand, the assessment of the attractiveness of an area in terms of ESs provision needs further analysis due to its volatile character. For this reason, this aspect should be addressed in a future research as well.

Finally, in terms of exportability, our methodological approach can be applied and tested in other European metropolitan contexts where the Natura 2000 network is in force, in order to define and complete, through the connections represented by the ECs, the spatial structure which is presently represented only by a system of disconnected nodes, and, consequently, not entirely consistent with the Habitats and Birds Directives.

Acknowledgements This essay is written within the Research Program "Natura 2000: Assessment of management plans and definition of ecological corridors as a complex network", funded by the Autonomous Region of Sardinia for the period 2015–2018, under the provisions of the Call for the presentation of "Projects related to fundamental or basic research" of the year 2013, implemented at the Department of Civil and Environmental Engineering and Architecture (DICAAR) of the University of Cagliari, Italy.

Ignazio Cannas, Sabrina Lai, Federica Leone and Corrado Zoppi have jointly made substantial contributions to the essay's conception, design, and conclusions. Sabrina Lai has taken care of Sect. 1 (Introduction); Federica Leone has taken care of Sect. 2; Sabrina Lai and Federica Leone have jointly taken care of Sect. 3; Ignazio Cannas has taken care of Sect. 4; Corrado Zoppi has taken care of Sect. 5.

References

Adriaensen F, Chardon JP, De Blust G, Swinnen E, Villalba S, Gulinck H, Matthysen E (2003) The application of 'least-cost' modelling as a functional landscape model. Landsc Urban Plan 64(4):233–247. https://doi.org/10.1016/S0169-2046(02)00242-6

AGRISTUDIO, CRITERIA, TEMI (2011) Realizzazione del sistema di monitoraggio dello stato di conservazione degli habitat e delle specie di interesse comunitario della Regione Autonoma della Sardegna. Relazione generale, Allegato 1b: Carta dell'idoneità faunistica [Implementation of the monitoring system concerning the conservation status of habitats and species of Community interest of the Autonomous Region of Sardinia. General Report, Attachment 1b: Habitat suitability map]. Mimeo

Ahern J (2011) From fail-safe to safe-to-fail. Sustainability and resilience in the new urban world. Landsc Urban Plan 100(4):341–343. https://doi.org/10.1016/j.landurbplan.2011.02.021

Arcidiacono A, Ronchi S, Salata S (2016) Managing multiple ecosystem services for landscape conservation: a green infrastructure in Lombardy region. Procedia Eng 161:2297–2303. https://doi.org/10.1016/j.proeng.2016.08.831

Battisti C (2004) Frammentazione ambientale, connettività, reti ecologiche. Un contributo teorico e metodologico con particolare riferimento alla fauna selvatica [Environmental fragmentation, connectivity, ecological networks. A theoretical and methodological contribution with particular reference to wild fauna]. Provincia di Roma, Dipartimento V, Servizio 1° [Province of Rome, 5th Department, 1st Office], Rome, Italy

Baudry J, Merriam HG (1988) Connectivity and connectedness: functional versus structural patterns in landscapes. In: Schreiber KF (ed) Connectivity in landscape ecology. Proceedings of the 2nd international seminar of the international association for landscape ecology (IALE), Munster, Germany 1987, pp 23–28

Beier P, Majka DR, Newell SL (2009) Uncertainty analysis of least-cost modeling for designing wildlife linkages. Ecol Appl 19(8):2067–2077. https://doi.org/10.1890/08-1898.1

Ben-Akiva M, Lerman S (1985) Discrete choice analysis: theory and application to travel demand (transportation studies). The MIT Press, Cambridge, MA, United States

Benedict MA, McMahon ET (2002) Green infrastructure: smart conservation for the 21st century. Washington, D.C., Sprawl Watch Clearinghouse Monograph Series. http://www.sprawlwatch.org/greeninfrastructure.pdf. Accessed 05 Oct 2017

Boitani L, Corsi F, Falcucci A, Maiorano L, Marzetti I, Masi M, Montemaggiori A, Ottaviani D, Reggiani G, Rondinini C (2002) Rete Ecologica Nazionale. Un approccio alla conservazione dei vertebrati italiani. Relazione finale [National Ecological Network. An approach to conservation of Italian vertebrates. Final report]. https://www.researchgate.net/publication/272436763_Rete_Ecologica_Nazionale_Un_approccio_alla_conservazione_dei_vertebrati_italiani_Relazione_finale Accessed 05 Oct 2017

Burkhard B, Kroll F, Müller F, Windhorst W (2009) Landscapes' capacities to provide ecosystem services—a concept for land-cover based assessments. Landscape Online 15:1–22. https://doi.org/10.3097/LO.200915

Burkhard B, Kroll F, Nedkov S, Müller F (2012) Mapping ecosystem service supply, demand and budgets. Ecol Indic 21:7–29. https://doi.org/10.1016/j.ecolind.2011.06.019

Cannas I, Zoppi C (2017) Ecosystem services and the Natura 2000 Network: a study concerning a green infrastructure based on ecological corridors in the Metropolitan City of Cagliari. In: Gervasi O, Murgante B, Misra S, Borruso G, Torre CM, Rocha AMAC, Taniar D, Apduhan BO, Stankova E, Cuzzocrea A·(eds) Computational science and its applications—ICCSA 2017. 17th international conference on computational science and its applications, vol 6, part IV. Springer International Publishing, Basel, Switzerland, pp 379–400

Chang Q, Li X, Huang X, Wu J (2012) A GIS-based Green infrastructure planning for sustainable urban land use and spatial development. Procedia Environ Sci 12:491–498. https://doi.org/10.1016/j.proenv.2012.01.308

Cherchi E (2009) Modelling individual preferences, State of the art, recent advances and future directions. Resource paper prepared for the workshop on "Methodological developments in activity-travel behavior analysis" at the 12th International Conference on Travel Behavior Research, Jaipur, India, 13–18 December 2009. Available on the Internet at http://citeseerx.ist.psu.edu/viewdoc/download?doi=10.1.1.464.2693&rep=rep1&type=pdf. Accessed 05 Oct 2017

CRITERIA, TEMI (2014a) Monitoraggio dello stato di conservazione degli habitat e delle specie di importanza comunitaria presenti nei siti della Rete Natura 2000 in Sardegna. Definizione della rete di monitoraggio. Volume 2: Piano di monitoraggio degli habitat e delle specie vegetali [Monitoring process of the conservation status of habitats and species of Community interest located in Sardinian Natura 2000 network sites. Definition of the monitoring network. Volume 2: Monitoring plan of habitats and vegetal species]. Mimeo

CRITERIA, TEMI (2014b) Monitoraggio dello stato di conservazione degli habitat e delle specie di importanza comunitaria presenti nei siti della Rete Natura 2000 in Sardegna. Elaborazione rapporto di sintesi sullo stato di conservazione di habitat e specie (Linea 4, 4.c.1) [Monitoring process of the conservation status of habitats and species of Community importance located in Sardinian Natura 2000 network sites. Elaboration of the synthesis report concerning the conservation status of habitats and species (Line 4, 4.c.1)]. Mimeo

D'Ambrogi S, Nazzini L (2013) Monitoraggio ISPRA 2012: La rete ecologica nella pianificazione territoriale [ISPRA Monitoring 2012. The ecological network in spatial planning]. Reticula 3. http://www.isprambiente.gov.it/files/pubblicazioni/periodicitecnici/reticula/Reticula_n3.pdf. Accessed 05 Oct 2017

D'Ambrogi S, Gori M, Guccione M, Nazzini L (2015) Implementazione della connettività ecologica sul territorio: il monitoraggio [Implementation of ecological connectivity at the spatial level: the monitoring process]. Reticula 9. http://www.isprambiente.gov.it/it/pubblicazioni/periodicitecnici/reticula/Reticula_n9.pdfAccessed 05 Oct 2017

European Commission (2012) The multifunctionality of green infrastructure. http://ec.europa.eu/environment/nature/ecosystems/docs/Green_Infrastructure.pdf. Accessed 05 Oct 2017

European Commission (2013) Communication from the commission to the European Parliament, the council, the European economic and social committee and the committee of the regions. Green Infrastructure (GI)—Enhancing Europe's natural capital. http://eur-lex.europa.eu/resource.html?uri=cellar:d41348f2-01d5-4abe-b817-4c73e6f1b2df.0014.03/DOC_1&format=PDF. Accessed 05 Oct 2017

European Environment Agency (1995) CORINE land cover. http://www.eea.europa.eu/publications/COR0-landcover. Accessed 05 Oct 2017

European Environment Agency (2014) Spatial analysis of green infrastructure in Europe. EEA Technical report 2/2014. Publications Office of the European Union, Luxembourg

Forman RTT (1995) Land mosaics. The ecology of landscapes and regions. Cambridge University Press, Cambridge, United Kingdom

Garmendia E, Apostolopoulou E, Adams WM, Bormpoudakis D (2016) Biodiversity and green infrastructure in Europe: boundary object or ecological trap? Land Use Policy 56:315–319. https://doi.org/10.1016/j.landusepol.2016.04.003

Graves T, Chandler RB, Royle JA, Beier P, Kendall KC (2014) Estimating landscape resistance to dispersal. Landscape Ecol 29(7):1201–1211. https://doi.org/10.1007/s10980-014-0056-5

Greene WH (1993) Econometric analysis. Macmillan, New York, NY, United States

Hansen R, Pauleit S (2014) From multifunctionality to multiple ecosystem services? A conceptual framework for multifunctionality in green infrastructure planning for urban areas. Ambio 43(4):516–529. https://doi.org/10.1007/s13280-014-0510-2

Horwood K (2011) Green infrastructure: reconciling urban green space and regional economic development: lessons learnt from experience in England's north-west region. Local Environ 16(10):37–41. https://doi.org/10.1080/13549839.2011.607157

Hosmer DW, Lemeshow S (1989) Applied logistic regression. Wiley, New York

Jordán F (2000) A reliability-theory approach to corridor design. Ecol Model 128(2–3):211–220. https://doi.org/10.1016/S0304-3800(00)00197-6

Kovács E, Kelemen K, Kalóczkai A, Margóczi K, Pataki G, Gébert J, Málovics G, Balázs B, Roboz A, Krasznai Kovács E, Mihók B (2015) Understanding the links between ecosystem service trade-offs and conflicts in protected areas. Ecosyst Serv 12:117–127. https://doi.org/10.1016/j.ecoser.2014.09.012

Lechner AM, Sprod D, Carter O, Lefroy EC (2017) Characterising landscape connectivity for conservation planning using a dispersal guild approach. Landscape Ecol 32(1):99. https://doi.org/10.1007/s10980-016-0431-5

Liquete C, Kleeschulte S, Dige G, Maes J, Grizzetti B, Olah B, Zulian G (2015) Mapping green infrastructure based on ecosystem services and ecological networks: a Pan-European case study. Environ Sci Policy 54:268–280. https://doi.org/10.1016/j.envsci.2015.07.009

Madureira H, Andresen T, Monteiro A (2011) Green structure and planning evolution in Porto. Urban For Urban Greening 10(2):141–149. https://doi.org/10.1016/j.ufug.2010.12.004

Madureira H, Andresen T (2014) Planning for multifunctional urban green infrastructures: promises and challenges. Urban Des Int 19(1):38–49. https://doi.org/10.1057/udi.2013.11

McFadden D (1978) Modelling the choice of residential location. In: Karlqvist A, Lundqvist L, Snickars F, Weibull JV (eds) Spatial interaction theory and planning models. The Netherlands, North Holland, Amsterdam, pp 75–96

McFadden D (1980) Econometric models for probabilistic choice among products. J Bus 53(3):13–29

McFadden D (2000) Disaggregate behavioral travel demand's RUM side. A 30-Year Retrospective. In: Hensher D (ed) Travel behavior research: the leading edge. Pergamon Press, Oxford, United Kingdom, pp 17–63

McRae BH, Kavanagh DM (2011) Linkage mapper connectivity analysis software. The Nature Conservancy, Seattle, WA. http://www.circuitscape.org/linkagemapper. Accessed 05 Oct 2017

Meerow S, Newell JP (2017) Spatial planning for multifunctional green infrastructure: growing resilience in Detroit. Landsc Urban Plan 159:62–75. https://doi.org/10.1016/j.landurbplan.2016.10.005

Millennium Ecosystem Assessment (2003) Ecosystems and human well-being: a framework for assessment. Island Press, Washington

Nerlove M, Press S (1973) Univariate and multivariate log-linear and logistic models. In: Report No. R1306-EDA/NIH. RAND Corporation, Santa Monica

Ortúzar JdD, Willumsen LG (2001) Modelling transport, 3rd edn. Wiley, Chichester

Sawyer SC, Epps CW, Brashares JS (2011) Placing linkages among fragmented habitats: do least-cost models reflect how animals use landscapes? J Appl Ecol 48(3):668–678. https://doi.org/10.1111/j.1365-2664.2011.01970.x

Snäll T, Lehtomäki J, Arponen A, Elith J, Moilanen A (2016) Green infrastructure design based on spatial conservation prioritization and modelling of biodiversity features and ecosystem services. Environ Manage 57(2):251–256. https://doi.org/10.1007/s00267-015-0613-y

Train K (2009) Discrete choice methods with simulation. Cambridge University Press, Cambridge

Tversky A (1972) Elimination by aspects: a theory of choice. Psychol Rev 79:281–299. https://doi.org/10.1037/h0032955

Wang YH, Yang KC, Bridgman CL, Lin LK (2008) Habitat suitability modelling to correlate gene flow with landscape connectivity. Landscape Ecol 23(8):989–1000. https://doi.org/10.1007/s10980-008-9262-3

Williams HCWL (1977) On the formation of travel demand models and economic evaluation measures of user benefit. Environ Plan A 9:285–344

Zeller KA, McGarigal K, Whiteley AR (2012) Estimating landscape resistance to movement: a review. Landscape Ecol 27(6):777–797. https://doi.org/10.1007/s10980-012-9737-0

Zoppi C, Lai S (2013) Differentials in the regional operational program expenditure for public services and infrastructure in the coastal cities of Sardinia (Italy) analyzed in the ruling context of the Regional Landscape Plan. Land Use Policy 30:286–304. https://doi.org/10.1016/j.landusepol.2012.03.017

Sustainability of Local Plans and Urban Sectors. Proposal for an Operating Procedure

Giuseppe Mazzeo

Abstract The city is the human structure with the largest consumption of energy. To achieve more sustainable levels of resource's usage and to reduce its environmental impact, the city needs to transform its way of working. For this purpose, it is necessary to act on the urban structure by ensuring that the essential resources become more and more efficient. To achieve this fundamental goal we can act by improving the functioning of both the existing urban structures, and of the new planning actions. For the latter, it is desirable to build a plan containing elements of sustainability that, once implemented, will enable the direct achievement of our objectives. It is appropriate to identify the right planning level to which we can apply these innovative procedures. Despite the fact that the extension of the planned area may vary from case to case, we can assume the scale of the neighbourhood/urban sector scale is the best thanks to its two basic features: the limited extension, and the immediate possibility to transform the indications of plan in physical and functional elements. The goal of the paper is to specify the characteristics of these plans and the size of their effects on the creation of urban areas that seeks to become neutral in the consumptions (Zero Impact Neighborhoods).

1 New Meanings of Sustainability

Improving the energy efficiency of buildings, mobility, and other urban functions it is a necessity for cities that want to achieve a sustainable growth: only to quote a figure, buildings account for about one-third of all greenhouse gas emissions (Lucon et al. 2014) and it is more like half in many big cities. Reducing the energy used to heat, cool and light buildings is an essential measure of the municipality's wider climate goals.

G. Mazzeo (✉)
ISSM, National Research Council of Italy, Naples, Italy
e-mail: mazzeo@issm.cnr.it; gimazzeo@unina.it

R. Papa et al. (eds.), *Smart Planning: Sustainability and Mobility in the Age of Change*, Green Energy and Technology,
https://doi.org/10.1007/978-3-319-77682-8_9

The building of traditional sustainability standards founds upon the thinking shaped by the Brundtland definition for sustainable development (1987), according to which the actual consumption's levels must not compromise those of the future generations. This meaning of sustainability is dated and it is overtaken by more recent evolutions, as that based upon the seventeen U.N. Sustainability Development Goals (2015),[1] which propose a far more proactive and ambitious approach.

These goals can be an interesting guide for the planning and the construction sectors. This means to revisit and to revise our thinking about the standards, so that they can be used for aiming at a socially just, culturally rich and ecologically restorative future.

One example in this direction is "USA Sustainable Cities Initiative—Baltimore".[2] The city's first sustainability plan was presented in 2009, and an update is underway. Likewise, a community-led initiative called "Baltimore 2030" already is in programme. Efforts are taking place to harmonize these plans with the aims of the UN-SDGs.

The core of the project is a system of locally relevant measures—or indicators—to track the progress toward the broad aims of the 17 SDGs. It uses 56 specific indicators for tracking the city's actions toward achieving various aspects of the SDGs. The used set is a customized version of a broader set of global indicators that the U.N. Statistical Commission is still pulling together (Table 1).

Another example is the report "Accelerating Building Efficiency—8 Actions for Urban Leaders", published by The World Resource Institute.[3] The study highlights strategies to maximize the efficiency in buildings. It involves actions of the public administrations, designer, builders, and users and it underlines that decisions made today on how to build, design and operate edifices will have lasting consequences owing to its long lifespan. The report cites eight main approaches:

- *Building codes and standards*. Defined requirements for new and existing structures must ensure at least a "minimum level of energy efficiency". They can cover design, construction, and operation, besides to promote long-term savings.
- *Energy efficiency targets*. Municipalities can establish mandatory energy-reduction goals for public buildings and voluntary benchmarks for private structures.
- *Measure and track performance*. Cities can use tools as audits, certification programs and disclosure requirements to measure the energy performance of buildings. The data can help building owners, managers and occupants to use energy in a smarter way.
- *Financial incentives*. Cities can introduce some economic approaches (grants, rebates, bonds, mortgage financing, loans, credit lines, tax incentives, and so on) to help projects to overcome economic hurdles. Funding mechanisms are particularly useful for initiatives with steep upfront expenses.

---

[1] https://sustainabledevelopment.un.org/sdg1.

[2] http://www.ubalt.edu/about-ub/sustainable-cities/.

[3] http://publications.wri.org/buildingefficiency/.

Table 1 Indicators for "USA Sustainable Cities Initiative—Baltimore" built on SDG goals

SDG goals (UN, 2015)	USA SCI—Baltimore indicators
Goal #1: End poverty	• Distressed communities index • Percent of children living in poverty • Number of homeless persons • Liquid asset poverty
Goal #2: Zero hunger	• Percentage of residents living in food deserts • Percentage of residents experiencing food hardship
Goal #3: Good health and well-being	• Number of drug- and alcohol-related emergency department visits per 1000 residents • Average life expectancy • Infant mortality rate per 1000 residents
Goal #4: Quality education	• High school graduation rate (includes equivalency), population 18–24 years • Percentage of students demonstrating readiness on the kindergarten readiness assessment • Percentage of students reaching advanced or proficient levels on the national assessment of educational progress in reading at grade 4 • Percentage of disconnected youth
Goal #5: Gender equality	• Gender wage ratio • Survivors of human trafficking per 10,000 • Percent of businesses owned by women
Goal #6: Clean Water and sanitation	• Lead in water of 90 percentile sample • Number of households to whom water service is unaffordable • Average residential daily water usage
Goal #7: Affordable and clean energy	• Total electricity consumption per capita • Total gas consumption per capita • Ratio of household utility expenditure to pre tax income
Goal #8: Decent work and economic growth	• Median household income, in 2014 inflation adjusted dollars • Labor force participation • Percentage of residents earning a living wage • Total number of jobs
Goal #9: Industry, innovation and infrastructure	• Annual hours of delay per auto commuter • Total value of city innovation fund awards • Number of utility patent grants

(continued)

Table 1 (continued)

SDG goals (UN, 2015)	USA SCI—Baltimore indicators
Goal #10: Reduced inequalities	• Percent children living in poverty, disaggregated by race • Percent residents living in food deserts, disaggregated by race • Infant mortality rates per 1000 residents, disaggregated by race • Percent students demonstrating readiness on the kindergarten readiness assessment, disaggregated by race • Percent students reaching advanced or proficient levels on the national assessment of educational progress in reading at grade 4, disaggregated by race • Percent firms owned by minorities
Goal #11: Sustainable cities and communities	• Number of days with air quality index "Good" • Number of affordable housing units • Housing and transportation cost as a percentage of income • Percentage of households commuting more than 45 min • Percentage of residential properties that are vacant or abandoned
Goal #12: Responsible consumption and production	• The recycling rate • Annual waste generation by category
Goal #13: Climate action	• Greenhouse gas emissions • Number of excessive Heat Code Red Days
Goal #14: Life below water	• Baltimore Harbor overall water health score • Impervious surface ratio • Cumulative stream restoration number
Goal #15: Life on land	• Tree canopy coverage rate • Number of urban agriculture projects • Number of species observed in BioBlitz
Goal #16: Peace and justice strong indicators	• State/local public funding for legal aid for eligible clients • Length of time in jail pretrial for misdemeanor offenses • Civil legal aid attorney ratio • Violent crime rate per 1000 residents • Percent population (over age of 18) who voted in the general election

- *Lead by example*. Municipal governments can start policies and projects that become an example for the community and foster greater acceptance for energy-efficiency solutions. They can involve public and private buildings.
- *Engagement strategies*. Competitions, awards and kiosks are potential means to spur dialogue among tenants, owners and building administrators.
- *Technical training*. Educational programs allow stakeholders to implement the latest energy conservation designs, to standardize investment terms, and to reduce transaction costs.
- *Partners with utilities*. Municipalities can widen their access to energy usage data. Alliances with local utility's providers can help city planners make smarter choices about energy efficiency goals.

2 New Processes of Assessment

Urban standards and certification systems are two distinct ways to interpret urban quality. While the standards that build our cities are mandatory and, usually, they are within the competence of public administrations, the certification systems, not yet required by law, are used on a voluntary basis and, therefore, developed by non-public institutions.

The elements of the established sustainability certification have well worked to uphold the principles of sustainability. However, they need to be replaced by a new generation of elements based on restorative or regenerative sustainability (Brown 2016).

To do this it is necessary to involve the productive system to a more advanced thinking, through the influence of certification standards. We don't have the luxury of time to do otherwise, also because the costs of addressing climate change will increase the longer we defer the current mode of action (Stern 2006).

This means, for example, that we need to stop to consider the green buildings as an occasional testimony to be proud. They should be the norm.

To achieve the aim the point of view must be changed. In other word, first the vision of the outcome, and then the path to get there. Normally, indeed, we focus on making current practice a little less harmful, possibly because of actual difficulties to envision a better future.

Built environment efforts need to become more responsible in addressing negative trends relating to health issues, social inequity, and environmental damage. The new standards must give these elements a similar emphasis to that of energy and resource management (Papa et al. 2014).

Living Building Challenge (LBC) is an alternative way to do certification. The system does not use a score system as LEED or BREEM: This make LBC a clean, easily understood tool. Furthermore, the obligations of this certification system must be fully meeting, whereby, in the case of energy performance, this means to demonstrate the achievement of the design aims over a 12-month period.

Despite their specificities, the development of LBC in the United States has led to a collaboration with LEED. In particular, the analysis of the sectors of Living Building Energy, Water and Material Requirements is carried out according to the methodologies of LEED certification. Another point of contact is that the LEED v.4 Materials Requirement is similar to the Living Building Challenge Red List, Pharos and Healthy Product Declaration thinking.

A prospective tendency in the future of sustainability standards is the development of interoperable, open-sourced approaches. This means that will be more dovetailed standards, assembled for obtaining certification standards best suited to meet specific values, requirements and sustainability aspirations. This means that in a next future it will possible to "Uberize" the sustainability, namely that new organizations, unlocked by the "traditional" way to do things, will offer alternative sustainability certification and assessment services tailored to the ever more complex needs of specific organizations (Elkington 2015).

A first example of this potential trend is Portico, the Google construction portal used for its own building projects and focused on healthy materials. In this case a major, powerful organization is developing its own standard of project and certifications. "Constructing buildings isn't our core business", Google notes on its Portico support pages,[4] "but creating work environments that inspire and energize our people is a major goal. The result: healthy Google buildings for healthy Google people". This will, of course, disrupt and challenge rules and criteria of established standards, as in the same way Uber want to do with taxi services, PayPal in the field of payments, or Amazon in that of trade (Brown 2016).

3 Methodology

The proposed methodology is closely operating. The paper presents the structure and the phases connected with it. A subsequent deepening, not present in this paper, will realize an application testing.

At the basis of the methodology is (1) the identification of action's areas structuring the local plan and (2) the use of the same areas for the assessment and certification of the local plan (Fig. 1).

The first part of the methodology presents a hypothesis about a structure of local plan. It want to shape a plan that is careful to the sustainable aspects and that can bring to the building of neutral urban sectors. Specific indicators delineate each area of action and measure the true results.

The second part of the methodology presents a logical structure for the evaluation and certification of plans and urban areas. It uses the same indicators of the first part, and, for this reason, it is useful to evaluate the results and the performances of operative plans. Besides, it can be useful to evaluate an existing urban area, whit the aim to define its sustainability level.

[4]https://portico.healthymaterials.net/login.

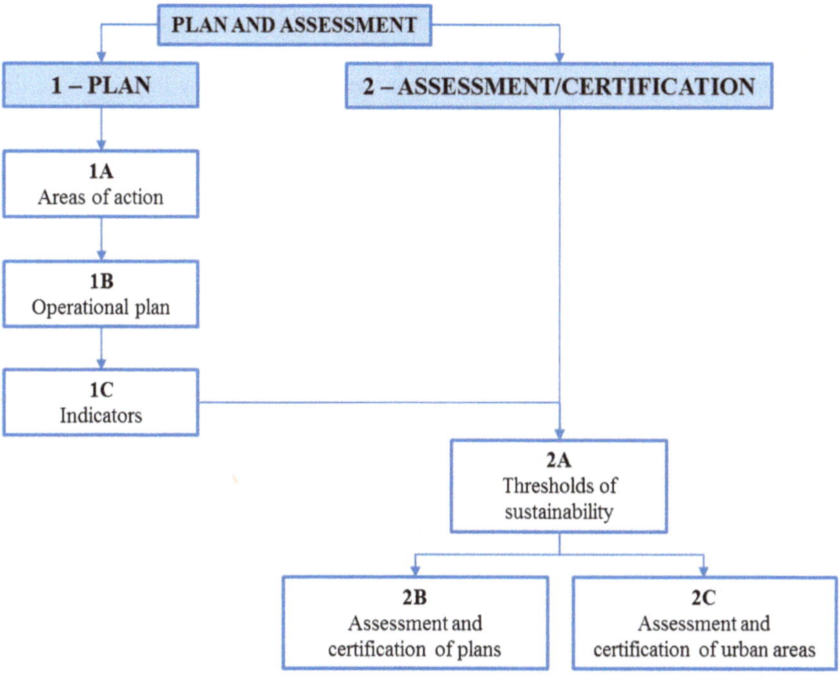

Fig. 1 Logical structure connecting planning and assessment

In this paper, the expression sustainability is used in a wider meaning and it encloses elements as resilience, vulnerability, and so on.

3.1 Areas of Action of Local Plans

The insights for new meanings of sustainability and new evaluation processes lead to the need to rethink contents and procedures. The contents are connected to the local transformation plan, while the procedures are related to the assessment of the results of these new planning structures.

In terms of planning the core elements to be considered lie in the need to insert new meanings and new elements within the plan. The purpose is to make possible the transformation of the plan in direction of more sustainable operations bringing to the construction of lighter cities. In this way can be useful the experiences that are under development in countries historically careful to the environmental aspects.

In UK, for example, the "Planning Policy Statement: Eco-towns" (Communities and Local Government 2009), now finished, had proposed eco-towns as urban structures that respond to the opportunities and challenges of their location and community aspirations. They should meet the standards "as set out in the Planning Policy

Statement or any standards in the development plan which are of a higher standard" (3).

Eco-towns respond to specific criteria for their location. They "should have the functional characteristics of a new settlement; that is to be of sufficient size and have the necessary services to establish their own character and identity and so have the critical mass necessary to be capable of self-containment whilst delivering much higher standards of sustainability" (3).

In particular, the obligations for to identify suitable locations for eco-towns were:

- The area should be able to make provision for a minimum of 5000 homes, scale that allows exploiting a number of opportunities and benefits.
- The area should be near to a higher order centre(s) where there is clear capacity for public transport links and other sustainable access.
- The area should be near to existing and planned employment opportunities.
- The eco-town can play an important role in delivering other planning, development and regeneration objectives.

The study had proposed four eco-town: Whitehill-Bordon, Rackheath, North-West Bicester, and St. Austell (China Clay Community).

The standard for the realization of these eco-town were the following:

- *Zero carbon in eco-towns (ET7).* The definition is that over a year the net carbon dioxide emissions from all energy use within all the buildings in the eco-town are zero or below. It excludes embodied carbon and emissions from transport.
- *Climate change adaptation (ET8).* Eco-towns should be sustainable communities. They must be resilient to and appropriate for the climate change, accepted as inevitable. Their planning aims to minimise future vulnerability in a changing climate, and to respect principles of mitigation and adaptation. The design must take account of the climate using the most recent available climate change scenarios. Eco-towns should realize a local environment of high quality and meet the standards on water, flooding, green infrastructure and biodiversity.
- *Homes (ET9).* Homes and buildings must reach zero carbon emissions, as part of the whole built environment. Furthermore they must meet some quality's level. In particular: (a) Achieve Building for Life Silver Standard[5] and Level 4 of the Code for Sustainable Homes[6] at a minimum. (b) Meet lifetime homes standards and space standards. (c) Have real time energy monitoring systems, real time public transport information, high-speed broadband access, a potential use of digital access to support assisted living and smart energy management systems. (d) Provide for at least 30% affordable housing (which includes social rented and intermediate housing). (e) Demonstrate high levels of energy efficiency in the construction of the building. (f) Achieve carbon reductions (from space heating,

[5]Building for Life—www.buildingforlife.org/.

[6]Code Level 4 contains standards for household waste recycling, construction waste, composting facilities, water efficiency measures, surface water management, use of materials, energy & CO_2, pollution, health & wellbeing, ecology & ongoing management of the development.

ventilation, hot water and fixed lighting) of at least 70% relative to current Building Regulations (Part L 2006).

- *Employment (ET10).* Eco-towns must be genuine mixed-use communities with minimum unsustainable commuter trips. Planning applications for eco-towns must be accompanied by an economic strategy that demonstrate how access to work will be achieved and easily reached by walking, cycling and/or public transport.
- *Transport (ET11).* Travel in eco-towns should support people's desire for mobility whilst achieving the goal of low carbon living. The design of the access to it and through it must give priority to options such as walking, cycling, public transport and other sustainable options. To achieve this, homes should be within 10 min' walk of (a) frequent public transport and (b) neighbourhood services. Other indications are in the case the eco-town is near a greater city, for the integration of ultra-low carbon vehicle options, and for the movement of children.
- *Healthy lifestyles (ET12).* The built and natural environments are important components to improve the health and well-being of people. Eco-towns should be planned to support healthy and sustainable environments and to enable residents to make healthy choices easily.
- *Local services (ET13).* Planning applications should include a good level of provision of services, proportionate to the size of the development. This should include leisure, health and social care, education, retail, arts and culture, library services, sport and play facilities and community and voluntary sector facilities.
- *Green infrastructure (ET14).* Forty per cent of the total area should be allocated to green space, of which at least half should be public and consists of a network of well managed, high quality green/open spaces which are linked to the wider countryside. Planning applications should demonstrate a range of types of green space. The space should be multifunctional, and support wildlife, urban cooling and flood management. Particular attention should be given to allow the local production of food.
- *Landscape and historic environment (ET15).* Planning applications should demonstrate that they have adequately considered the consequences on the local landscape and historic environment.
- *Biodiversity (ET16).* Eco-towns should demonstrate a net gain in local biodiversity. Planning permission may not be granted for proposals which have a significant adverse effect on internationally designated nature conservation sites or sites of special scientific interest.
- *Water (ET17).* Water efficiency must be a primary aim of the proposals, particularly in areas of serious water stress. It should contribute towards improving water quality in these localities.
- *Flood risk management (ET18).* Eco-towns should reduce and avoid flood risk wherever practicable. They should not increase the risk of flooding elsewhere and should use means to address and reduce existing flooding problems.
- *Waste (ET19).* Eco-town should include a sustainable waste and resources plan, covering both domestic and non-domestic wastes.
- *Master planning (ET20).* All planning applications should include an overall master plan and supporting documentation to demonstrate how the eco-town standards

ET7-ET19 will be achieved. It is vital to the long-terms success of eco-towns the sustainment of the standards. In developing the master plan, there should be a high level of engagement and consultation with communities.

- *Transition (ET21)*. To support the transition process, planning applications should set out: (a) The detailed timetable of delivery, employment and community facilities and services. (b) Plans for operational delivery of priority core services to underpin the low level of carbon emissions. (c) Progress in and plans the provision of health and social care. (d) How will support the initial formation and growth of communities, which enhance well-being and provide social structures. (e) How will provide information and resources to encourage environmentally responsible behaviour, especially as new residents move in. (f) The specific data which will be collected annually to monitor, support and evaluate progress in low carbon living, including those on zero carbon, transport and waste. (g) A governance transition plan from developer to community. (h) How carbon emissions resulting from the construction will be limited, managed and monitored.
- *Community and governance (ET22)*. A long term approach is necessary to ensure the integrity of structures as an eco-town, and is able to manage change in a planned way. Planning applications should be accompanied by long term governance structures to ensure that: (a) Appropriate governance structures are in place to ensure that standards are maintained and evolved to meet future needs. (b) The community is continually involved to develop social capital. (c) Sustainability data are agreed and monitored. (d) Future development continues to meet the eco-town standards. (e) Community assets are maintained.

Figure 2 relates the areas of standards according PRS 2009 with the areas identified as invariants of a potential new local plan (Mazzeo 2016). Even if the elements are different in type and number it is possible to identify narrow connections between the two lists. The relations allow us to say that the areas in which to deepen local planning in a sustainable perspective are outlined with a remarkable precision and that one of the major tasks of the near future will be to fill of meanings these areas. Another consideration is that the two lists outline new meanings of urban standards, more sustainable, more flexible and more modern.

3.2 Assessment and Certification

Five steps structure the proposed methodology:

- (a) Identification of the areas of action of the local plan.
- (b) Connection among areas of action and design elements of the specific local plan.
- (c) Selection of indicators for to outline the evolution of an urban sector and identification of their connection with the areas of action. For this step can be of help national and international reference standards.

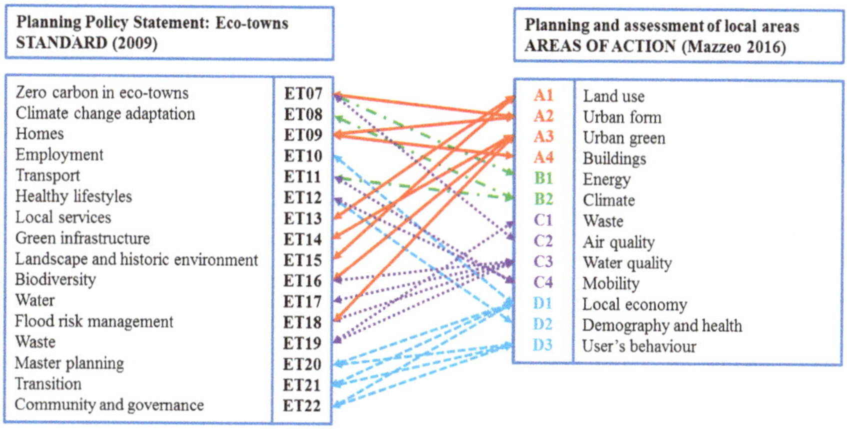

Fig. 2 Connections between standard in Eco-towns foreseen by UK Planning Policy Statement (Communities and Local Government 2009) and planning and assessment of local areas proposed by Mazzeo (2016)

- (d1) Use of the indicators for assessing a local plan assigning to it a level of sustainability within a predetermined scale.
- (d2) Use of the indicators for assessing an existing urban sector/neighbourhood assigning to it a level of sustainability within a predetermined scale.

The success of the method is largely based on the choice of a system of quantitative and qualitative indicators able to lead to a clear and shared assessment. The choice results by a combination of vulnerability and resilience criteria (Mazzeo 2014a, b; Galderisi et al. 2016).

Meaningfulness of this assessment is based on the building of value's classes useful to identify the level of sustainability. The value reached by an indicator put it in one of the value's classes. The combination of the values defines the overall class reached by the area of action.

The certification system can be applied either to a local plan or to an urban neighbourhood. Table 2 shows the areas of action. Each area is associated with one or two codes: the LP code indicates the use of the same in the local plan certification, while the N code indicates the use in the neighbourhood certification.

Areas from A1 to C4 are related to measurable elements by means of physical and functional indicators that, for the most part, are quantitative; those from D1 to D3 are related with socio-economic characteristics of urban space and with anthropic behavior, once the area is transformed or in the case of existing neighbourhoods.

Moreover, indicators representing D1÷D3 are both quantitative and qualitative and they are more complex to define since they are related to social welfare and to the use of public spaces done by citizens and city users. They are connected with the knowledge of how to use anthropic structures, with the processes of development of the activities carried out there, and with the changes occurring over time and that can be linked to persistent or change factors in the use of the city (EEA 2013) (Fig. 3).

Table 2 Areas of action and their use for planning and for assessment

Areas of action	Use for local planning	Use for local planning (LP) assessment	Use for urban sector (N) assessment
A1—Land use	Yes	Yes	Yes
A2—Urban form	Yes	Yes	Yes
A3—Urban green	Yes	Yes	Yes
A4—Buildings	Yes	Yes	Yes
B1—Energy	Yes	Yes	Yes
B2—Climate	Yes	Yes	Yes
C1—Garbage	Yes	Yes	Yes
C2—Air quality	Yes	Yes	Yes
C3—Water quality	Yes	Yes	Yes
C4—Mobility	Yes	Yes	Yes
D1—Local economy	No	No	Yes
D2—Demography and health	No	No	Yes
D3—User's behaviour	No	No	Yes

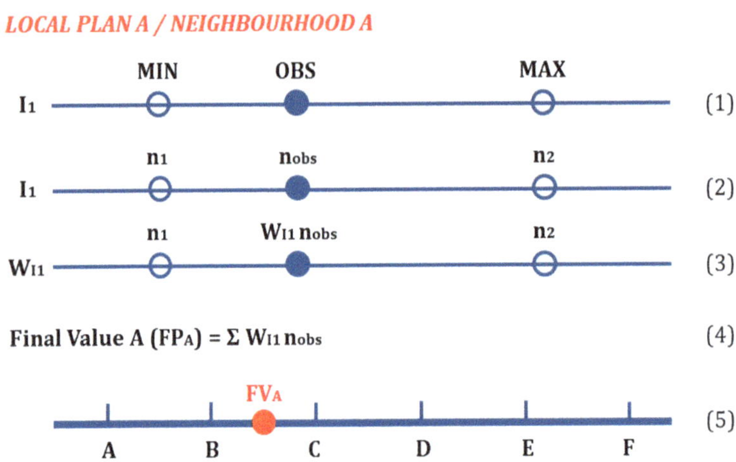

Fig. 3 Steps for the use of the indicators in the operations of assessment of local plans or neighborhoods

A system of indicators characterizes and measures the results of each area. They can be formed by qualitative or quantitative data usable to determine the sustainability of a local plan (LP) or of a neighborhood (N). For each indicator is defined a quality range on the basis of two limit values: the first is the minimum quality value (MIN), the second is the maximum quality value (MAX). Within this range the observed datum (OBS) is the actual value related with case study A (Fig. 1). This value (foreseen or observed) represents the state of LP/N for that indicator (step 1).

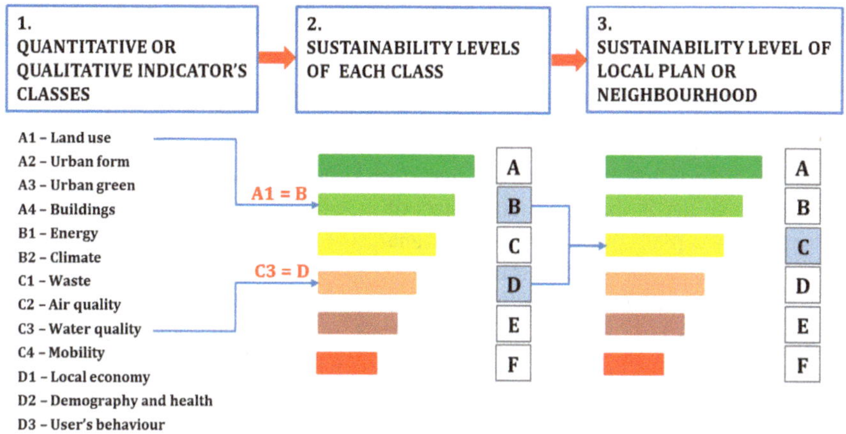

Fig. 4 Certification system of a local plan or a neighborhood as a combination of areas of actions and indicators

Each indicator is quantified in its measuring system. Consequently, the necessity of to reach a final value makes necessary the homogenization of the measure's units. For this aim, each observed datum is transformed in a standardized value using one of the available methods (step 2). The following passage is the association of a weight W to each area (step 3). This passage is necessary for defining the relative significance of an area compared to the other. The sum of the points represents the final value (FV) (step 4) that is used to associate the LP/N(A) to a level in the certification process (step 5).

Physical and functional indicators represents the areas of action A1÷A4, B1, C1÷C4, while social and behavioural indicators represents D1÷D3. For their characteristics, A1÷A4, B1, C1÷C4 can be used to define the certification of a plan, while A1÷A4, B1, C1÷C4, D1÷D3 can be used to define the certification of an urban area.

The application of this procedure makes it possible to define a certification system for local plans and/or urban areas based on values related to the consumption of resources and to the impact on environmental elements. This certification system associates to each operative plan or neighborhood a synthetic index defining its level of sustainability (Fig. 4).

3.3 Results and Discussion

The outlined operative structure can be used for three different aims:

- To assess the sustainability level of a LP/N at a given time. If we consider a specific time point, the use of the operative structure allows to define the level of sustainability at that time. The object of the analysis can be an operative plan (analysis of a plan) or a neighborhood (analysis of an urban sector).
- To assess the evolution of an urban area over time. The application of the method to the time points t_0 and t_1 allows to investigate possible positive and negative changes in the level of sustainability. Two cases can occur: (a) plan in t_0; urban sector in t_1; (b) urban sector in t_0, urban sector in t_1. In both the cases, the method allows to define two sustainability levels and to confront the changings among the two time points.
- To assess the level of sustainability of an urban structure as a sum of more areas, each of which characterized by specific environmental sustainability level. Even this assessment can be diachronic. A city can be considered as a sum of parts. The implementation of the procedure to each urban sector brings to a complete classification of the performances and, consequently, to the possibility of defining a synthetic index for the whole city. The analysis can be realized using a specific time point, or the evolution among two specific time points.

The third aim is of great interest: We consider at time t_0 an urban structure formed by N neighborhoods, existing or in planning. A certified sustainability level in A÷F scale can be associated to each neighborhood, for which the N system may be characterized by a vector of values formed by the sum of the N areas associated to their k certified levels. The resulting vector defines the level of sustainability of the entire urban structure (Fig. 5).

Repeating the operation at a time t_1 there is the high probability that one or more neighborhoods have changed their level of sustainability affecting, in this way, the final level of the entire urban structure.

The proposed methodology is characterized by being an expeditious procedure that certifies the sustainability of current or planned urban areas. Also if all the phases are important in the development of the procedure, the choice of the indicator system is a crucial moment for the success of the operation. In addition, we can assume that the system of the indicators is not closed, but may vary with the changing of the basic conditions created by the necessity of an urban transformation.

The same things can be said as regards the areas of action: They seem to be well defined and representative of the categories of transformation affecting urban transformations, but the system is open to the future contribution of other elements.

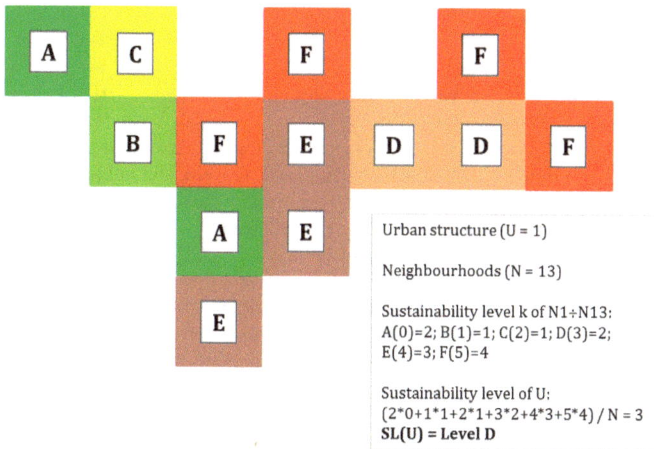

Fig. 5 From the sustainability of a neighborhood to the sustainability of the entire urban structure

4 Conclusions

The city is a human organized structure that often ends up under indictment for its unsustainability. Among the phenomena affecting the urban load we mention the urban diffusion and the functional concentration, negative especially in relation to the high levels of energy consumption and greenhouse gas emissions. Not forgetting the social phenomena, such segregation and exclusion.

75% of Europeans lives in towns and cities, 85% of GDP is created in urban areas, towns account of 80% of energy use and they face special social cohesion and environmental challenges. These numbers create the necessity to rethink towns and cities as pleasant and environmentally sustainable places where to live and work.

Over the past years, a series of innovation applied to urban and regional planning have proposed. The result is a changing in the approach to the city and a different affection on the performances of the urban and territorial systems. The examples are various and, among them, it is possible to mention the urban regeneration tools, designed to improve the use of urban space, or the mobility planning tools that want to increase the extension of the public network connection and its efficiency, or to experiment new techniques based on the control of the energy processes in the planning.

The developed methodology wants to put a further step in this process because it highlights the elements that must be part of a sustainable plan and it defines a procedure to quantify the sustainability reached by that plan.

It is therefore possible to arrive at an overall sustainability labelling, namely a categorization of the plans based on their contribution to the reduction of the climate change processes. This process of labelling can be extended to the city as a whole, characterizing it in relation to the advances actually realized.

References

Brown M (2016) FutuREstorative: working towards a new sustainability. RIBA Publishing, Newcastle Upon Tyne

Communities and Local Government (2009) Planning Policy Statement: eco-towns. A supplement to Planning Policy Statement 1. Communities and Local Government Publications, London

EEA (2013) Achieving energy efficiency through behaviour change: what does it take? Technical report 5/2013, European Environment Agency, Copenhagen

Elkington J (2015) Let's 'Uberize' sustainability. September 14, 2015. https://www.greenbiz.com/article/lets-uberize-sustainability. Accessed 31 Mar 2017

Galderisi A, Mazzeo G, Pinto F (2016) Cities dealing with energy issues and climate-related impacts: approaches, strategies and tools for a sustainable urban development. In: Papa R, Fistola R (eds) Smart energy in the Smart City. Springer International Publishing, Cham, pp 199–217. https://doi.org/10.1007/978-3-319-31157-9_11

Lucon O, Ürge-Vorsatz D, Zain Ahmed A, Akbari H, Bertoldi P, Cabeza LF, Eyre N, Gadgil A, Harvey LDD, Jiang Y, Liphoto E, Mirasgedis S, Murakami S, Parikh J, Pyke C, Vilariño MV (2014) Buildings. In: Edenhofer O, Pichs-Madruga R, Sokona Y, Farahani E, Kadner S, Seyboth K, Adler A, Baum I, Brunner S, Eickemeier P, Kriemann B, Savolainen J, Schlömer S, von Stechow C, Zwickel T, Minx JC (eds) Climate change 2014: mitigation of climate change. Contribution of working group III to the fifth assessment report of the intergovernmental panel on climate change. Cambridge University Press, Cambridge, United Kingdom. http://www.ipcc.ch/report/ar5/wg3/

Mazzeo G (2014a) Urban labelling: resilience and vulnerability as key concepts for a sustainable planning. INPUT 2014, Smart City: planning for energy, transportation and sustainability of the urban system. Tema J Land Use Mobility Environ S.l.:671–682. https://doi.org/10.6092/1970-9870/2483

Mazzeo G (2014b) Labelling urbano come programma di lavoro sulla città. Eyesreg 4(1):27–30

Mazzeo G (2016) La città leggera. Smart City e urbanistica attuativa. FedOA Press (Federico II Open Access University Press), Napoli. https://doi.org/10.6093/978-88-6887-008-9

Papa R, Gargiulo C, Zucaro F (2014) Urban systems and energy consumptions: a critical approach. Tema J Land Use Mobility Environ S.l.:784–792. https://doi.org/10.6092/1970-9870/2552

Stern N (2006) The economics of climate change: the stern review. Cambridge University Press, Cambridge

Conflicts and Sustainable Planning: Peculiar Instances Coming from Val D'agri Structural Inter-municipal Plan

Giuseppe Las Casas, Francesco Scorza and Beniamino Murgante

Abstract This paper examines experiences coming from the process of Inter-Municipal Structural Plan focussing on the evidences coming from conflicts assessment among major stakeholders' groups compared with the strategic development framework which represents the core of the plan. The composition of these conflicts makes the case study an example of the complexity challenge that we faced through a rational planning approach. Conflicts as factors of liquid society: we link our research in planning disciplines to this social debate because when we identified conflicts among stakeholders and social groups in a territory demanding for planning and more effective governance, we assume these factors as an indicator of the liquefaction of local community bringing to the instability of agreements among the components of that society. Among the proposed planning strategies, we discuss the Energy Museum project as the most characterizing context based perspective for the study area addressing the request to increase the knowledge and shared awareness on energy sector in order to enhance community awareness and reinforce social agreement in the area.

1 Introduction

This paper examines territorial organization choices defined by LISUT[1] research group in the framework of the scientific activities developed supporting the Regional Administration of the Project Val d'Agri, with reference to the design of Inter-Municipal Structural Plan (Las Casas et al. 2016, 2017). In particular, we focus on the evidences coming from conflicts assessment among major stakeholders' groups compared with the strategic development framework which represents the core of

G. Las Casas · F. Scorza (✉) · B. Murgante
Laboratory of Regional and Urban Territorial Systems Engineering (LISUT), School of Engineering, University of Basilicata, Potenza, Italy
e-mail: francesco.scorza@unibas.it

[1]Laboratory of Urban and Regional System Engineering—University of Basilicata.

R. Papa et al. (eds.), *Smart Planning: Sustainability and Mobility in the Age of Change*, Green Energy and Technology,
https://doi.org/10.1007/978-3-319-77682-8_10

the plan. The planning procedure is still not closed and the results discussed in this paper are still to be discussed with local decision makers (in particular Majors and other public institutions involved in the process).

Therefore, even the presented results are not to be considered the final planning scenario, what We propose as a disciplinary contribution is the technical and methodological instances at the base of this work: accountability, systematic and independent frameworks of knowledge, sharing of uncertainty and transparency remain the basis of our research project.

Val d'Agri is a very peculiar place: a territorial context where natural and water landscapes represent structural and identity heritage and evoke a development model based on agriculture and tourism, while the presence of the biggest continental oil field in Europe, in the last decades, started an alternative industrial development model generating—of course—significant impacts.

The result is that on one hand we find people that consider hydrocarbon revenues as a chance to reverse impoverishment and economic and socio-demographic trends (characterized by depopulation and unemployment) and, on the other hand, there are ranchers, farmers and part of the scientific community that invoke at least the precautionary principle in order to compare the oil industrial development with alternatives scenarios reducing environmental impacts.

The composition of these conflicts makes the case study an example of the complexity challenge that we faced through a rational planning approach.

It corresponds with the research for the three safeguard principles that we consider as the logical basis of our proposal:

i. efficient allocation of resources
ii. equity in the distribution of opportunities
iii. protection of non-renewable resources

and it expects to test the research of an a priori rational logic, in which the connection between targets, products, activities, means, and inputs become clear through the Logical Framework Approach (LFA) implementation.

This methodology is based on an operative vision of the Faludi's "proceduralist approach" (Faludi 1987) that—besides the desired scenario—proposes a process to define and to monitor objectives and strategies and promotes synergies in order to ensure the concentration of efforts on a few well-defined directions.

The strengths of this approach are the three above-mentioned principles and the explicit sharing of objectives: this represents Popper's demarcation principle that, by following Faludi (1987) identifies what can be called "a good plan."

We consider the evidences of the disciplinary debate that since the end of the seventies, the "classical" period of the so-called *systemic approach*,[2] relevant elements of dissatisfaction about the transition from analysis to project persisted.

[2]Among the authors, besides the best known McLoughlin (1969) and Chadwick (1971), we find a very interesting reconstruction of Michel Wegener (1994), proposing a survey 10 years later Lee's article (1973).

Such transition remained predominantly linked to the optimization attempt connected to Operative Research (Friend and Jessop 1969) and to the flourishing production of simulation models (cfr. Wilson 2016).

According to our approach (Las Casas et al. 2008, 2016), the rationality of decisions about citizens' needs and aspirations and the use of common goods and non-renewable resources must be considered as a citizen's right and so a prerequisite in the development of plan proposals. An approach whose method focuses on:

- collective learning processes that feed themselves through the awareness of the interaction system complexity connected to social fabric, economy and environment[3];
- governance processes that could be applied after the definition of objectives, means and activities, logical links between the achievement of the desired scenario and available means, an adequate system of indicators measuring effectiveness and efficacy.

The references of this approach are:

- from the technical point of view, the Logical Framework Approach, included SODA, about whom we'll talk at the point 3;
- from the point of view of the normative feasibility, the GPRA (Government Performance and Result Act of 1993 of the United States) (Archibugi 2002);
- from a theoretical point of view, our main reference is Faludi (1987), whose proposal considers the transition from a static concept of planning, that adopts technical knowledge to the development of a desired future scenario, to a dynamic vision, focused on the decision as a process.

In the next section, we present the general features of the case study area: Val d'Agri. Than we discuss the "conflicts" identification coming out from territorial analysis and strategy development.

The strategic framework is described in the Chap. 4 and an example of program structure for the implementation of a specific integrated development project is proposed in the Sect. 5. Conclusions regard persistent need for new rational tools supported by robust territorial models to support planning and decision making in a context of social conflicts and "liquid society" (Bauman 2013).

2 The Implementation Context: A Synthetic Description of Strategic Sectors

The application context of this research is represented by an inner territory of Basilicata Region. Val d'Agri is a mountainous area whose settlement system is placed

[3]Cfr. the conspicuous production of Roy Benard's group with reference to what was produced in the field of understanding and modeling of decision processes. Among others, see B. Roy (1985); Ostanello and Tsoukias (1993); Las Casas (2010).

along "Agri's valley" where the S.S. 598, a road infrastructure that connects the Valley with the provinces of Potenza and Matera and Vallo di Diano, represents the backbone of the settlements.

This context consists of 23 municipalities, grouped in our development perspective into four sub-areas of specialization. It is a territory characterized by very high environmental values: approximately 50% of territory is included in the National Park of Appennino Lucano and there are about 12.300 ha of SIC areas and 5.400 ha of ZPS.

The analysis of land use shows that the most of the territory is characterized by woods, pasture lands and agricultural areas.

The area presents a lot of physical and intangible endogenous resources linked to the forms of historical villages, to agricultural techniques and products, to cultural traditions such as Lady Maria cult or arboreal rites.

The physical component of territorial heritage is widespread on the whole area. Moreover we can find unique landscapes and villages, historical and monumental riches that are not yet enhanced by a structural policy of local development.

The rural and pastoral vocation is proved by local gastronomy: Sarconi's beans, protected by IGP, Moliterno's 'Pecorino Canestrato', 'Caciocavallo podolico', Missanello's oil, cured meats, Roccanova's 'Grottino' IGT, horseradish. All of these products represent a territorial identity that connects geo-climatic features with traditional practices and knowledge. This is a structural dimension that must be considered by prospects of economic development based on innovations and on the qualification of land products.

Since the beginning of fossil energy resources exploitation, Val d'Agri has obtained an "energetic vocation" that is shown in recent regional planning documents (especially in PIEAR Basilicata).

This development dimension has not an endogenous foundation but seem to be as a strategic area of interest that can attract investments, infrastructures, job, financial resources deriving from royalties and compensation forms.

It should be underlined that the identification of environmental impacts, a consequence of industrial plants and processes, is not entirely known and it depends on monitoring systems which are planned and managed at a regional level.

Today we can say that the expectations of local communities are only partially satisfied in terms of public resources invested in requalification, in support of local production activities (including agro-zootechny), in territorial promotion. Employment represents a weakest link in terms of local development since it delays a cross growth in economic sectors different from the industrial one and the migration of people in productive age is still persisting.

The "Green Economy" can be considered as a prospect for the territorial system: public and private interventions concerning the use of plants and technologies with low environmental impact are evaluated.

Renewable Energy Sources (RES) have been adopted according to the Regional Programme Framework contained in the Regional Environmental Energy Plan (PIEAR) that considers the Valley as a Regional "Energy District".

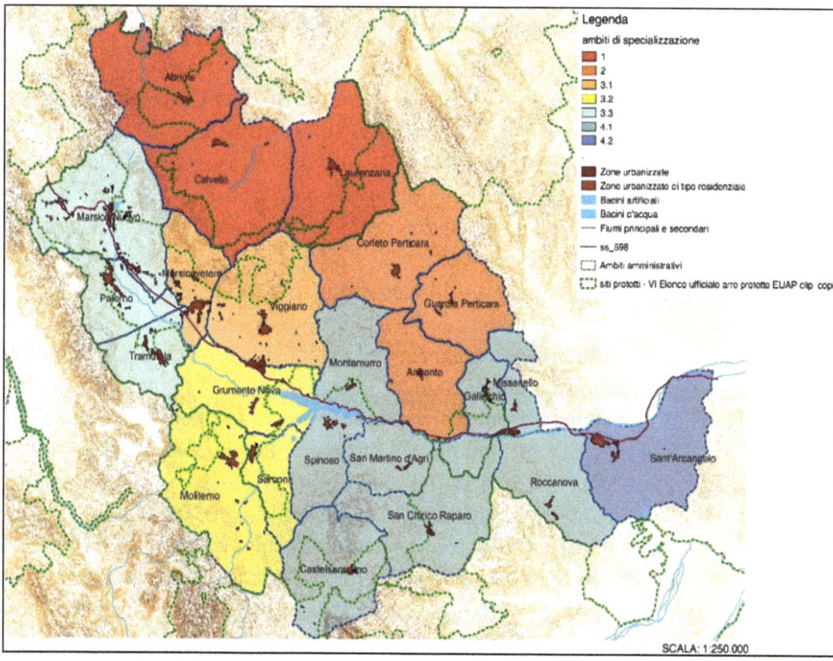

Fig. 1 Val d'Agri strategic sub-areas

Currently this forecast is not configured in terms of systemic interventions that are able to combine production processes and technological development with installations and implementations.

As regards oil, even if the discovery of oilfields dates back to the 80s, mining activities have been started during the 90s. The Oil Center of Viggiano was built in 1996 and in subsequent years an oil pipeline was devised in order to transport crude oil to the refinery of Taranto. The growth of mining activities and the hydrocarbon processing have brought several concerns about environmental safeguard, that is a precondition for all other forms of development.

There is the need to develop a knowledge framework that allows to define an important system of choices that, despite the decline of some areas (from an environmental point of view), could reinforce the defense of other areas through the strengthening of hiking, cultural and gastronomic tourism, agriculture livestock and dairy industry, small local industries that could benefit from the low cost of energy.

The following figure shows four sub-areas in which the implementation context has been divided (Fig. 1).

3 Conflicts as Evidence of Liquid Society Paradigm

Starting for Bauman (2013) position concerning Liquid Society Umberto Eco (2015) comments: *Is there a way to survive to liquidity? Yes there is. We have to realize we live in a liquid society and in order to understand it and also—hopefully—to overcame it we need new tools. The problem remains in the politic and in the most part of the intelligentsia that still not understood the extend of the phenomenon. Baumann remains a "vox clamantis in deserto".*

We link our research in planning disciplines to this social debate because when we identified conflicts among stakeholders and social groups in a territory demanding for development and more effective governance, we assume this factor as an indicator of the liquefaction of local community bringing to the instability of agreements among the components of that society, from which depend territorial planning choices.

The experience of the PSI (Inter-municipal Structural Plan) of the Val d'Agri (Las Casas et al. 2016, 2017) has offered us the opportunity to verify how the conflicts' overlap are related to the most dramatic problem: the conflict between nature conservation and industrial activities characterized by significant emissions and dramatic socio-economic effects. In the case of Agri Valley, it is possible to describe a high level of complexity in conflicts overlapping among the groups of actors, among individuals inside each group and, finally, any individual incoherence. Probably in this complex structure of conflicts lies the complexity that, according to Herbert Simon (1982), limits the possibility of knowledge about the planning context and imposes limitations to plan rationality (Fig. 2).

The conflicts examined in the case of the Val d'Agri can be identified on two levels: those emerging between "codifiable" groups—that is to say those categories of decision makers interacting within plan decision making process; and internal conflicts to those macro groups marking of disaggregated social tissue (i.e. "liquidity"). This happens due to the absence of appropriate means of knowledge on sensitive issues: pollution, health risks, loss of identity resources, a persistent disadvantage of opportunities for residents (in terms of work, services, infrastructures) in front of a territory facing an "announced disaster".

Our disciplinary position is based on the statement that the new tools, referred by Eco (2015), are those of a renewed approach to the plan rationality in order to reduce the impact of the lack of knowledge that determines the limits of rationality according to Simon (1972, 1982).

To answer Eco's questions, in our experience of PSI development we experimented two operational levels:

- Anti-fragile strategies to be confronted with communities;
- Procedures that clarify an implementation accompanied by monitoring of transformations (tactics).

This approach (Las Casas 1995) is based on the application of the following three principles, as a solid anchor point for the identification process of the hierarchical structure of problems based on a cause-effect relation and the consequent formulation of a hierarchical structure of the objectives (we call it "Program Structure").

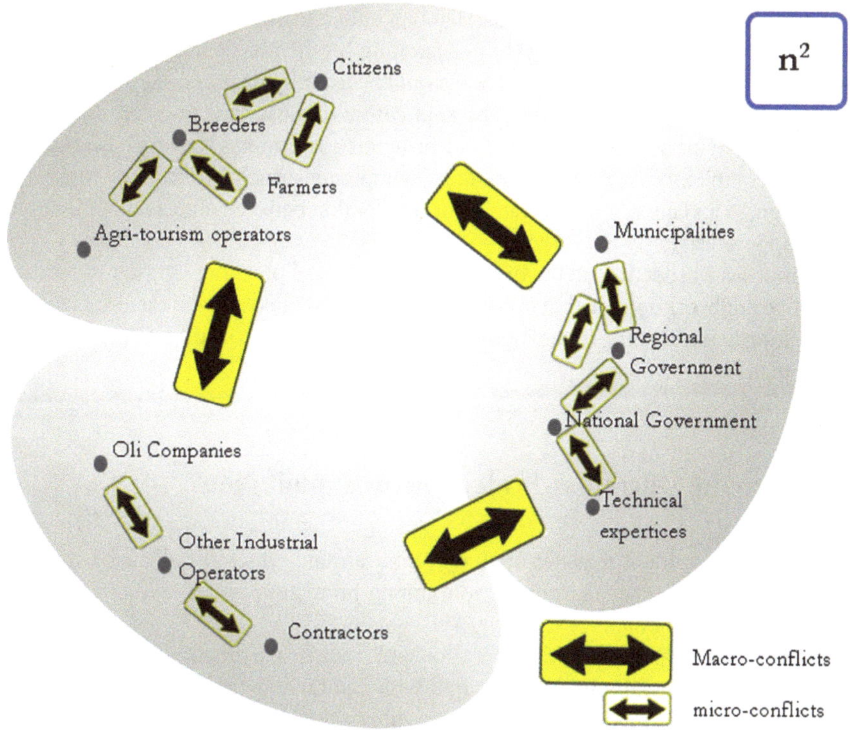

Fig. 2 Conflicts between groups and actors—The Val d'Agri case (*Source* Las Casas and Scorza 2017)

Following A. Faludi (1985, 1987), in fact, the question arises whether the problem is identified before of the objective according with the assumption that "the problem is what is meant to achieve an objective and the objective exists in order to remove the problem" and the answer was given by the three planning principles (Las Casas et al. 2018):

- equity;
- efficacy;
- un-reproducible resource conservation.

Their acceptance is given as universal as they correspond to the essence of the social contract that would ensure the cohabitation of members of a community. The constitutions of many countries rest on them. So, in applying these basic principles we affirm that in all that cases in which one or more of these principles are un-respected "problems" emerge, and to remove those problems is aimed objectives elicitation and planning strategies.

It seems that the need for planning secure territory represents a statement of aggregative intentions that arise with the natural disasters increase, international agreements, widespread literature.

In our view, it clearly appears that Bauman's liquefaction is precisely about tampering with these principles in favor of searching for the prevalence of an individual over another. Consequently, any strategy could be usefully be proposed in the field of public decision, without defending the application of those fundamental principles.

In addition, the acceptance of the three principles determines in every case another level of complexity: the need to manage competing principles such as equity vs effectiveness and ending with creating further conflict between ethics and individual or group interests.

It imposes a reflection on the disciplinary renewal of planning in order to face an operative and working role that oscillates between two opposite dimension: idealism and compromise bending to the interests of emerging powers leading to the negation of the plan.[4]

4 Dramatic Dilemma: Protection or Exploitation?

It seems to be evident that the dramatic point around which the decision-making process is tangled up is the connection among protection issues, exploitation and endogenous development of territory.

In the case of Val d'Agri, this aspect concerns research, extraction, transportation, processing of hydrocarbons and their link with natural and agricultural system.

We recorded the highest levels of uncertainty on which the debate, or better the dispute, became livelier about this key topic. These regard:

- objective, updated and forecasting data on the location and extent of activities;
- impacts on the air;
- effects on surface and underground hydrology system;
- effects on the quality of water transported by water schemes or reintroduced into surface water bodies;
- impacts on inhabitants' health;
- the allocation of revenues;
- future of the area after the depletion of deposits;
- direct and indirect impacts on employment, for activities related to hydrocarbon processing but also for agriculture and tourism;
- desires of a population that leaves the valley with a dramatic trend and achieves aging rates which, for some centres, do not foretell the permanence of inhabitants in the near future.

As for these and other problems, we cannot ask the question in a radical way: yes oil/no oil. Conversely, we have to attempt the research of appropriate context or place-based compromises, based on the sharing of information and uncertainty.

[4]The cases of decisions that in the name of acquired rights, or the need to cover public debts, or to accomplish instances of flexibility are available to introduce variants of the plan which, in fact, deny its robustness.

Only through awareness and sharing of uncertainty sources, debate and negotiation can develop a process whose main product is information. Financial resources from oil production could support such difficult research even if the total amount is not fully predictable for the future.

The proposed *vision* is based on the hypothesis of an area where the cultivation of oil fields and subsequent processes can coexist with the preservation of the most important share of natural features and the reinforcement of traditional activities and other innovations connected with local peculiarity.

We propose in fact a territory where the main mountain crests that enclose the valley and more internal ones, that include the greater and more significant natural areas, crown a valley where, conversely, the oil centre, pipelines and other innovative activities are dominant.

The hydrological system of the highest quotas will be protected and subject to monitoring procedures and aquifers will have to be effectively defended. The water collected in Pertusillo reservoir will be conveyed in a suitable pre-treatment plant before being introduced into the aqueduct.

Above all the idea to explore, produce and process oil in Val d'Agri must correspond to the exclusion of all other specified locations, avoiding the dissemination of plants without limits whose negative impact to land degradation has proved to be very serious.

The price paid in terms of environment is certainly high and it will be balanced by a development through which the activities of the valley may be partially sacrificed to the inevitable impacts (but not only those) derived from the exploitation of hydrocarbons.

Therefore, the measures of prevention, mitigation and precaution will be precondition for the reinforcement of an image, only dreamed nowadays, in which the weakest economic asset based on niche products, such as certified productions, or on the highly sensitive landscape natural elements and cultural heritage can be exploited through an integrated offer that links hiking, culture, food and wine to proposals such as the widespread museum of energy, widespread welcome and for the elderly (Table 1; Fig. 3).

5 "Energy Museum" Project

Among the proposed strategies, the one that characterizes the development perspective of the study area and which addresses the request to increase the knowledge and shared awareness on the underway processes in the energy sector is the "Energy Museum".

The project intends to integrate the conspicuous cultural heritage and environmental heritage through a system enhancing the knowledge dissemination of the historical relationship between man, the environment and energy, including future scenario. It is also oriented to link the undeniable sacrifices in environment to concrete prospects of replacing fossil fuel energy with clean energies, promoting a wide use

Table 1 Strategic functions and types of intervention

Function	Typology				
	a	b	c	d	e
Nature tourism	Walking tours	Horse	Cycle tourism	Environmental education	Equipment
Cultural and religious tourism	Museum	Fruition of energy museum	Festivals	Religious celebrations	Libraries and homeland history
Tourist accommoda-tion	Hotel	B&B	Farmhouse		
Health and old age	Senior housing	Health Aid			
Recovery and re-use for residences and services	Recovery and re-use for Leisure and socialization	Recovery and re-use for culture, entertainment and training	Education in the field of recovery and re-use and energy efficiency		
District/energy museum	Oil refining and monitoring	Energy history and economy	Technological park and innovative R.E.S. systems	The water mills	
Industry	Energy production	Electrochemistry	Metallurgy	Electronics	
Agro-Zootechnics	IGP bean	Ovine slaughter, preserves and dairy products	Podolica slaughter, preserves and dairy products	Restaurants	Logistics
Network: energy, water supply, transport and communica-tions	Freight transport	Purifiers and water schemes	ICT for people transport	Organization LPT	Connection with tourist demand

of the economic resources deriving from the extraction of hydrocarbons (royalties) in research and large scale RES (renewable Energy Sources) experiments considered as a relevant compensation form.

To this end, the museum will exhibit the technologies, showing their evolution and the evolution of the relationship with the daily life of man, with the economy and with the environment.

Following the well-known examples of the "territorial museums", it is a widespread museum that consists of some exhibition nodes and examples of techno-

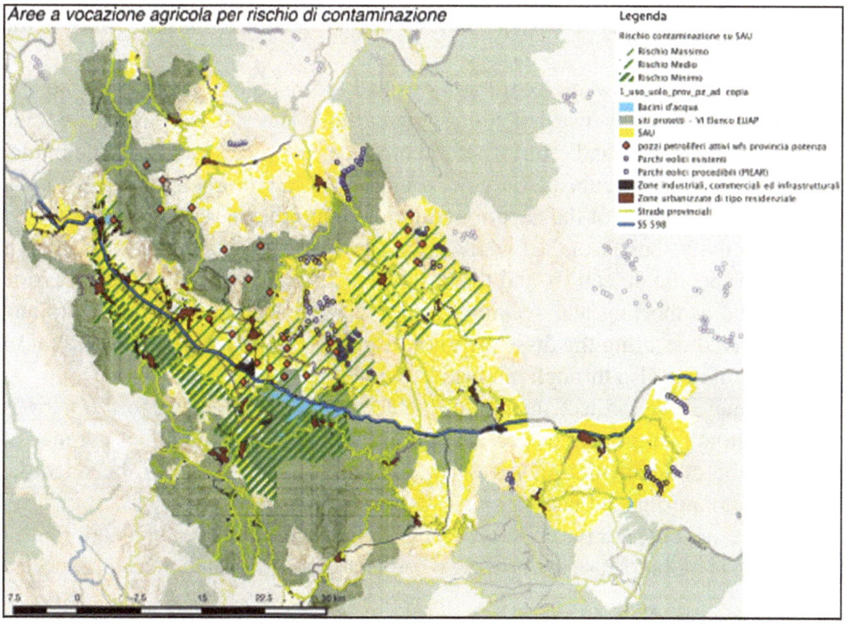

Fig. 3 Agricultural land per contamination risk

logical plants and will propose itineraries that represent a form of integrated tourist supply.

Key elements of the "Energy Museum" project are the energy plants on the territory, in particular those related to the oil extraction process as well as those related to energy production from renewable sources, intangible studies, research and experimentation, examples of innovation and experimental facilities and training/information activities aimed at local communities and visitors.

It is a project aimed at generating repercussions in terms of awareness of the processes underway from an environmental and social standpoint, assessments regarding specific conflicts, in-depth studies and research projects geared towards sustainable development, mitigation of environmental and human risks, to the formulation of strategies to safeguard and enhance the identity of characters within a process of exploitation of fossil resources.

The objectives concern actions from which local communities will benefit [O1]. Such actions are aimed at an understanding and awareness of the processes underway in the energy sector, accompanied by the establishment of a center of expertise [O2] aimed at the rigorous study of phenomena and the screening of territorial development scenarios. Suitable mean is the implementation of inter-institutional, interregional and transnational projects that involve local research and excellence centers. These transversal activities are enriched through the implementation of exploitation processes for the Widespread Energy Technology Park [O3] which includes the traditional forms of fossil resources' exploitation, the most recent ones linked to RES and innovative energy efficiency projects to be supported and disseminated on the territory.

The identity link between local traditions and future scenarios is represented by the network of old mills distributed throughout the territory. These artifacts—to be cataloged, redeveloped and refurbished according to the precise needs of the implementation of the "Museum of Energy" project—represent architectural historical emergencies and nodes of the territorial development schema [O4].

In this activity, the role of the Local Administrations will be of highly relevant. For each of the specific objectives and themes, it will set up an institutional discussion group aimed at meeting and seeking synergies and integration by proposers and actors involved ensuring the dissemination of results and the wider involvement of local communities also through innovative online tools.

Furthermore, this project intends to create a cooperation between the various research actors that operate or can operate in the region, with the dual purpose of achieving the dissemination of knowledge initiatives and to deepen the scientific aspects of monitoring and innovation.

This role and cooperative approach between the institutions is a precondition for the realization of the "Museum of Energy" project and, at the same time, an expected outcome.

6 Conclusions

The implementation of this renewed approach to plan rationality can take place on two levels:

- *"a tavolino" (on the desk)* it means as a theoretical exercise of planning developed in a technical dimension avoiding the debate with decision makers and relevant stakeholder comparing different steps of the program structure with appropriate MOVs
- on the field i.e. developing appropriate participation (Table 2; Fig. 4).

In the first case uncertainty remains, while working on field and so: accepted the ethical principle of sharing uncertainty and in accordance with the most recent international guides (UN Habitat 2015, 2016, 2017), comes out the usefulness of promoting participation.

Table 2 Energy museum program structure and M.O.V.

Energy museum program structure	
Overall objective	To disseminate knowledge of technologies and their relationship with man and the environment and to promote RES
Specific objectives	O1 Valorisation of knowledge about oil exploitation and treatment processes and environmental monitoring
	O2 Disseminate knowledge of the history of energy and its contribution to the economy
	O3 Valorization of the widespread technological park and experimentations of innovative R.E.S.
	O4 Knowledge enhancement of water mills and integration in the tourist offer system
Products	P 1.1. Spaces and equipment for reception and exposure
	P 1.2. Multimedia dissemination programs in the history of energy
	P 1.3. Short dissemination courses
	P 2.1. Study and project center
	P 2.2. Collections of ancient technological equipment
	P 3.1. Best practices: itineraries and guides
	P 3.2. Realization of innovative and demonstration plants
	P 3.2. Spaces and equipment for reception and exposure
	P 4.1. Routes of the mills equipped for the visit
	P 4.2. Mills and artifacts recovered

Participation represents a necessary condition for urban and territorial governance in a creative direction but should be extended beyond the only stakeholders that traditionally have the possibility to decide. It has to include the other groups that traditionally remain the "object" of the decisions.[5] Of course, it should include an extensive use of suitable knowledge-structuring tools as in Las Casas and Scorza (2016).

Since the Sixties[6], participation has been seen as a conquest of democracy and has resulted in the spread of assemblies, neglecting the effect, far more productive, which could be produced by a synergy of brains.

[5]In french: "Les agis" (B. Roy 1985)

[6](cfr. Davidoff 1965; Friedmann 1987)

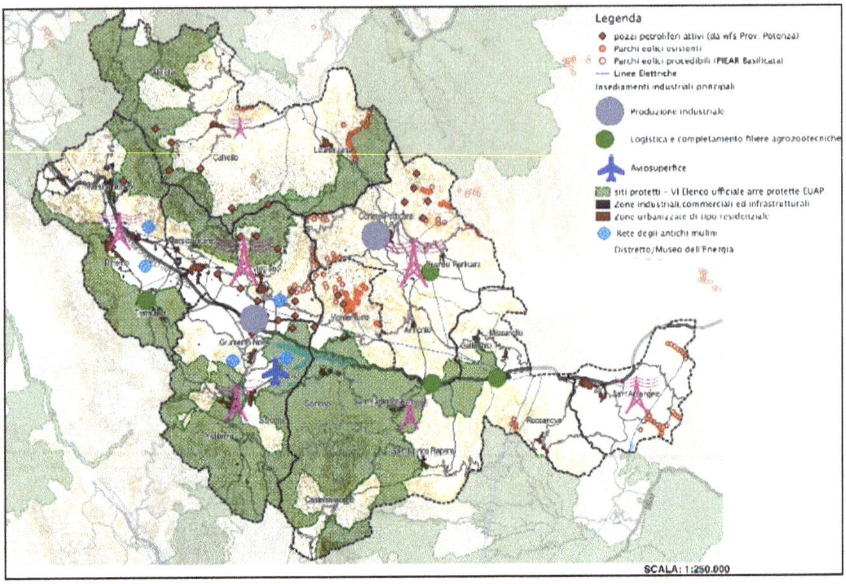

Fig. 4 Energy museum project

Now, if participation is certainly a conquest of democracy, it must also be taken into account that participation is a precondition for the development of a collective learning process which is at the basis of a decision-making process that invests the public sphere.[7]

References

Bauman Z (2013) Liquid modernity. Wiley, Hoboken
Chadwick G (1971) A systems view of planning. Towards a theory of urban and regional planning process. Pergamon Press, Oxford
Davidoff P (1965) Advocacy and pluralism in planning. J Am Inst Planners 31(4):331–338
Eco U (2015) La società liquida. Con questa idea Bauman illustra l'assenza di qualunque riferimento "solido" per l'uomo di oggi. Con conseguenze ancora tutte da capire. Repubblica L'Espresso
Faludi A (1985) A decision-centred view of environmental planning. Elsevier, Amsterdam
Faludi A (1987) A decision-centred view of environmental planning. Elsevier, Amsterdam
Friedmann J (1987) Planning in the public domain: from knowledge to action. Princeton University Press, Princeton

[7]"*participation and participatory process remain a central element of modern society, representing a prerequisite and a democratic right in Western nations. However, although their importance is underlined at the international level, the participatory practices has resulted in some criticism and problematic aspects due to the ambivalent nature and concept of participation. .... Governments, some times, implement inclusive procedure in order to reinforce the existing power relations...*". Leone e Zoppi (2016)

Friend JK, Jessop WN (1969) Local government and strategic choice: an operational research approach to the processes of public planning. Tavistock Publications, London
Las Casas GB (1995) L'etica della Razionalità. In: Urbanistica e Informazioni, vol 144
Las Casas GB (2010) criteri multipli e aiuto alle decisioni nel processo di piano. Available from: https://www.researchgate.net/publication/268262221_CRITERI_MULTIPLI_E_AIUTO_ALLE_DECISIONI_NEL_PROCESSO_DI_PIANO. Accessed 29 Apr 2016
Las Casas G (2011) Governo del territorio: innovare la ricerca per innovare l'esercizio professionale. In: Francini (eds) Modelli di sviluppo di paesaggi rurali di pregio ambientale. Franco Angeli, Milano. ISBN: 9788856840308
Las Casas GB, Sansone A (2004) Un approccio rinnovato alla razionalità nel piano. G. Depilano (a cura di), Politiche e strumenti per il recupero urbano, EdicomEdizioni, Monfalcone
Las Casas G, Scorza F (2016) Sustainable planning: a methodological toolkit. In Gervasi O et al (eds) Computational science and its applications—ICCSA 2016: 16th international conference, Beijing, China, July 4–7, 2016, Proceedings, Part I. Springer International Publishing, Cham, pp 627–635. http://doi.org/10.1007/978-3-319-42085-1_53
Las Casas GB, Scorza F (2017) I conflitti fra lo sviluppo economico e l'ambiente: strumenti di controllo. In: Atti della XIX Conferenza nazionale SIU, Cambiamenti, Responsabilità e strumenti per l'urbanistica a servizio del paese. Catania 16–18 Giugno 2016, Planum Publisher, Roma-Milano
Lee DB (1973) Requiem for large-scale models. J Am Inst Planners 39:163–178
Leone F, Zoppi C (2016) Participatory processes and spatial planning. The Regional Landscape Plan of Sardinia, Italy, Territorio, governance e sostenibilità. Franco Angeli, Milano. ISBN: 9788891740984
McLoughlin JB (1969) Urban and regional planning: a systems approach. Faber and Faber, United Kingdom, p 329
Ostanello A, Tsoukias A (1993) An explicative model of public interorganizational interactions. Eur J Oper Res70:67–82
Roy B (1985) Méthodologie Multicritère d'Aide à la Décision, Economica, Paris
Simon HA (1972) Theories of bounded rationality. Decision and organization 1(1):161–176
Simon HA (1982) Models of bounded rationality: empirically grounded economic reason, vol 3. MIT press, Cambridge
UN Habitat (2015) International guidelines on urban and territorial planning. UN-Habitat, Nairobi
UN Habitat (2016) New urban agenda. UN-Habitat, Nairobi
UN Habitat (2017) Action framework for implementation of the new urban agenda. UN-Habitat, Nairobi
Wegener M (1994) Operational urban models: state of the art. J Am Plan Assoc 59:17–29
Wilson A (2016) New roles for urban models: planning for the long term. Reg Stud Reg Sci 3(1):48–57. https://doi.org/10.1080/21681376.2015.1109474

Linking Knowledge to Action with Geodesign

Elisabetta Anna Di Cesare, Roberta Floris, Chiara Cocco and Michele Campagna

Abstract This chapter presents the novel geodesign framework methodology, as proposed by Steinitz (A framework for geodesign. Changing geography by design. Esri, Redlands, 2012), and relevant enabling technology as viable way to innovate spatial planning. Through the detailed description and critical review of a case study on the future development scenarios of the Cagliari Metropolitan City (Italy), the authors aims at demonstrating the geodesign approach may contribute to address some of the most urgent issue of the contemporary planning, as required in Europe by the introduction of Strategic Environmental Assessment. To this end the paper shows how the geodesign framework may link planning knowledge to action, support collaboration for pluralist and democratic decision-making, and ensure that impact assessment is considered during the design. Possible limitations and issues for further research are also discussed aiming at suggesting improvements of the current practice.

1 Introduction

Strategic Environmental Assessment (SEA), introduced by the European Directive 42/2001/EC, promotes a significant methodological innovation in the plan-making process aiming at enriching it with environmental considerations and public participation. Important conditions for SEA to be effective are represented by its inclusive and incremental attitude in defining the objectives of the policies which need to be assessed (Fisher 2003), and the effective participation of all the key actors in the process (Zoppi 2012), both during the preliminary and the *in-itinere* evaluations (Brown and Thérivel 2000). However, many difficulties are often found by professionals on the proper implementation of these principles (De Montis et al. 2014), especially in

E. A. Di Cesare (✉) · R. Floris · C. Cocco · M. Campagna
DICAAR, Università Degli Studi Di Cagliari, Cagliari, Italy
e-mail: elisabetta.dicesare@unica.it

R. Papa et al. (eds.), *Smart Planning: Sustainability and Mobility in the Age of Change*, Green Energy and Technology,
https://doi.org/10.1007/978-3-319-77682-8_11

setting a democratic process, in finding compromises during the participation phase and in consensus building (Zoppi, ibidem).

Geodesign may offer a viable contribution to address these pitfalls. It can be defined as a process for integrating methods, techniques and geo-information technologies, with the goal of supporting planning the physical evolution of the territory. It entails an integrated, collaborative and participatory approach that starts from the conceptualization of the project and continues with data analysis, process simulation, suggestion of alternatives, evaluation of impacts towards the final choice (Steinitz 2012).

Geodesign is currently receiving growing international attention in the field of landscape and urban planning, thanks to the advances and diffusion of cutting-edge information and communication technology introducing new potential for knowledge building and decision support.

Against this background, and in order to explore the potential of the geodesign methodological approach for bringing innovation into planning and SEA practices, we present the results of the international "Geodesign Workshop on Future Scenarios for the Cagliari Metropolitan Area", held in May 2016 at the University of Cagliari, from the workshop preparation phase to the elaboration of alternative design proposals. This case study shows the potential of the geodesign methodological approach to develop both environmental savvy decision-making and efficient collaboration processes, as required by the SEA.

The paper is structured up as follows. The next section gives a brief overview of the geodesign methodological approach and its application within the workshop workflow. The third paragraph describes all the pre-workshop activities: from the definition of the development objective for the metro area and the underlying scenario, to the representation of the territorial context and the evaluation of its dynamics. Paragraph four briefly describes the main steps in the workshop, and the last section concludes with a discussion on the geodesign potential to bring innovation into planning practices.

2 The Geodesign Methodological Approach

Geodesign, intended as a methodological approach to decision making informed by territorial knowledge, allows promoting multidisciplinary collaboration and participation. Geodesign workflows can be applied to urban and regional planning in order to understand how the context should be transformed in the future, according the geodesign framework (GDF) proposed by Steinitz (2012), whose core consists of six models. The first three models describe the study area before the implementation of the plan: based on a detailed description of the study area (Representation Model—RM), the evolution trends of the main territorial dynamics are investigated (Process Model—PM) and then assessed in order to evaluate possible opportunities and risks for proposed transformations (Evaluation Model—EM). The last three models consist of a design stage in which, starting from the identification of possible alternatives for change (Change Model—CM), and their impact assessment (Impact Model—IM), it is possible to choose a final agreed plan (Decision Model—DM). As such, the initial three models represent the assessment phase while the last three models constitute the intervention phase (Steinitz, ibidem).

With these respect, the GDF shows a consistent logic with SEA, which should be run since the early stages of the planning process in order to inform decisions at any step, and it may contribute to address many current SEA pitfalls encountered in the urban and regional planning practices (Campagna and Di Cesare 2016), and relating to the shift from knowledge to action, including the frequent lack of collaborative processes and plan alternatives design.

In order to test the potential of the geodesign framework to a planning case study, the GDF was applied during all activities of the Cagliari metro area futures scenario design, and it served as a guide for the workshop workflow since the very beginning of the process (Table 1).

The workshop is enabled by an advanced technology called Geodesignhub, a web-based collaborative planning support system (https://www.geodesignhub.com) which allows to implement the intervention models of the Steinitz' geodesign framework. In Geodesignhub projects and policies can be drawn in form of geo-referenced vector diagrams (i.e. line and polygons) so creating a matrix of possible transformation options. The latter, which are shared among the participants involved in the design collaboratively, can be then selected by stakeholder design groups to compile complex change proposals, or *syntheses*, which represents alternative plans proposals. The input data for a Geodesignhub project supporting the workshop are depicted with GIS procedures with which the models of the assessment phase are prepared.

3 The Workshop Preparation

While the core workshop activities involving the collaboration of the stakeholder groups involved in the geodesign study were carried-on in the two intensive days focusing on the intervention phase of the study, the local coordination team, responsible for the workshop preparation and consisting of senior and young researchers of the UrbanGIS Laboratory (i.e. experts in architecture, planning, environmental engineering and geo-informatics) worked beforehand with the conductor focusing on the assessment phase.

The Cagliari metropolitan city, located in the southern coastal part of Sardinia (Italy), was established through the Sardinian Regional Law n. 4/2016 including 17 municipalities, for a total population of approximately 420,000 inhabitants. The intrinsic economic vocation of the area include a rich variety of agricultural and fishery activities and, thanks to its natural and cultural landscapes together with improvement in the air-transport and accommodation offer, the area recently became one of the most important tourist destination of Sardinia.

One of the first stages during the workshop preparation was the definition of underlying scenario for the development of the Metropolitan City within a 25-year

Table 1 The geodesign methodological approach applied to the workshop workflow

GDF model	Process phase	Actors	Activities
RM	Workshop preparation	Local coordination team	Legislation framework review Choice of the 10 systems to be represented as design support structure Data collection and analysis
PM	Workshop preparation	Local coordination team	Analysis of territorial dynamics for each system
EM	Workshop preparation	Local coordination team	Creation of the evaluation maps of the 10 selected systems
CM	Workshop	WS participants	Design of conceptual projects and policies proposals related to each system as georeferenced diagrams in Geodesignhub Selection of diagrams and creation of 6 different development synthesis in Geodesignhub
IM	Workshop	WS participants	Real-time impact assessment of the six synthesis in Geodesignhub
DM	Workshop	WS participants	Negotiation towards a final agreed scenario with a sociogram and Geodesignhub

time frame, pursuing the overall goal to create new job opportunities and to promote a new sustainable tourism model. The objectives entailed in this scenario include: development opportunities in tourism, agribusiness, and the creation of an ICT industry pole, together with the essential sustainable transport system improvement, risk prevention, and residential and commercial land-use patterns balancing in the shift of the development scheme from the municipal to the metropolitan scale.

In addition, the local coordination team dealt with the territorial knowledge acquisition and evaluation, including data collection and integration, interpretation and analysis, in order to represent the territorial context in a consistent format with the Geodesignhub input requirements.

3.1 The Context Representation: The Choice of the Ten Systems

The representation of the study area commonly constitutes the first stage of a planning study and it constitutes the first model of the geodesign framework. In the Cagliari case study, the territorial knowledge was structured in ten systems, which were analyzed with the aim of providing common background information to all participants as a base for the design. The choice to describe the main opportunities and risks for the transformation of the land using ten systems was a reasonable compromise: ten is a number big enough to satisfactory represent the main features of the study area and limited enough to enable to handle the territorial system complexity.

The choice of the themes to be represented and the main characteristics to analyze, was strongly influenced by the scale of the study. Accordingly, the first step was the screening of the main national and regional legislation framework and in particular:

- The Sardinian Regional Spatial Planning Law (L.R. n. 45/1989), aiming at planning the use of the regional territorial resources at all levels and at regulating land-use modification interventions.
- The Legislative Decree on organization and responsibilities for Local Authorities (D.Lgs. 267/2000).
- The Sardinian Regional Landscape Plan (PPR-NTA Art. 106, 2006), which pursues the aim of preserving, protecting and enhancing the environmental, historical, cultural and local identity of the whole Sardinian territory, and to promote sustainable forms of development, thus it represents a framework for local land-use planning.
- National law on metropolitan areas, provinces and union of municipalities (L. Delrio 56/2014).

The main objectives affecting the future development of the metropolitan area, defined by regulatory framework, are summarized in Table 2. They are primarily related to the coordination of activities, as well as works, of major supra municipal interests in both the economic, productive, commercial and tourist sectors, as well as in the social, cultural and sport sectors.

The choice of the 10 themes was strongly affected by development goals previously defined. Three of them assess vulnerability elements: cultural heritage (CULTH), ecology (ECO), hydrogeological hazard (HYDRO); the last seven systems devise opportunities for development, or attractiveness, elements: tourism (TOUR), agrifood (AGRI), transports (TRASP), low density housing (LOW-H), high density housing (HIGH-H), commerce and industry (COMIND), smart services (SMRT).

Within the workshop context, that was a research study, the choice of the ten systems was made by the coordination team, playing the role of local expert. In a real case study the choice could have been made according to the SEA regulation requirements, by representatives of the local authorities and of the relevant authorities in environmental matters (i.e. scoping phase).

Table 2 The selection of the systems to be represented according to the legislation framework review

	Systems	Reference regulations	Related environmental systems	Vulnerability (V) or Attractiveness (A)	Objective	Selection
1	Hydrogeological hazard	D.Lgs.267/2000 NTA-PPR Art. 106	Soil, water, settlements, climate change	V	Soil protection, settlement safety in relation to floods and landslides	✓
2	Water bodies	D.Lgs.267/2000 NTA-PPR Art. 106	Water	V	Water resource protection (water body quality)	
3	Desertification	NTA-PPR Art. 106	Soil, climate, land-use	V		
4	Fire hazard	NTA-PPR Art. 106	Soil Flora, fauna and biodiversity Settlements	V	Fire prevention, with particular reference to forests and urban areas	
5	Industrial accident hazard	NTA-PPR Art. 106	Economic and productive sector Settlements	V	Settlements safety/human health	
6	Ecology/bio-Habitat	D.Lgs.267/2000 NTA-PPR Art. 106 L. Delrio 56/2014	Flora, fauna and biodiversity	V/A	Identification of protected areas, natural habitats, flora and fauna species of community interest Identification of ecological corridors in order to allow the connection among protected areas, habitats and natural areas, rivers and wetlands	✓

(continued)

Table 2 (continued)

	Systems	Reference regulations	Related environmental systems	Vulnerability (V) or Attractiveness (A)	Objective	Selection
7	Cultural heritage	L.R. n. 45/1989 D.Lgs.267/2000 NTA-PPR Art. 106	Landscape, historic and cultural resources	V/A	Protection and/or promotion of the historic and cultural resources	✓
8	Transportation	L.R. n. 45/1989 D.Lgs.267/2000 NTA-PPR Art. 106 L. Delrio 56/2014	Mobility	A	Transport Infrastructures and services development	✓
9	Tourism	NTA-PPR Art. 106	Economic and services sector	A	Coordination and promotion of economic and social development, ensuring support for innovative business initiatives, consistent with the metropolitan area vocation	✓
10	Housing	L.R. n. 45/1989 NTA-PPR Art. 106	Settlements Economic and productive sector	A	Coordination of municipal initiatives aimed at residential settlements identification	✓
11	Industrial/commercial	L.R. n. 45/1989 NTA-PPR Art. 106	Economic and productive sector	A	Coordination of municipal initiatives aimed at productive districts identification	✓

(continued)

Table 2 (continued)

	Systems	Reference regulations	Related environmental systems	Vulnerability (V) or Attractiveness (A)	Objective	Selection
12	Social/cultural/sports facilities	L.R. n. 45/1989 NTA-PPR Art. 106	Settlements Economic and productive sector	A	Integrated metropolitan area services management	✓
13	Energy/renewable energy	D.Lgs.267/2000	Energy	A	Renewable energy resource exploitation	
14	Waste management	D.Lgs.267/2000	Waste	A	Ensure an efficient waste disposal system	
15	Health services/schools	D.Lgs.267/2000	Settlements	A	Management of health and education services	✓
16	Land capability classification (agricultural productive soils)	L.R. n. 45/1989	Soil Economic and productive sector	A	Coordinated management of agricultural land uses	✓

3.2 From the Representation to the Evaluation Model

This section reports on how the maps representing the EM of two of the ten systems were created in a GIS environment. Each system was analyzed starting from the description of its current condition (i.e. RM) to the assessment of its territorial opportunities for change or conservation (i.e. EM), in order to give to the workshop participants ten evaluation maps of all the selected systems, as a base for the design activities (i.e. CM). All the EM maps were created in a GIS environment integrating several geographic datasets as input for a land suitability or risk analysis.

The construction of the first system reported here concerns the "Cultural Heritage" (CULTH), which identifies the most vulnerable areas (i.e. risk) in relation to the concentration of the most significant historical assets. The second system is the "Tourism" (TOUR), which identifies the most attractive areas (.i.e. suitability) to undertake tourism development. The data sources used for the evaluation maps creation includes official datasets, or Authoritative Geographic Information (A-GI), retrieved from the regional Spatial Data Infrastructure (SDI), as well as social media geographic information or SMGI (Campagna 2014) such as Panoramio.com georeferenced posts, and Booking.com hotels customers preferences. As a matter of fact SDIs faced prosperous development worldwide in the last decade and in many European regions give accessibility to spatial data for the wider public in order to support informed decision-making (Campagna and Craglia 2012). However, the use of SMGI may enrich the insight potential of AGI for, if properly analysed, (i) it supplies real-time proxy data on citizens movement in and usage of places, and (ii) it gives hints of user interests and preferences.

The integrated analysis of AGI and SMGI was carried on applying the SMGI analytical framework proposed by Campagna (2016) (Massa and Campagna 2016). As argued also by Briassoulis (2002) a relevant knowledge regarding economic, socio-cultural and environmental activities in destinations, for better investigating tourism phenomenon, is necessary. In this perspective, SMGI could provide a meaningful information and discloses opportunities for building analytical scenarios related to urban and regional planning.

CULTH, as a vulnerability system, identifies the areas affected by the major spatial distribution, density and proximity to the cultural heritage sites and artefacts to be protected for their historical value, according to the Sardinian Regional Landscape Plan (RLP). The data used to define the RM of the CULTH system (Fig. 1) is retrieved from the regional SDI in digital formats, and it represents the cultural and historical heritage assets in the area.

Specifically, these areas include:

- historic city centres;
- cultural goods (i.e. the combination of historic architectures and the archaeological sites);
- archaeological industrial areas related to the production processes of historical relevance (e.g. the Geological Mining Park and the historic saltworks).

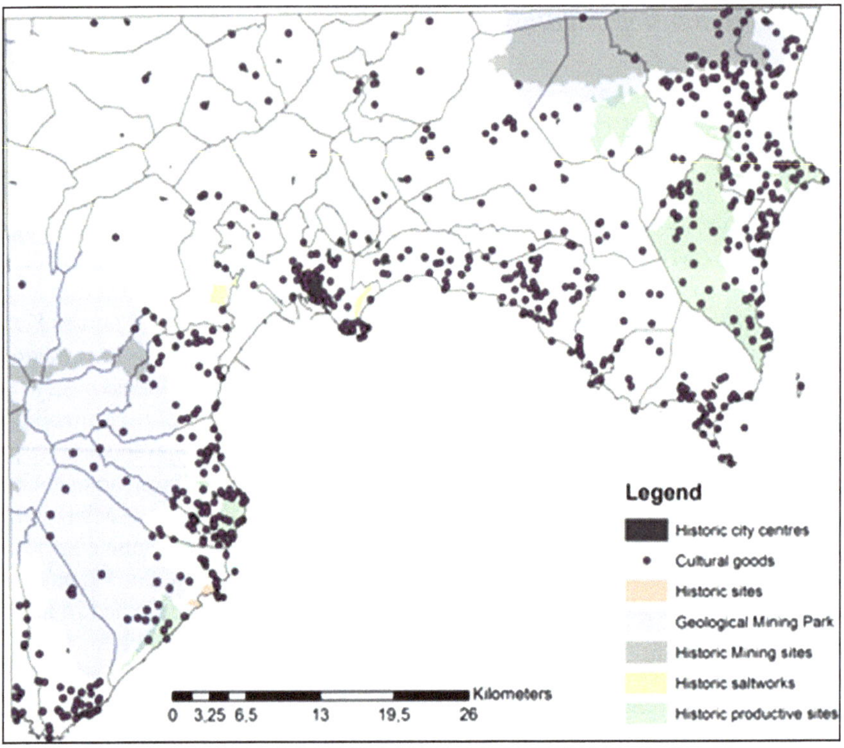

Fig. 1 Representation model of the CULTH system

In order to obtain an EM map of the CULTH system, each dataset is considered as a locational criterion. The CULTH map is implemented in order to describe spatial distribution of historical areas to be protected for future preservation strategies within the Metro area. Firstly, the historic city centers are given the highest vulnerability score, while a decreasing weights are assigned to two buffer zones of influence around them: the first buffer zone extending up to 300 m away from the site/artefact and the second one up to 1500 m. Secondly, a kernel density is implemented for points representing the cultural goods distribution, in order to identify the areas affected by their highest concentration. Lastly, the historical industrial sites are identified and assigned a vulnerability score (Fig. 2).

Each criterion was assigned a weight depending on its susceptibility, and consequently its need for protection: historic city centers and cultural assets have the highest weights, while the historical industrial sites the smallest one, for their lesser vulnerability. The final map was generated combining all these criteria together (Fig. 5a), also on the basis of the context knowledge of the local team, playing the role of local expert.

TOUR represents an attractiveness system, which depicts the spatial distribution of tourists' preferences regarding existing Tourism Lodging Services (TLS) as well as

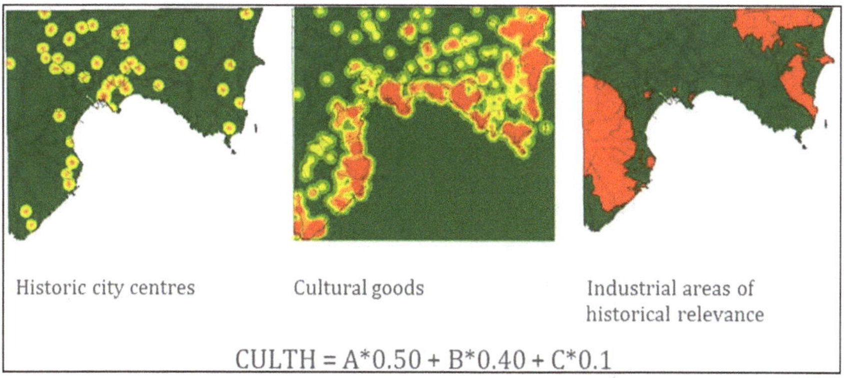

Fig. 2 Process model of the CULTH system

natural and non-natural resources. The innovative aspect of this map is the fact that it includes and represents, thanks to the use of SMGI, tourists' and local communities' perceptions and opinions, spontaneously generated by users (Goodchild 2007).

In fact, understanding the tourists' perceptions and opinions, and integrating this information with traditional A-GI, may represent an opportunity of great potential to enrich sustainable tourism goals with a broader, deeper and more multifaceted understanding of tourist destinations. With an improved awareness of the users' preferences, decision making can be simplified (Leslie et al. 2007) by emphasizing the strengths of tourist destinations for past and potential visitors.

In the light of these considerations, the RM of the TOUR system (Fig. 3) includes the spatial patterns of the following three key elements:

- The existing TLSs and their relative perceived quality, as retrieved from TripAdvisor.com and Booking.com. This dataset includes quantitative information concerning the TLSs scores based on rankings, divided into several categories, such as value/price, rooms, location, cleanliness and sleep quality.
- The already planned tourist areas, or "F areas" as defined by the 2266-U/83 Regional Decree. Data comes from the Municipal Master Plans of the 17 municipalities included in the Metro area and from the Sardinian RLP.
- The georeferenced users' image posts on Panoramio.com, considered as points of interest, from which it is possible to elicit the local landscape, and the visitors' perceptions of natural and historic and cultural resources.

The TOUR system evaluation map was built in order to describe spatial patterns of tourists' preferences and to identify locations of interest for future tourism development strategies within the Metro area. In order to obtain an evaluation map of the areas suitable for tourism development, three different criteria were defined, relying on the three spatial criteria described above. Firstly, a kernel density is implemented for points representing the spatial distribution of tourists' preferences, in order to identify the areas affected by their highest concentration. Secondly, the existing F areas are identified for the 17 municipalities comprising the Metro area and treated

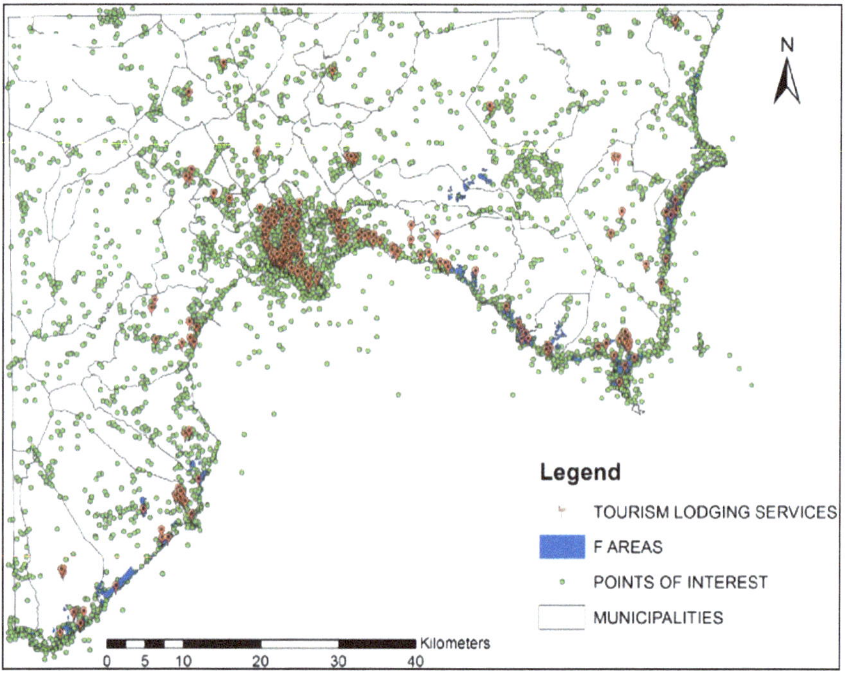

Fig. 3 Representation model of the TOUR system

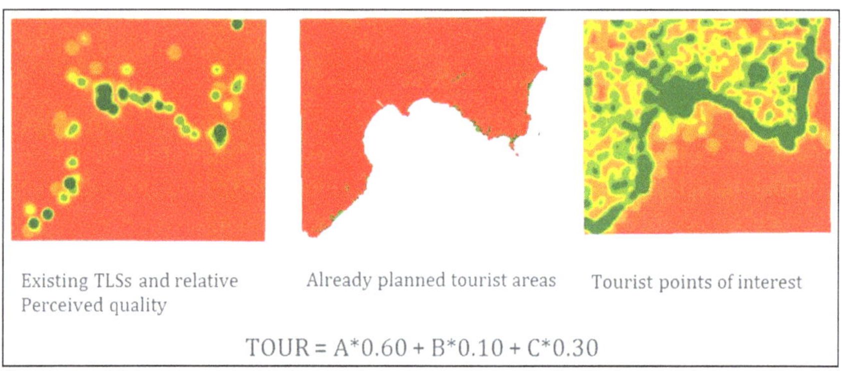

Fig. 4 Process model of the TOUR system

as a boolean variable. Finally, a kernel density is implemented for points concerning the users' contributions on landscape, natural and non-natural resources perception (Fig. 4).

Applying this model, the evaluation map was generated (Fig. 5b), by assigning different weights to each of the three criteria, considering their relative importance, and combining them together. More specifically we consider the presence of tourism

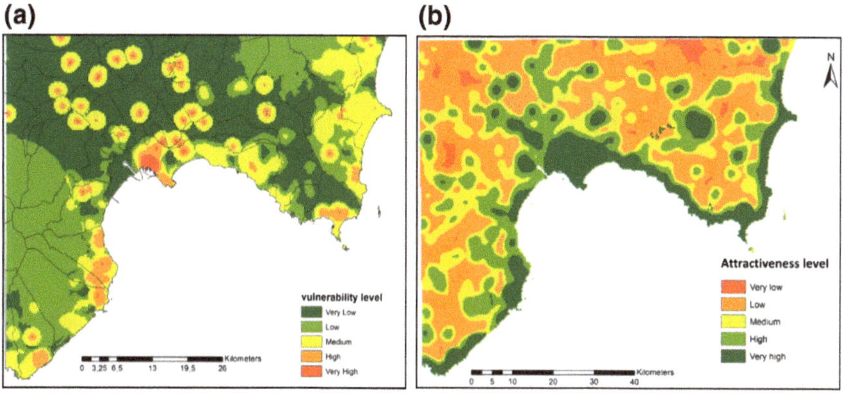

Fig. 5 Evaluation models of the CULTH (**a**)and TOUR (**b**) systems

facilities, accommodation and high tourists' satisfaction level with regards to location, as the most important factors for determining the attractive areas to implement appropriate tourism strategies. For this reason we assign the highest weight to the spatial distribution of tourists' preferences, which indicates the high tourism vocation of a specific area, while the users' perceptions on landscape and resources and existing F areas were assigned a medium and the smallest weight, respectively. Indeed, Municipal Master Plans regulations on the F areas, may not correspond to the real people' perception of places.

The result of the analyses of the EM of the CULTH system is a thematic map which classifies the study area in 5 vulnerability levels, where red areas indicate those characterized by a very high vulnerability, in which only actions aimed at preserving and promoting these sites can be permitted. To the contrary, the dark green areas are the less vulnerable ones, in which do not persists any restriction in use. Whereas in the TOUR system, the final evaluation map classifies the territory into 5 levels, with a color ramp where green color identify very high attractiveness areas for developing tourism development actions, thanks to the presence of existing tourism facilities, accommodations, scenic values and high users interest level for the area. Conversely, areas affected by very low attractiveness, due to the lack of tourism facilities, users' interest and very low accessibility, are depicted with the red color.

Applying the same methodology and the same classification and color code, the local coordination team produced the other eight evaluation maps related to the remaining systems. In Fig. 6 the whole set of evaluation maps given as input to the workshop participants is shown.

According to McHarg (1969) each place is a sum of natural processes to which corresponds social values. In order to respect these values it is important to identify the intrinsic vocation of a territory. The EM pursues this objective, but it is strongly influenced by the cultural and scientific knowledge of the those who create the model and by their role in decision making. As a matter of fact, maps can vary considerably

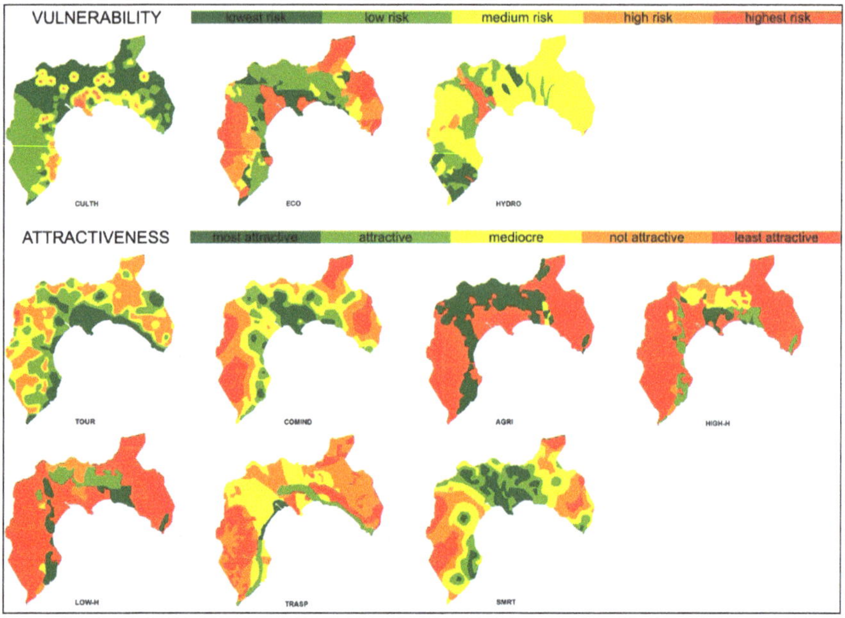

Fig. 6 The ten evaluation maps in the geodesign Workshop

in function of the data collected to describe a specific phenomenon, the criteria analyzed and their respective weights, the suitability/risk function and the modeling tools implemented and, inevitably, all these criteria relies on the planners' expertize. Since the output of the EM, in a geodesign study, provides the knowledge base for the design of alternative plans (CM), the decision-making process is strongly influenced by its results. Considering a real-world planning studio, a subjective perception of phenomena may affect deeply the decision making stage, influencing the shape of the final plan. Hence, an inclusive, participatory and multidisciplinary approach is fundamental in order to ensure a more democratic and transparent process during the definition of the EM. This position is clearly promoted by the 2001/42/EC Directive, which identifies as a primary goal of the SEA the definition of the more appropriate way to represent all the interests of involved parties, and especially to find as many agreement as possible so that all the key-actors' needs are represented in the decision-making processes.

Lastly, the local coordination team compiled and uploaded as input in the Geodesignhub platform a cross-systems impact matrix in order to qualitatively identify the impacts, from the most positive (value +2) to the most negative (value −2), of each single change action over the ten systems (Fig. 7), allowing the calculation and the visualization during the workshop in order to have real-time feedback of each design proposal.

System #	SYSTEM CHANGE	1 CULTH	2 ECO	3 HYDRO	4 TOUR	5 COMIND	6 AGRI	7 HIGH-H	8 LOW-H	9 TRASP	10 SMRT
1	CULTH	0	0	0	2	-1	0	-2	-1	-2	-1
2	ECO	0	0	1	0	-1	-1	-2	-1	-1	1
3	HYDRO	-2	1	0	1	1	2	1	1	1	1
4	TOUR	-2	-2	-2	0	2	1	-1	-1	2	0
5	COMIND	-2	-2	-2	-2	0	-1	1	0	-2	-1
6	AGRI	0	-1	-1	2	1	0	-2	-1	-1	0
7	HIGH-H	-2	-1	-2	-1	2	-2	0	-2	-2	1
8	LOW-H	-1	-2	-1	-1	1	-2	-1	0	-1	0
9	TRASP	-2	-2	-2	2	2	1	2	2	0	2
10	SMRT	-1	-1	-1	0	2	-1	2	1	-2	0

(Column headings: SYSTEM EXISTING CONDITION, System #)

IMPACT MODEL	
2	most positive, best
1	positive, good
0	neutral
-1	negative, bad
-2	most negative, worst

Fig. 7 Cross-systems impact matrix

4 The Workshop Workflow

The 2 days geodesign workshop was attended by the 32 participants with diverse backgrounds: researchers from several universities, students from architecture and civil engineering, public administration officials from the Sardinia Regional Government's Planning department and independent professionals (i.e. engineers, architects, agronomists). During the workshop the conductor was responsible for managing the timeline of all the activities and encouraging much as possible an effective communication among all the participants.

At the very beginning of the workshop each participant had a computer available to login and get familiar with the online platform Geodesignhub. The local coordinator introduced the study area and the earlier preparatory work for the workshop and explained the building of the ten systems aimed at representing a common knowledge basis for the participants on the base of which they could start design. As a first stage the participants, grouped in 10 teams, one for each system, were asked to producing a set of geo-referenced diagrams, representing change proposals (i.e. a project or a policy) related specifically to the system they were in charge of. To this end, the platform offers a sketch planning tool for drawing within the study area lines or polygons and for visualizing these changes in the geographic space in real time. The diagrams created in this first stage were around 200 and they were systematically organized by the software in the chronological order of creation into a matrix, where each column represent a territorial system (Fig. 8).

By the end of the morning, participants were divided into six teams, each one with a specific strategic role within the decision process, in order to consider the

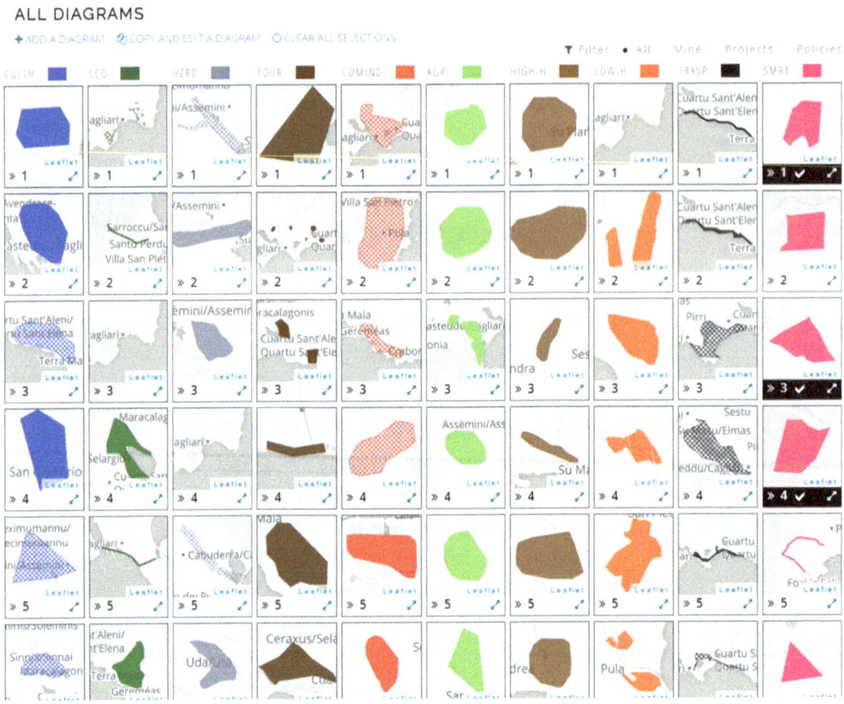

Fig. 8 Project and polices diagrams

outlook of diverse stakeholders from the Institutions, the private sector, NGOs, and other groups of interests, as listed below:

- METRO: Metropolitan government
- RAS: Regional Government of Sardinia
- GREEN: Green (NGO)
- CULTH: Cultural Heritage Conservation
- DEV: Developers
- TOUR: Tourism Entrepreneurs

They were asked to prioritizing the ten systems according to their specific role, expertize, and preferences and to selected a collection of projects and policies among the ones previously designed, in line with their development goals and interests. This way the first plan alternative draft (or syntheses) were created. Hence, the first 6 different syntheses, one for each group, were created. The online platform not only supports rapid syntheses creation, but it also computes real time impacts with a series of maps and histograms showing the direct impact of the proposed changes in each of the systems on a three-classes ordinal scale from positive (i.e. purple) to neutral (i.e. yellow), and negative (i.e. orange). The possibility of visualizing the performance of each alternative synthesis represents one of the central advantages of using digital

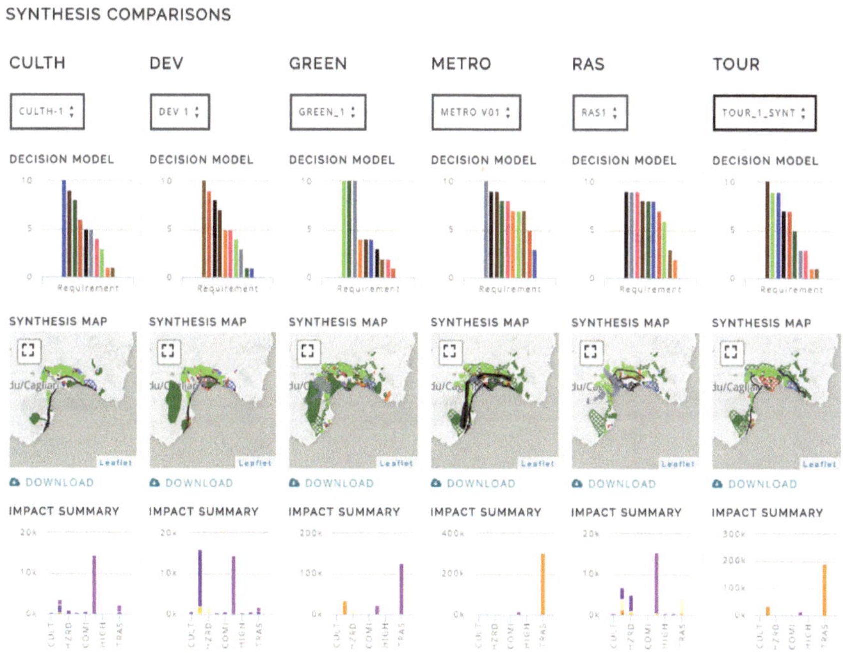

Fig. 9 The scenarios comparative tool showing the impacts performance of the six designs

geodesign technologies for it enables the dynamic revision of the change models through agile trial-and-error iterations. Moreover, Geodesignhub makes available tools for effective comparison of the alternative scenarios in form of maps and graphs (Fig. 9), facilitating the participants analyze differences and affinities between the designs.

After the first plan proposal were consolidated through impact assessment iterations, a sociogram for negotiation agreement was created with the aim of finding compatibility, hence possible alliances, between the groups. The sociogram is a matrix where each team vote the syntheses of the other groups, giving them a judgment, from positive to negative, depending on their compatibility with the goals of the team (Fig. 10).

The sociogram clearly showed the teams having higher potential for collaboration. On this basis, two different coalitions of teams were formed (i.e. TOUR, CULTH and RAS, and GREEN and METRO) reaching, after a first stage of negotiation, two combined design solutions (Fig. 11). The DEV group, which initially could not join any coalition for strong actual divergences, during the second stage of negotiation, decided to collaborate with the strongest coalition on the definition of a third negotiated scenario. Lastly, the final Cagliari metro area agreed scenario was reached at the end of a third stage of negotiation among all the teams (Fig. 12). This was the most crucial moment of the workshop, where the discussion not without difficult moments and animated debates, eventually led to the final agreement through negotiation.

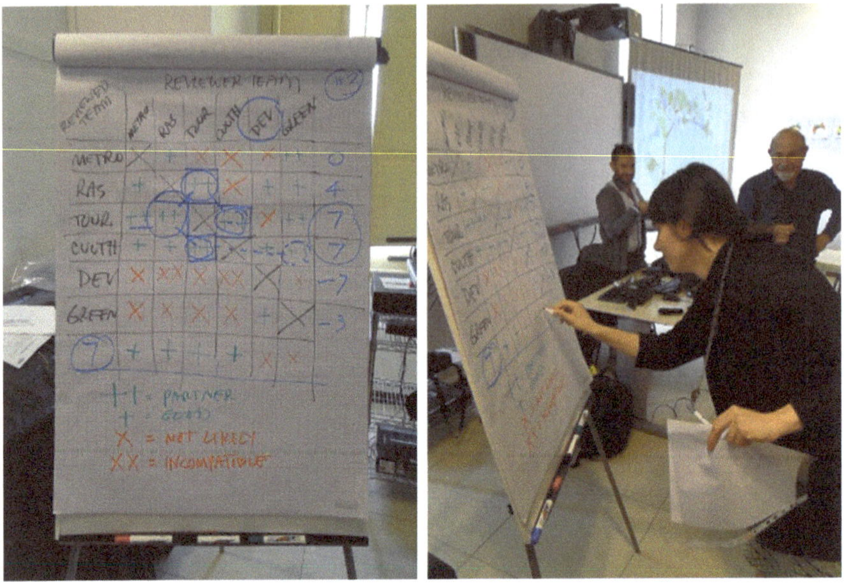

Fig. 10 The sociogram for negotiation agreement

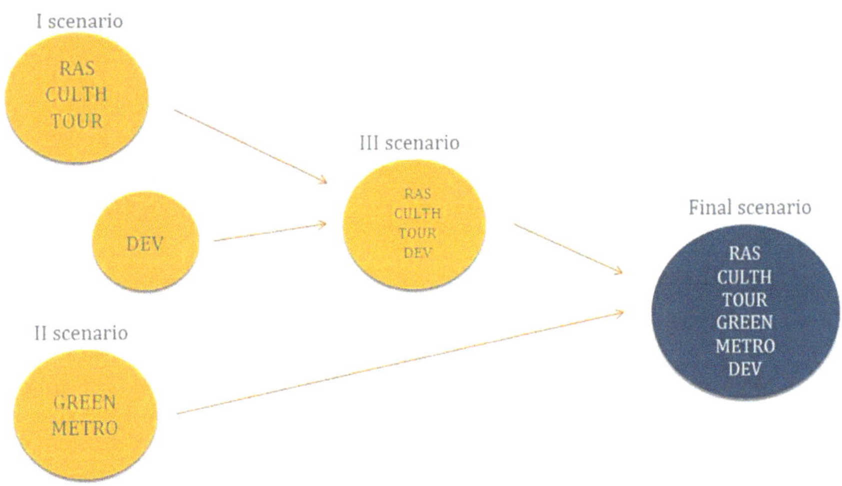

Fig. 11 Re-grouping during the negotiation phase

5 Results and Discussion

Indeed, the workshop represented an opportunity to reflect on the potentiality of the geodesign methodological approach to bring innovation in spatial planning. As a matter of fact, such a dynamic methodology, favored reasoning on the possible

Fig. 12 The negotiation process among the stakeholders and the final agreed design

strategic scenarios for the metropolitan area in a collaborative manner where all the participants could contribute actively, and allowed to attain a conceptual plan of development for the Cagliari metropolitan city after two only days of work.

The analyzed case study confirmed that geodesign might be well applied for the management of a planning process of remarkable complexity, which involves numerous actors and foresees and face the necessity to create and evaluate heterogeneous design alternatives. Thus it appears particularly suitable in the context of the landscape and urban planning process innovation, as introduced by the Directive 2001/42/EC.

6 Conclusions

This chapter presented the application of geodesign methods and techniques to spatial planning. After discussing the innovation of the current planning season introduced by Strategic Environmental Assessment, the authors reported the geodesign study on the future scenario for the metropolitan city of Cagliari (Italy). The chapter focused on the link between knowledge and actions in plan design and decision-making. On the base of the results of the workshop the authors demonstrated how the geodesign approach can support pluralism in design through collaborative decision-making.

Acknowledgements Elisabetta Anna Di Cesare gratefully acknowledges Sardinia Regional Government for the financial support of her Ph.D. scholarship (P.O.R. Sardegna F.S.E. Operational Programme of the Autonomous Region of Sardinia, European Social Fund 2007–2013—Axis IV Human Resources, Objective l.3, Line of Activity l.3.1.).
Chiara Cocco gratefully acknowledges Sardinia Regional Government for the financial support of her Ph.D. scholarship. (P.O.R. Sardegna F.S.E. Operational Program of the Autonomous Region of Sardinia, European Social Fund 2014–2020—Axis III Education and training, Thematic goal 10, Priority of investment 10ii.).

References

Briassoulis H (2002) Sustainable tourism and the question of the commons. Ann Tourism Res 29(4):1065–1085

Brown AL, Thérivel R (2000) Principles to guide the development of strategic environmental assessment methodology. Impact Assess Proj Appraisal 18(3):183–189

Campagna M (2014) The geographic turn in social Media: opportunities for spatial planning and Geodesign. In: Murgante B et al (eds) Computational mscience and its applications–ICCSA 2014. Springer International Publishing, Cham, pp 598–610

Campagna M, Craglia M (2012) The socioeconomic impact of the spatial data infrastructure of Lombardy. Environ Plan B: Plan Des 39(6):1069–1083

Campagna M (2016) Social media geographic information: why social is special when it goes spatial. European handbook of crowdsourced geographic information, pp 45–54

Campagna M, Di Cesare EA (2016) Geodesign: lost in regulations (and in practice). In: Papa R, Fistola R (eds) Smart energy in the smart city. Springer International Publishing, Cham, ISBN 978-3-319-31155-5

De Montis A et al (2014) SEA effectiveness for landscape and master planning: an investigation in Sardinia. Environ Impact Assess Rev 47:1–13

Fischer TB (2003) Strategic environmental assessment in post-modern times. Environ Impact Assess Rev 23(2):155–170

Goodchild MF (2007) Citizens as voluntary sensors: spatial data infrastructure in the world of Web 2.0. Int J Spat Data Infrastruct Res 2:24–32

Leslie E et al (2007) Walkability of local communities: using geographic information systems to objectively assess relevant environmental attributes. Health & place 13(1):111–122

Massa P, Campagna M (2016) Integrating authoritative and volunteered geographic information for spatial planning. European handbook of crowdsourced geographic information, pp 401–418

McHarg IL (1969) Design with nature. American Museum of Natural History, New York

Steinitz C (2012) A framework for geodesign. Changing geography by design. Esri, Redlands

Zoppi C (2012) Governance, pianificazione e valutazione strategica: sviluppo sostenibile e governance nella pianificazione urbanistica. Gangemi Editore spa

Part II
Mobility

“Smart” Planning to Enhance Nonmotorised and Safe Mobility in Today’s Cities

Maurizio Tira, Michela Tiboni, Silvia Rossetti and Michelle De Robertis

Abstract The objective of “smart” planners is to improve the quality of life in today’s cities. That quality is not only a matter of land use, urban design and density, it is also largely linked to mobility patterns. Urban mobility should take place in comfortable and safe conditions for all users, satisfying people’s expectations (especially those of vulnerable users) and the right to move. The paper aims to merge safety and quality of public spaces as an inextricable scenario for a better quality of life in our cities. Past research developed on nonmotorised mobility will be reviewed, in order to address the topic of vulnerable users’ safety from a more comprehensive perspective. Designing streets to balance the needs of diverse users is proposed as the key to shape an enticing environment that ensures access, safety, comfort, and enjoyment for everyone.

1 Planning and Designing Cities for Cars: Historical Milestones

Mobility has grown consistently since the end of the Second World War. In the industrialised regions of the world, travelled kilometres per person per day have boomed from 12.3 in 1950 to 45.5 in 1997, resulting in a staggering increase in total passenger-km travelled from 2628 to 14,951 billion between 1950 and 1997

M. Tira · M. Tiboni (✉) · S. Rossetti · M. De Robertis
DICATAM, Università degli Studi di Brescia, Brescia, Italy
e-mail: michela.tiboni@unibs.it

M. Tira
e-mail: maurizio.tira@unibs.it

S. Rossetti
e-mail: silvia.rossetti@unibs.it

M. De Robertis
e-mail: m.derobertis@unibs.it

R. Papa et al. (eds.), *Smart Planning: Sustainability and Mobility in the Age of Change*, Green Energy and Technology,
https://doi.org/10.1007/978-3-319-77682-8_12

Fig. 1 Motor Mania and the road rage of Goofy. *Source* Walt Disney (1950)

respectively (WBCSD 2001: 36). In 2015, only in the EU-28, the total number of passenger-km travelled was 6602 billion (WBCSD 2017). In the meantime, motorisation in Europe (and elsewhere) has shown an impressively fast growth, and development planning has been dictated primarily by car users' demand.

Massive car ownership has gradually reduced the quality of public spaces, now invaded by motor vehicles, and has impacted the behaviour of motorists who have become affected by so-called "road rage".[1] Road rage is not a recent phenomenon, but a recurring theme which affects motorists. This was ironically evidenced already in 1950 by "Motor Mania", a short Walt Disney cartoon, where Goofy, an animated character from Walt Disney's Mickey Mouse Universe, undergoes a metamorphosis due to a Dr. Jekyll and Mr. Hyde split personality: when he goes on foot he acts as Mr. Walker, a very kind pedestrian, but while when driving a car he becomes Mr. Wheeler, an aggressive car-driver (Fig. 1).

Automobile dependence (Newman and Kenworthy 1989; Dupuy 2002) was the inevitable consequence of municipal transport planning which was clearly biased towards cars. During the '50s, transport planning was particularly based on a modelling process that combined land use-demand and transport. The purpose of those plans was to anticipate the growth of population, activities and traffic flows for a

[1]Road rage is defined by the Oxford Dictionary as "a situation in which a driver becomes extremely aggressive or angry with the driver of another car because of the way they are driving".

period of twenty years, in order to ensure a balance between transport supply and the emerging demand generated by urban development. However, those plans were heavily geared towards accommodating the latent demand for driving and thus the planning of road infrastructures, rather than balancing the different transport modes (Tiboni 2010: 25). That process combined with the development of low density residential areas inevitably increased the dependence on the automobile for daily movements (Newman and Kenworthy 1999).

In the '50s and '60s the number of cars in Europe grew almost exponentially, and the traffic caused a strong pressure on the road network. In the context of economic growth, many efforts were made to adapt road infrastructure to the needs of individual motorised traffic. In that period, managing traffic congestion was a much greater concern than safety. Therefore, existing roads and new streets were widened, often by drastically reducing the public space available for pedestrians and cyclists (Tiboni 2010; Muhlrad 2000: 7).

In 1963, a team of technicians, coordinated by Prof. Colin Buchanan, produced a report for the UK Ministry of Transport titled "Traffic in Towns. A study of the long term problems of traffic in urban areas". The team was appointed by the UK Minister of Transport to study and analyse the problems arising from the rapidly increasing car use. "Traffic in Towns" is the first official document to recognise that the growth in traffic is a threat to the quality of life in the city (Ewing 1999: 11). Their results were interesting: Buchanan and his team, already in 1963, were able to identify and describe in detail the main difficulties connected to car use. In Buchanan's view, "*it is impossible to spend any time on the study of the future of traffic in towns…without at once being appalled by the magnitude of the emergency that is coming upon us….we are nourishing at immense cost a monster of great potential destructiveness, and yet we love him dearly….to refuse to accept the challenge it presents would be an act of defeatism*" (Buchanan 1963). In his opinion, congestion (what he calls "frustration in the use of vehicles"), accidents and deterioration of the environment were the three major difficulties, but it is possible to say that nowadays (50 years after the Buchanan's report) they are still the main concerns connected to car use.

At the core of Buchanan's approach was the belief that, in a traditionally designed town or city, there is an irreconcilable conflict between vehicle-based movement and a high quality local urban environment, which can be solved only through a physical separation among the different road users.

As a consequence Buchanan made a fundamental distinction between two kinds of roads: those for traffic distribution (movement) and those for local access to buildings and activities adjacent to the road, within protected "environmental areas" (see, among others, Bresciani 2007; Jones et al. 2008; Tiboni and Rossetti 2011). Therefore, the Buchanan report is often credited to have enhanced the modern traffic calming movement (Ewing 1999: 11, Fig. 2).

Even if Buchanan's approach was thought to best serve the needs of urban areas with a growing number of cars, its actual implementation revealed some problems: many new roads crossing urban areas were built, and city streets were often redesigned as high capacity roads with restrictions on footways (Tiboni 2010) and several negative side-effects (Tira 2003).

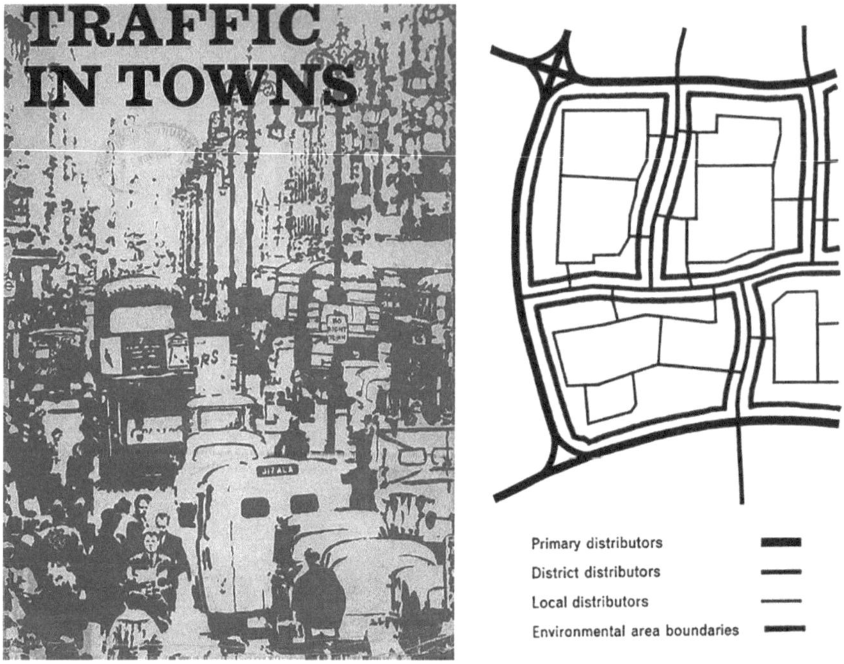

Fig. 2 Traffic in towns report and environmental areas layout. *Source* Buchanan (1963: cover and page 44)

Although part of the population as well as urban planners became aware that new courses of action were needed, policies focusing on car traffic were seen as "modern" and legitimized by technical progress, and lasted well into the '70s. Intentions were clear, as stated, for example, by the former French President George Pompidou in a famous speech in 1971: "*Cities must adapt to cars*" (Muhlrad 2000: 7).

At the end of the '70s the global energy crisis set in and on July 15th, 1979 U.S. President Jimmy Carter gave a nationally-televised speech, that came to be known as his "malaise" speech. In this speech, Carter started to promote the use of environment-friendly transportation modes, like car-pooling and public transport, in response to the energy crisis: "*I'm asking you for your good and for your nation's security to take no unnecessary trips, to use carpools or public transportation whenever you can, to park your car one extra day per week, to obey the speed limit, and to set your thermostats to save fuel.... I have seen the strength of America in the inexhaustible resources of our people. In the days to come, let us renew that strength in the struggle for an energy-secure nation*" (Carter 1979).

In the meantime, road accidents also started to be perceived as a major problem: the rapid growth of the traffic intensity both on arterial roads and local roads led to a growth of the accident rates, especially involving vulnerable road users. Road accidents had been a concern since the very beginning of the "car age". On August

Fig. 3 The "black widow" poster of 1946. *Source* Buchanan (1963: 16)

17th 1896, Bridget Driscoll was the very first pedestrian to be hit by a car (CROW 2009), so becoming the first known traffic fatality. The accident happened in London during a car show. Witnesses reported of the car colliding with the victim at extreme speed: the driver was going 4 miles per hour (Fleury 1998, Fig. 3)!

Furthermore, the "traffic segregation" approach started to show some difficulties in its application, especially within the city centres. The first examples of "traffic integration" were applied in the Netherlands, through the so called "*Woonerf design*" (Kjemptrup and Herrstedt 1992): residential roads were redesigned and organised as leisure areas for pedestrians and vulnerable road users, but allowed cars if driven at "walking speed". Following the Dutch example, new measures started to be implemented in European cities to calm traffic (chicanes, road humps, raised pedestrian crossings, pedestrian platforms, narrowings, pedestrian streets etc.) and to improve existing neighbourhoods (what later will be called shared spaces and liveable streets) (Fig. 4).

2 Land Use, Mobility and Automobile Dependence

Land use and mobility are strictly interrelated, and planning has a far-reaching effect on people's travel modes. That is a clear lesson from history.

Nevertheless, the integrated approach between urban and mobility planning is still nowadays an opportunity and a problem. In Italy, for example, already in 1935, Cesare Chiodi, a prominent professor of urban planning, described the importance of traffic issues for modern cities, stating that transport development should come first in a good urban policy, as a mean to control urban development (Chiodi 1935). One can easily recognize this as a precursor of Transit Oriented Development schemes. Similar ideas were then reaffirmed by Luigi Dodi in 1972. In Dodi's view, transport

Fig. 4 The Dutch "Woonerf Design". *Cited by* Tiboni (2010: 28)

networks and physical urban structure cannot be separated, rather they should be the object of a single plan (Dodi 1972: 358).

In the '80s, Newman and Kenworthy (1989) found a strong link between urban densities and oil consumption. Analysing 32 major cities in four continents, they found a correlation between urban density and the annual fuel per capita consumption, thus concluding that more compact cities are less dependent on gasoline. It clearly emerges from their study that the lower the density, the higher the motorised transport demand (Fig. 5). They also found a threshold of urban density of around 40 residents per hectare: at lower densities the gasoline use per capita grows very rapidly.

In subsequent research, Newman and Kenworthy (1999) defined the unsustainability of current urban transport patterns as "automobile dependence". When the accessibility level for cars increases, the degree of their use increases as well, while public transport use decreases and, consequently, society becomes car-dependent (Dupuy and Bost 2000). Therefore, car dependency must be addressed both as a social problem and from an economic perspective (Gorham 2002; Newmann and Kenworthy 1999).

The Victoria Transport Policy Institute (VTPI) describes automobile dependency as the result of a self-reinforcing cycle of increased automobile ownership, reduced travel options and more dispersed automobile-oriented land use patterns (Fig. 6). The challenge for sustainable urban transport and land use planning becomes that of achieving urban development with no, or at any rate, with as little as possible, increase in car use (Bertolini and Le Clercq 2003: 575).

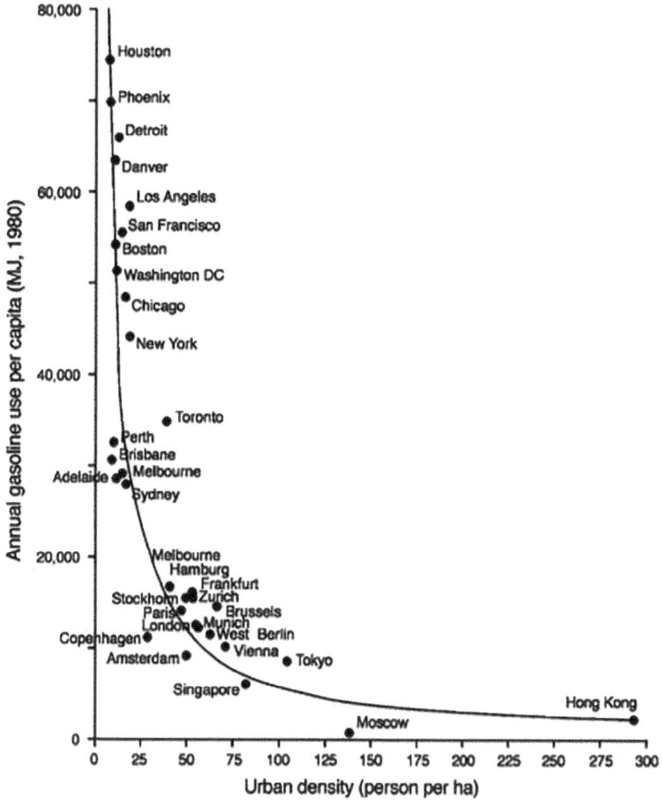

Fig. 5 Gasoline use per capita versus urban density (1980). *Source* Newman and Kenworthy (1989)

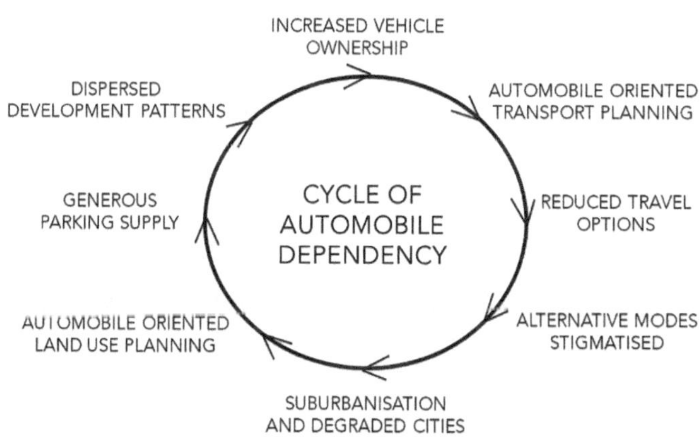

Fig. 6 The cycle of automobile dependency. *Source* Adapted from VTPI (2013)

Fig. 7 Urban Sprawl and road layout lead to land consuption and traffic growth

As represented in the circle, suburbanisation and, hence, urban sprawl, have increased travel distances and thus more reliance on motorised transport. The era of the private car has completely changed town design worldwide. Most new urban settlements have been planned explicitly assuming the use of the private car. In addition, urban sprawl causes several negative impacts to the environment, the main consequences being land consumption and traffic growth. When analysing sprawl developments, it is often difficult to distinguish causes and consequences since several factors interact and mutually amplify their contribution, causing greater pressures on the environment (Fig. 7).

Among others, Lord Roger's report "Towards an Urban Renaissance" (1999) recommended improvements to design standards to encourage people as well as facilities and activities to return to the cities. According to Lord Roger, "*...a dispersed city has large areas of low density development which are quite remote from the urban "hub" or centre. While the hub contains the core functions that support urban life – public transport, civic services, commercial and retail facilities – the lower density areas are mainly occupied by residential accommodation, with a limited mix of uses...*" (DETR 1999: 23–24). Therefore, he advocates for compact and well-connected city plans. The compact-city policy is based on a strategy of reducing the expansion of urbanised areas and filling the empty derelict spaces, now known as infill development.

3 How to Counteract Automobile Dependency? A Planning Point of View

The key urban planning approaches to curb automobile dependence and promote environment-friendly means of transport start by reducing the need for cars. According to Newman and Kenworthy (1999), four main policies are considered the basis to overcome automobile dependence:

(1) Implementing traffic calming measures as a means to slow car traffic and create more liveable environments;
(2) Improving the quality of public transport and facilities for cycling and walking to provide genuine alternatives to the car;
(3) Designing urban villages and managing growth to create multimodal centres with dense and mixed land uses thus reducing the need for travelling;
(4) Introducing transportation tax policies (e.g. adopting parking pricing and road pricing), to cover externalities related to individual motor vehicle use.

Land use policies are then one of the four legs playing a crucial role in a long-term perspective to change current trends and move towards sustainable cities. These policies include densification, transformation of former industrial areas and derelict land, and the concentration of development around public transport corridors.

Urban containment aims at restricting land development beyond a defined urban area, while encouraging infill development and redevelopment inside the urban area. The objective of urban containment policies is to reach and maintain a minimum urban density and a compact urban form. The redevelopment of brownfields, the set of boundaries to urban growth and the preservation of greenbelts rank among the most important containment policies.

The New Urbanism movement, which emerged in the '80s in the United States, aims at enhancing the quality of life, and at promoting an urban design that can contribute to curb automobile dependence. The movement incorporates the need of reducing automobile use, trying to achieve through town planning rules and practices: it attempts to create walking-friendly environments and neighbourhoods through denser and more mixed land uses, characterised by good connectivity and by a high-quality architectural and urban design, essentially reinventing the types of city layouts found in many European cities (Fig. 8).

New Urbanism particularly addresses the layout of streets and on the orientation of buildings, as well as with density and mixed activities (see, among others, Calthorpe 1993; Katz 1994; Congress for the New Urbanism 1999, Fig. 9).

More recently, relevant studies were carried out by Bertolini and Le Clercq (2003), who summarised some policy implications of a sustainable mobility approach (Fig. 10), that include dealing with daily activities: without travelling, by walking and cycling, by public transport, and, only as a last resort, by car.

4 Recent Development in Planning Research: Smart Planning for Nonmotorised Mobility

Automobile dependence is still a problem and safety of vulnerable road users a main concern. In 2013, 7600 people died in road traffic while cycling or walking in the European Union (source ETSC 2015). Furthermore, it has been argued that between

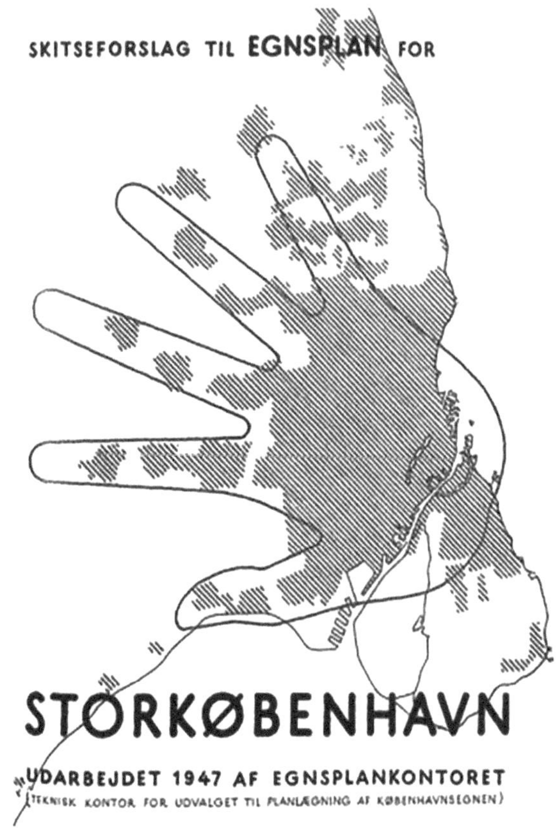

Fig. 8 The "Finger Plan" of Copenhagen (1947) formed an important basis for all further development. Train and bus lines service as "fingers", characterised by high densities, and the inhabitants are assured short distances to green areas between "fingers". *Source* Knowles (2012: 252)

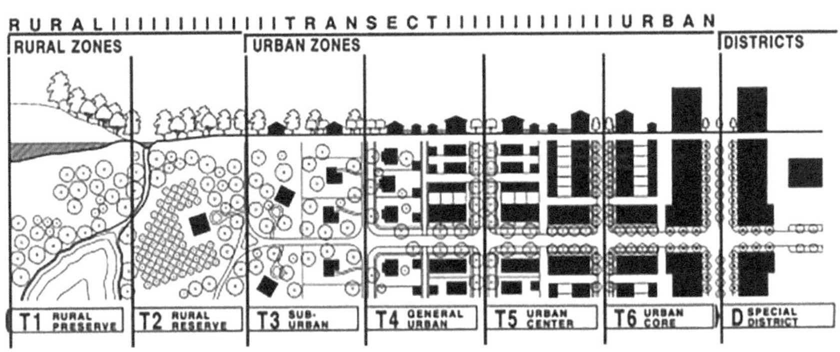

Fig. 9 The "transect" as schematised by the New Urbanism, with highest densities at town centre and progressively less dense towards the edge. This urban-to-rural transect hierarchy has appropriate building and street types for each area along the continuum. *Source* Duany and Talen (2002: 248)

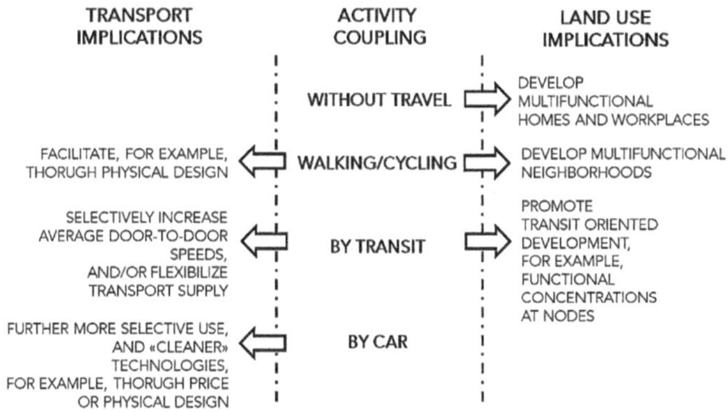

Fig. 10 Policy implications of sustainable mobility patterns. *Source* Adapted from Bertolini and Le Clercq (2003: 577)

20 and 30% of the variation in travel patterns can still be attributed only to land use and physical characteristics[2] of our cities (Banister et al. 2000).

Nonmotorised road users, especially walkers, are nowadays the indicator species for quality of life in our cities. Traditional research focused on infrastructure provisions and technical measures for vulnerable road users' safety (see, among others, Rossetti et al. 2014), while 21st century research should indeed focus on walking and nonmotorised mobility as a transportation mode: network, intermodality and integrated policies but also placemaking, living streets and Active Design approach (see, among others, Arenghi et al. 2016).

"*First life, then spaces, then buildings – the other way around never works*": 'life between buildings' should be the focus of city design (Gehl 2008). The buildings, which frame the spaces, need to contribute to the urban walking experience.

For "Smart Planners", it is now the time to place (once again and finally) people back at the heart of our cities and drive a human-focused approach to the design of the built environment.

The newly published Global Street Design Guide is a manual that aims at setting a global baseline for designing streets and public spaces while redefining the role of streets in a rapidly urbanizing world. The Guide broadens how to measure the success of urban streets to include access, safety and mobility for all users, environmental quality, economic benefit, public health and overall quality of life (Global Designing Cities Initiative 2016, Fig. 11).

The guide stresses the role of people: people use urban streets for mobility or for stationary activities, for leisure or for work, out of necessity or by choice. People of all ages and abilities experience streets in different ways and have many different needs. Whether sitting, walking, cycling, using collective or personal transport, moving goods, providing city services, or doing business, the various activities that streets

[2]The remaining 70–80% is accounted for by the socio-economic characteristics of people, including car ownership.

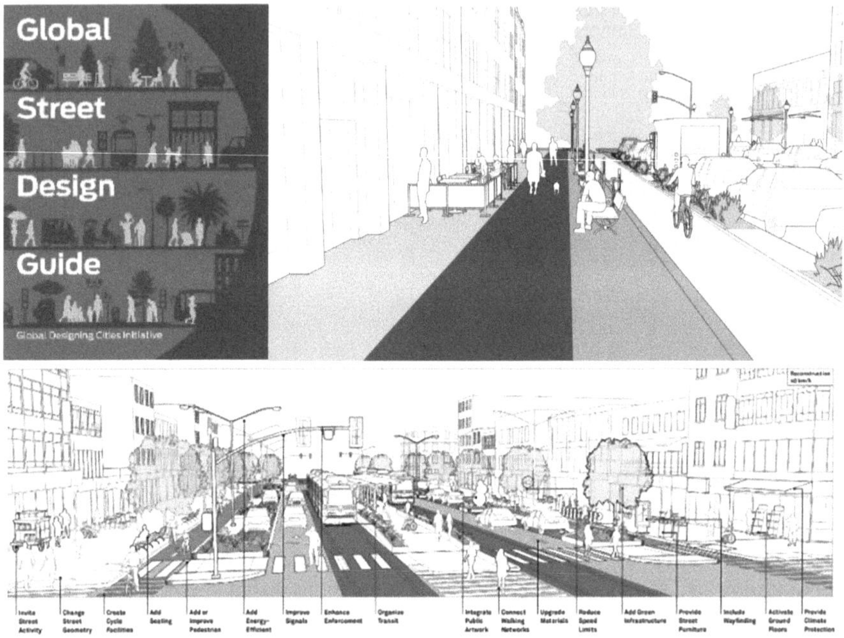

Fig. 11 The global street design guide. *Source* Global Designing Cities Initiative (2016)

accommodate and facilitate shape the accessibility and liveability of the city (ARUP 2016).

References

Arenghi A, Garofolo I, Laurìa A (2016) On the relationship between 'Universal' and 'Particular' in architecture. Studies in health technology and informatics 229:31–39

ARUP (2016) Cities alive. Towards a walking world

Banister D, Stead D, Steen P, Akerman J, Dreborg K, Nijkamp P, Schleicher-Tappesser R (2000) European transport policy and sustainable mobility. Routledge (Taylor & Francis), London and New York

Bertolini L, Le Clercq F (2003) Urban development without more mobility by car? Lessons from Amsterdam, a multimodal urban region. Environ Plan 35(A 2003): 575–589

Bresciani C (2007) Urbanistica e mobilità: regole e innovazione. Aracne editrice, Roma

Buchanan C (1963) Traffic in towns: a study of the long term problems of traffic in urban areas, reports of the Steering group and Working group appointed by the Minister of transport, Her Majesty's stationery office, London

Calthorpe P (1993) The next American metropolis: ecology, community, and the American dream. Princeton Architectural Press, New York

Carter J (1979) Crisis of confidence speech, televised speech delivered on July 15, 1979

Chiodi C (1935) La città moderna. Tecnica Urbanistica, Ulrico Hoepli, Milano

Congress for the New Urbanism (1999) Charter of the new urbanism: region/neighborhood, district, and corridor/block, street, and building. McGraw Hill, New York
CROW (2009) Road safety manual. Crow, The Netherlands
Detr (1999) Towards an urban renaissance. Final Report of the Urban Task Force chaired by Lord Rogers of Riverside. Spon Press, London
Dodi L (1972) Città e territorio. Urbanistica Tecnica, Tamburini editore, Milano
Dupuy G (2002) Cities and automobile dependence » revisité: les contrariétés de la densité, Revue d'Économie Régionale & Urbaine, n.1, Armand Colin
Duany A, Talen E (2002) Transect planning. J Am plann Assoc 68:245–266
Dupuy G, Bost F (2000) L'automobile et son monde. Edition de l'Aube, Paris
Ewing R (1999) Traffic calming. State of the practice. ITE, Washington
Fleury D (1998) Sécurité et urbanisme. La prise en compte de la sécurité routière dans l'amanagement urbain, Presses Ponts et Chausses
Gehl J (2008) Life between buildings: using public space. The Danish Architectural Press, Denmark
Global Designing Cities Initiative, and National Association of City Transportation Officials (2016) Global street design guide. Island Press, Washington, DC
Gorham R (2002) Car dependence as a social problem: a critical essay on the existing literature and future needs. In: Black WP, Nijkamp P (eds) Social change and sustainable transport. Indiana University Press, Bloomington, pp 107–115
Jones P, Boujenko N, Marshall S (2008) A comprehensive approach to planning and designing urban streets, Association for European Transport
Katz P (1994) The new urbanism: toward an architecture of community. McGraw-Hill, New York
Kjemptrup K, Herrstedt L (1992) Speed management and traffic calming in urban areas in Europe: a historical view. Accid Anal Prev 24(1):57–65
Knowles (2012) Transit oriented development in Copenhagen, Denmark: from the Finger Plan to Ørestad. J Trans Geogr 22(2012):251–261. https://doi.org/10.1016/j.jtrangeo.2012.01.009
Muhlrad N (2000) A short history of physical speed reduction measures in European urban areas. In: Proceedings of ICTC extraordinary workshop on traffic calming, ICTC Secretariat, New Delhi
Newman P, Kenworthy J (1989) Cities and automobile dependence: an international sourcebook. Gower Publishing Brookfield, VT USA
Newman P, Kenworthy J (1999) Sustainability and cities: overcoming automobile dependency. Island Press, Washington, DC
Rossetti S, Tiboni M, Tira M (2014) Road safety in Italy: An assessment of the current situation and the priorities of intervention. Period Polytech Transp Eng 42(2):159–165
Tiboni M (2010) La storia della moderazione del traffico in Europa e le sue implicazioni urbanistiche. In: Maternini G, Foini S (eds) Tecniche di moderazione del traffico. Linee guida per l'applicazione in Italia, Egaf, Forlì
Tiboni M, Rossetti S (2011) The role of urban planning into a long-term climate friendly transport vision. In: Radmilovič Z, Čišic D (eds) REACT "Shaping climate friendly transport in Europe: key findings and future directions". University of Belgrade Faculty of Transport and Traffic Engineering, Belgrade, pp 161–167
Tira M (2003) Safety of pedestrians and cyclists in Europe: the DUMAS approach. In: Tolley R (ed) Sustainable transport. Woodhead Publishing, Cambridge (UK), pp 339–350
Tira M, Rossetti, Tiboni M (2016) Managing mobility to save energy through parking planning. In: Papa R, Fistola R (eds) Smart energy in the smart city. Green Energy and Technology. Springer, Cham. https://doi.org/10.1007/978-3-319-31157-9_6
VTPI (2013) Defining and measuring automobile dependency. TDM encyclopedia victoria transport policy institute
Walt Disney (1950) Motor mania. Short animation comedy, USA
World Business Council for Sustainable Development (2001) Mobility 2001. World mobility at the end of the twentieth century and its sustainability, Atar Roto Presse, Geneva
World Business Council for Sustainable Development (2017) Passenger transport demand. Atar Roto Presse, Geneva

Territorial Governance and Mobility Management. A Smart Perspective for an Alpine City

Bruno Zanon

Abstract A rapid change has traversed local territories, which have been involved in globalization dynamics, in supra-local connections, and in wider processes of governance rescaling. These latter have implied the shift of competencies from national states to the European Union, on the one hand, and to regional authorities on the other; from the public domain to a variety of market actors. Mobility is part of this process, due to the larger scale involved, which breaks down the consolidated administrative borders, and because the traditional public transport systems, state-owned, have been substituted by a plurality of companies, both public and private. In regard to mobility, the stress is usually placed on the infrastructure; yet management is the key issue, because the governance of a complex system, responding to a variety of demands and composed of a plurality of actors, must be flexible and based on cooperation. And this requires a sound information basis and control mechanisms. The contribution addresses these issues by referring to the case of the Alpine regions along the Brenner corridor, and in particular to the city of Trento. Different scales interact, from the European one to the national, regional, and local ones. A smart perspective can capitalize on a variety of good practices developed in the regions involved, but it must overcome certain obstacles which hinder innovation. The appropriate use of knowledge and ICT technologies can be pivotal in supporting the re-organisation of mobility systems, appropriately connecting the different scales involved and stimulating the diverse actors to cooperate.

B. Zanon (✉)
Dipartimento di Ingegneria Civile, Ambientale e Meccanica, Università degli Studi di Trento, Trento, Italy
e-mail: bruno.zanon@unitn.it

R. Papa et al. (eds.), *Smart Planning: Sustainability and Mobility in the Age of Change*, Green Energy and Technology,
https://doi.org/10.1007/978-3-319-77682-8_13

1 Mobility Between Infrastructure and Service. The Role of Information

Mobility is a central factor in the dynamics of cities and territories. It interweaves personal, social and economic connections on a variety of spatial scales, from the local to the global, giving places specific opportunities and roles. Currently, a multiplicity of transport modes operated by private and public operators coexist, with a prevalent role of the automobile for people, and of trucks for freight. This is due to the freedom of choice and the flexibility of use provided by cars and trucks, to an unsatisfactory organisation of public transport, and to the uneven balance between the freight service by rail and that by road.

The consequences concern the direct and indirect costs of such a modal split in terms of high levels of fossil fuel consumption and greenhouse gas (GHG) emissions, as well as other environmental impacts and human health threats.

There is consensus that action is urgently needed, and many initiatives have been launched in terms of regulations, investments, and reorganisation of mobility systems. Some experiences have shown that results can be important, but generally they regard the local dimension (cities, regions) or single infrastructures or transport technologies. More complex is the connection among different scales and operators; and even more difficult is change in the behaviours of people and enterprises. Even price leverage is only partially effective in re-orientating the ways in which people move and goods are transported.

On addressing such issues, the stress (and the political rhetoric) is in general placed on the infrastructure, yet management plays the key role because the governance of a complex system, composed of a plurality of actors responding to a variety of demands, must be flexible and based on cooperation. What is required are a sound information basis and appropriate control mechanisms. Therefore, both 'hard' interventions—i.e. new infrastructures—and 'soft' initiatives based on coordination of actors, innovation of management systems, and better information, are required. In this regard, information and communication technologies—ICT—can provide key instruments with which to innovate practices.

The focus here is on the perspective of the application of ICT to mobility control in small-medium sized cities which are inserted in larger networks extending from the regional to the international levels. In general, they are not the most dynamic cities, but they are nonetheless involved in innovation processes, due to new initiatives—such as the allocation of educational and cultural institutions or innovative economic activities—but also due to their urban quality, which is attracting people and enterprises.

The urban dimension is a controversial aspect concerning mobility because medium sized cities suffer from congestion, but mass transportation systems cannot be adopted. Consequently, the quality of the service is low and the urban environment is negatively affected by car traffic. Moreover, new challenges must be faced, in terms of both supra-local connections (high speed railroad lines, freight re-organisation, etc.) and the effects of unexpected transport modes (low cost flights,

FlixBus, BlaBla Car, etc.). Other challenges concern uncertainty over the spread of new mass behaviours, such as cycling, car sharing, e-commerce, etc.

A territorial approach is adopted in order to address how spatial organisation is changing in relation to the extensive use of information technologies. The term 'territory' refers to the complex relationships among people, activities and places. Such a meaning is progressively extending from some languages and countries (France and Italy, in particular) to many European documents (CEC 1999; Territorial Agenda 2011), and can be helpful in framing the role of the new technologies in the management of a changing spatial organisation.

The paper develops a case study on the alpine regions along the Brenner corridor. It focuses on the city of Trento. Different scales interact, from the European one to the national, regional, and local ones. A smart perspective can capitalize on a variety of good practices developed in the regions involved, but it must overcome some obstacles which hinder innovation. The appropriate use of knowledge and ICT technologies can be pivotal in supporting the re-organisation of mobility systems, appropriately connecting the different scales involved and making the diverse actors cooperate.

2 Territory and Mobility: Distance, Speed, and Behaviours

Cities are dense places whose dynamics and success depend on the ability to provide easy relationships among people, activities and places, by means of an appropriate infrastructure which allows the mobility of people and goods, as well as the exchange of information and knowledge. The key features of historical cities consisted in their central role in a territorial network and in their physical structure characterized by diverse places and functions in close connection. Proximity was the factor supporting urban life. It enabled people to meet, markets to operate, and information to circulate.

The industrial revolution brought about dramatic changes as concerns both socio-economic processes and the spatial organisation of activities. Innovation of transport technologies made it possible to overcome physical obstacles and shortened distances, shrinking space and time. Proximity was no longer necessary for different production factors to be combined, while other communication technologies allowed a rapid and easy diffusion of information, regardless of distance. As Wegener (2012, p. 275) declared: "The rise of the modern city is built on mobility".

The contemporary socio-economic system is being affected by another radical change. This is produced by the rapid diffusion of ICT technologies, which provide new connections among places, people and activities by means of the instantaneous transmission of signals and information.

Material and immaterial exchanges between places and people are supported by physical means. This is the role of the infrastructure that forms the 'backbone of the territory', in particular of the urbanized space (Zanon 2015). It supports the relationships among communities, economies and places and shapes the spatial organisation and the urban form. It has been argued that the infrastructure is a 'sociotechnical' construction (Graham 2001) because it builds the territory by enabling communication, transport and relationships. But effects are not automatic. In fact, it is necessary to distinguish the infrastructure from the use made of it: 'mobility', 'transport', 'traffic', 'communication' are different terms denoting diverse processes and different issues. Hence 'mobility' means freedom of movement, while 'traffic' refers to flows (and often to congestion), 'transport' to specific systems, and 'communication' also to immaterial exchanges.

The increasing attention paid to mobility depends on the form assumed by the contemporary urbanized space, which extends across broad territories, and by the effects of the globalization of economy and society. Such aspects require easy connections on different scales: local/regional, and regional/global. But the spatial dimension is increasingly measured in terms of travel time, not in physical distance. In fact, "Distance travelled has grown dramatically over the recent past, and all activities depend on travel and movement to get people and goods to where they are needed" (Banister 2011, p. 950). The increasing speed is enlarging the spatial dimension, so that what used to be regional is becoming local, while the nodes connecting the local and the global levels are being displaced from the cities (railroad stations, in particular) to external points (airports, high speed train stations, multi-modal centres).

The consolidated approaches and the usual practices concerning mobility management must therefore be renovated. Banister (2008) stressed that transport planning is traditionally based on two principles: first, travel is a derived demand, not "an activity that people wish to undertake for its own sake"; second, the main concern is the minimisation of travel costs (Banister 2008, p. 73). Also Wegener (2012, p. 276) stresses that mobility theories are based on strict connections among costs, speed and distance. Consequences regard reduced attention to the experience of travel and a predominant effort to increase the speed, which means longer distances covered in less time. In fact, in transport planning time is considered a cost, and what is forgotten is the experiential side of travelling, for people do not decide simply on the basis of time or cost saving, but on the information at their disposal, on the quality of the service, or on the ease of access. And people travel not only to access workplaces or specific points of interest but often simply to move around (Mokhtarian and Salomon 2001).

3 Mobility Governance: Places and Flows

Mobility flows reflect the spatial allocation of human activities and places of interest within wider connection networks. A rapid change has recently affected local territories, which have undergone reorganization due to local dynamics and globalization processes. The resulting—often unexpected—supra-local connections generate both opportunities and threats.

In parallel, wider processes of governance rescaling have taken place, overcoming the consolidated institutional borders and redefining the traditional politico-administrative competencies. Responsibilities have scaled up from the nation states to the European Union, on the one hand, and scaled down to regional authorities on the other, and from the public domain to a variety of market actors (Zanon 2013).

As regards public transport, in the European countries the traditional—state-owned—public systems have given way to a plurality of companies, both public and private, some of which propose innovative solutions leveraging on use of the Internet to access information and to select the service. Mobility of people and goods is particularly affected by the use of ICT technologies and by management methods taking full advantage of the potential of the new information systems, in particular reducing operational costs, first of all ticketing. Also road pricing and access control in many areas are already automatized. On the customers' side, it has rapidly become common behaviour to make use of the Internet to obtain information on travel alternatives, to buy tickets, and to combine different travel means in order to reach a destination.

As regards freight transport, it reflects the epochal change in industrial production, which is currently based on the assembling of different components produced in distant places, following the 'just in time' manufacturing system. At the same time, e-commerce is changing access to the market place. Both phenomena produce increasing levels of traffic.

Such rapid and profound changes connected to the diffusion of the new technologies affect not only mobility but also the nature itself of regions and cities and how they operate. Consequences regard a new spatial organisation of activities and new modes of access—physical and immaterial—to an increasing number of opportunities, changing the consolidated form and organisation of the inhabited space, cities in particular (Delponte 2012).

Both opportunities and threats are therefore present, and they must be appropriately conceived and managed. In particular, an increase in transport capacity means generating environmental problems, while more efficient mobility levels must be pursued, together with a more efficient use of fossil fuels and the abatement of GHG emissions. Such goals imply, as regards people mobility, a more balanced modal split, discouraging useless journeys (dematerializing many exchanges), a more efficient collective transport, as well as more cycling and walking. As concerns freight transport, it requires transferring a good part of the traffic from road to rail. What is needed is not only the use of the best available technologies as concerns vehicles (in terms of energy and emission performances, safety control, etc.) but also the integrated organisation of mobility and transport systems.

Public transport can play an important role. It is aimed at providing freedom of movement and equal access to urban opportunities to all citizens, regardless of their personal circumstances and place of residence. It aims also at reducing traffic and abating emissions by providing shared means of transport. Such goals must be pursued by providing freedom of choice at reasonable costs for the customers. Therefore, the quality of the service must be comparable with—or higher than—that of private car in terms of comfort, safety, and speed of travel. Not only the quality of the service and of the fleet must be considered, but also the quality of the places involved (bus stops, stations, etc.), assuming an urban approach to mobility, which is a social phenomenon influenced by the perception of the environment and of the experience.

Another crucial point regards the fragmentation of the transport system among different companies and a variety of means, which solicits the activation of an appropriate management, providing a friendly access system for the user: single information portals, combined tickets, easy payment, etc. Many experiences are proving that there are feasible methods and solutions which produce an increase in public transport use. Such results are dependent not only on heavy technological investments but also on actors' coordination and integration, as well as on an extensive use of the technology we all share: our smartphones. In fact, they make it possible to access information, purchase services, and pay small (and, on certain conditions, large) amounts of money.

Physical aspects must not be forgotten, however, because coordination and integration start at the local scale. Therefore, mobility and parking demand must be appropriately connected to the urban form and land use decisions, considering that the allocation of activities implies intervening in traffic attractors.

4 Governance and ICT in Mobility Management

The consolidated urban form and functions are challenged by new spatial organisation models which make use of innovative methods and devices developed within the ICT disciplines and technologies. In this regard, a well-known proposal goes under the name of 'smart city'. Its definition is far from being univocal, but the idea mobilizes public policies and private investments in view of more efficient and effective ways to manage the urban space and activities. Data management and computer science experts played a central role in the first elaboration of this proposal, which is aimed, in short, at dealing with complexity. Urban planners came later, placing less stress on technologies and more on due attention to the effects of ICTs on the urban form and functions. 'Smart city', in their view, often overlaps with the concept of 'sustainable city'. In general, doubts arise as to the 'urban' meaning of the concept and of the actions undertaken, with the suspicion that there are some marketing strategies designed to strengthen the role of technologies in order to create some dependence on the new devices (Kunzmann 2014).

The new perspective is stimulating, anyhow, and many initiatives, at different levels, have been activated to support local authorities in shifting towards the smart city. To be cited, in particular, are the documents of the European Union, UN Habitat III, and, as regards Italy, the "Italian National Smart City Observatory" (Osservatorio Nazionale Smart City) promoted by the Association of the Italian Municipalities.

A number of definitions of 'smart city' have been provided, marking an evolution of the concept over a short time (Nam and Pardo 2011; EUP 2014; Albino et al. 2015; Papa et al. 2015). 'Smartness' can be conceived in different ways, with stress not only on the potential effects of the use of technologies but also on the efforts required by the governance of complex systems, and on the diverse abilities of urban communities (people and institutions) to learn and adapt.

The debate has seen a progressive evolution, with a shift of the focus from hardware (technologies and infrastructure) in the early 2000s ('smart city' as a 'digital city') to software (human and social capital, participation) around the middle of the decade ("social and inclusive city"), then to the integration of such aspects ("city with a high level of life") (Papa et al. 2015, p. 29). It has been argued that the problems of urban agglomerations have been positively addressed by making use of creativity, human capital, cooperation, as well as brilliant scientific ideas. These are, in reality, the 'smart solutions' not necessarily connected to ICT technologies which enabled modern cities to develop by supporting quantitative and qualitative improvements of productivity (Caragliu et al. 2011, p. 66).

The stress placed on the city as a self-organizing system has underlined the role of the (collective) ability to learn, to valorise the adaptability and flexibility of the urban system, shifting the stress from the technological factors to the social and human ones (Papa et al. 2015). In other words, availability and quality of the ICT infrastructure are not the only ways to define the smart city, because human capital and education are pivotal, considering that the 'creative class' tends to converge on some specifically characterized places (Caragliu et al. 2011, p. 67).

Stress has also been placed on the integration between 'smartness' and 'sustainability'. In 2015, the working group of ITU-T, a UN organisation engaged in definition of technological standards concerning telecommunication, proposed the following definition: "A smart sustainable city is an innovative city that uses information and communication technologies (ICTs) and other means to improve quality of life, efficiency of urban operation and services, and competitiveness, while ensuring that it meets the needs of present and future generations with respect to economic, social, environmental as well as cultural aspects" (ITU-T 2015).

A 'working definition' has recently been provided by a European document (EUP 2014, p. 9): "A Smart City is a city seeking to address public issues via ICT-based solutions on the basis of a multi-stakeholder, municipally based partnership". In short, a socio-technical approach must be applied, considering the city as a government unit which must attend to developing and coordinating technology factors (the infrastructure, ICT), human factors (human and social capital) and institutional factors (governance, policy and regulations) (Nam and Pardo 2011).

A smart model, therefore, cannot be limited to the use of technological systems, which "must be seen as an evolutionary stage of the city" (Fistola 2013, author's translation). For sure, technologies must be used appropriately so that many urban phenomena can be monitored and governed, also with a view to activating citizens as 'anthropic sensors' able to feed the flows of data and information (Fistola 2013). The smart perspective does not regard solely the metropolitan areas; it also concerns many medium-sized cities (50–200,000 inhabitants) which are characterized by a manageable size and by upper-level functions (Fistola 2013).

What are the key issues for a smart city? A well-known survey (Giffinger et al. 2007) conducted comparison among a number of European cities on the basis of various items which must be qualified as 'smart': economy, people, governance, mobility, environment, living. This survey marked a step forward by operationally integrating ICT issues with others which have emerged in recent years as pivotal for dynamic cities. As usual, problems regard how to measure the levels reached by each city, making use of appropriate criteria and indicators. In particular, different aspects must be considered, of a physical nature on the one hand, and concerning politico-administrative actions, socio-economic levels, and citizens' perception of the urban environment, on the other.

Another study, focused on the same features in a large number of European cities, observed that smart city initiatives intervene mainly in some areas because they "are spread across all six characteristics, but most frequently focus on Smart Environment and Smart Mobility" (EUP 2014, p. 9). Moreover, most of the actions remain at the planning or pilot testing stage, highlighting that much has still to be done to achieve the expected results.

5 A Methodological Framework for a Smart Perspective

The definition of future prospects requires a methodological basis. In this regard, planning methodologies have undergone progressive development. Banister and Hickman (2013) observed that, in general, the tasks and commitments of the different approaches have shifted from forecasting sectorial trends to a more comprehensive view of social, economic and environmental factors, "so that there is a more fundamental understanding of socio-technical change" (Banister and Hickman 2013, p. 291). Other steps have been made towards a "backcasting" method, "to look at preferred futures over the longer term". This is a normative approach concerned "not with what futures are likely to happen, but with how desirable futures can be attained", "in order to determine the physical suitability of that future and what policy measures would be required to reach that point" (Robinson 1990, pp. 822–823).

In this perspective, the use of information methods and technologies can be pivotal in constructing innovative paths so as to attain the integration of different processes, the coordination of actors, an active role of citizens and customers. Such integration requires the definition of a framework clarifying what issues are to be addressed and what actors are responsible for what actions.

Concerning mobility, crucial issues in the future will centre on the use of resources (and their cost), considering that we are depleting fossil fuels, and environmental problems. Such aspects will have important effects on the localization of activities and on accessibility, thus influencing the spatial organization and the urban form. On applying urban modelling, it has been observed that ambitious change goals can be achieved; but not technology alone must be used, because also a change in behaviours is required (Wegener 2012, p. 281). This means that most of the efforts should be devoted to helping people, communities and institutions to improve their knowledge and know-how, and to adopt responsible behaviours.

As concerns mobility and ICT, a preliminary step requires identifying what their reciprocal relationships are. Some authors have maintained that controversial effects are generated (Lyons 2009; Mokhtarian 1990; Salomon 1986):

- substitution of telecommunication for travel;
- stimulation of more travel because of ICT use;
- improvement of the transport system using ICT;
- indirect effects connected to change in spatial organisation caused by ICT.

Other relationships and effects have been noted, considering that information technologies can redistribute mobility, enrich travel, reconnect fragmented activities (Lyons 2009), and complement different systems (Salomon 1986).

Therefore, the expected effects regard not only the organisation of mobility but also the behaviours of people and companies, the urban form and the allocation of activities, the physical and virtual connections of people and places, on their different scales. There have been many schemes applying ICT methods and devices to the management of urban mobility, but the effects in general have been far from a massive reduction in private car use, accompanied by an abatement of fuel consumption and gas emissions, as well as by an increase in accessibility, personal safety and urban quality.

A second step regards clarification of the issues, of different kinds, which are interrelated in the discourse on cities, mobility, and ICT. In short, they can concern effects on:

- spatial organisation—the city as a central place in a wider space and as a node in mobility networks;
- infrastructure and mobility—connecting different spatial scales;
- accessibility and urban form—mobility as a factor of concentration/diffusion;
- information for accessing mobility services;
- governance of mobility flows, actors, and systems.

At the centre there are, obviously, the communities and the institutions, which are characterized by different motivations and abilities to learn, to change and to innovate. In fact, ICTs can only have an instrumental role, but an appropriate use of their potentials can attain important goals.

Assuming the point of view of a city, at stake is the interaction with supra-local policies and projects in order to maintain the role of mobility hub or to become one. A major factor is the new infrastructural organisation, but information and coordination can be crucial in connecting different mobility systems and in attracting new operators. Urban competitiveness and attractiveness are the goals.

The urban form is simultaneously the outcome of mobility systems and the determinant of specific modal splits, reflecting urban concentration or diffusion. Yet cities are changing rapidly due to the re-settlement of key activities and the formation of new nodes. Anyhow, the efficiency of urban systems is affected by how mobility is managed, and this is a specific field of application for ICT, allowing the deployment of information and stimulation of a change in behaviours. Cycling and walking must be a common way to approach basic services, while public transport efficiency can be increased by furnishing updated information to users and operators.

Governance—i.e. the coordination of actors and systems—is a specific field of application for ICT. What is expected is not simply a saving on costs, as usually achieved by replacing operators with informatics. It implies providing a single service—i.e. mobility—by means of a variety of systems. Problems regard not technology but overcoming the 'decisional silos' segmenting the provision of the service and the provider/user interface.

On the basis of the above-described considerations, the interpretative framework which can be applied in the case study to assess how 'smart' mobility systems are, and what can be the role of ICTs, can be summarized as the combination of the following factors:

- spatial scales: different territories are involved, as well as diverse mobility flows;
- mobility systems: corresponding to diverse services and territorial levels;
- actors: specific responsibilities and interests are involved.

As a consequence, actions regard:

- coordination: of the different services, systems, flows, on the diverse spatial scales;
- governance: of balancing contrasting interests;
- management: of the infrastructure, the mobility systems, the information.

Coordination of actors and scales, appropriate use of information and extension of the experiences concerning traffic control, road pricing, access to pedestrian zones control, etc., can be the main goals.

6 Mobility in an Alpine Region

The area considered in the case study is centred on the city of Trento, and involves the Province of Trento, the Trentino-Alto Adige Region, and the adjacent ones along the "Brenner corridor". Diverse scales and territories traversing the Alps are connected and by-passed by mobility systems and flows. The medium-sized cities of Verona to the south, Trento and Bolzano-Bozen at the centre, and Innsbruck to the north, are the central places of territories which gravitate on the services and activities furnished by such urban areas. At the same time, important merchandise and passenger flows are supported by the corridor connecting Northern and Central Europe and the Mediterranean, crossing the Brenner Pass, which is the most trafficked gateway of the Alps. In particular, the motorway is often congested, due the number of trucks. In short, data provided by Monitraf!, the traffic monitoring system of the Alpine regions, highlight that the Brenner corridor absorbs most road flows: "in 2015, 42% of vehicles crossed the Brenner", with a share of heavy vehicles close to 25%. (Monitraf! 2016). After a short period of decline, due to the economic crisis, flows are increasing, also for freight traffic. Rail is playing a minor role, with a share of around 28%. Problems connected to the gradient of the line and to the different railroad technologies adopted in Italy and Austria make rail traffic improvements complicated.

An important project is under way to construct a long tunnel lowering the maximum height of the railroad and its gradient. Anyhow, all the line from Munich, in Germany, to Verona, at the crossroad of the North-South corridor (Scandinavian-Mediterranean, in the EU terms) with the East-West line (Mediterranean corridor) needs to be improved, and this requires huge-scale works, in particular tunnelling.

Other issues concern the changing roles of mobility actors. The Italian state-owned railroad company has altered its commitment to internal high-speed lines, and runs local traffic only on request of administrations. The result is that the international passenger service along the Brenner corridor is left to other companies. Freight traffic on the Italian side is supported by some inter-modal centres, in particular the one based in Verona, but the long sought-after switch from road to rail is far from being achieved. Also the interoperability of the different railroad system is undergoing improvement through implementation of the ERTMS/ETCS (European Rail Traffic Management System/European Train Control System).

The Brenner corridor project is a challenge for the local territories because accessibility will change dramatically. It will be easy to reach—and to be reached from—the main European metropolitan areas, but the points of connection between the local system and the larger scale one have not yet been defined. Different attitudes are emerging in this regard (Zanon 2015), because Verona aims at becoming a hub of supra-local flows by leveraging on its position at the crossroads of two main European interest corridors, and on the multi-modal centres already in operation. Bolzano is searching for the containment of nuisances, and Trento is uncertain whether to maintain and refurbish the existing passenger station or to build a new one. Also the future of a multi-modal centre, which is not performing as expected, must be better defined.

On the local scale, some interesting schemes have been undertaken, in particular in the Province of Bolzano-Bozen (Alto Adige - Südtirol), where public transport is particularly efficient and local railroads have been improved to serve the local population's needs as well as those of tourists. A major effort has been made to provide information and to give tourists the opportunity to use public transport for free or at low prices. Also cycling is important, both in the urban area of Bolzano and in the rest of the province, where some bike trails have become tourism attractors. But the city of Bolzano is under pressure because of its huge number of commuters, and because the air quality is affected by the motorway skirting the urban area on a viaduct.

7 Trento, a Smart City in the Making

On entering the city of Trento, some billboards placed in the middle of certain roundabouts claim that Trento is a "smart city". In fact, since some years this is a commitment of the municipality, which activated a number of concrete actions, and has given space to some political rhetoric. The project is connected to the European framework and to specific EU funded projects, as well as to the above-mentioned Italian local authorities association's initiative.

The definition provided by the municipal administration's website states that "A smart city is a city which makes use of technology as a means to improve the services furnished to citizens and enterprises and the quality of life" (author's translation). A participatory process involves citizens, institutions and stakeholders. Some results can be already appreciated, both as a change in the administrative mind-set—more outcome-oriented—and as a number of innovative services, from opening the access to municipal data to some smart tools. A number of apps can be downloaded from the municipal website to interact with the municipal administration by accessing information and services. In particular, some apps can be used to consult public transport schedules and buy tickets, pay parking fees, or use the bike sharing service.

Mobility management, anyhow, involves decisions and control actions at different spatial and administrative levels. The city of Trento has adopted a mobility plan, which is oriented at managing traffic flows, at defining circulation and parking regulations and at planning some new infrastructure. Trento has around 110,000 inhabitants, but nearly half of them live in the suburbs, and the role of the city as the provincial capital attracts city users mainly heading to the historic centre and the business district. The municipal administration is monitoring traffic, which on a normal weekday amounts to nearly 100,000 vehicles entering the urban area. Only 20% of them do not stop in the city, while the rest requires a parking lot. Balancing the different mobility flows is a real challenge, in particular if a more sustainable modal split in favour of public transport and cycling is sought. In fact, the municipal mobility plan aims at a 20% reduction of traffic by supporting inter-modality in favour of public transport, cycling and walking.

A number of provisions have already been activated. They range among a large pedestrian zone in the city centre, the extension of toll parking in order to discourage the misuse of public space, the provision of a bike sharing service, and, as said, apps to spread information on the different opportunities for moving around in the city. More has to be done to improve information concerning public transport and the alternatives for accessing the various areas of the city, and this endeavour must involve diverse actors responsible for the infrastructure and public transport. A major problem regards the form of the city, which in the last decades has extended across the territory, reflecting the growth of car ownership and use, with a resultant mobility difficult to control.

Another level concerns the connections between Trento and the rest of the territory at the provincial-regional and national-international levels. This is a complex issue because a number of projects are being proposed, concerning in particular the new railroad line along the Brenner corridor, which will by-pass the city. It is not yet clear how the interconnection between the existing line and the new one will be realized, providing access to upper level mobility hubs—such as airports and multi-modal centres. Different companies will operate, both as concerns passengers and freight, soliciting effective governance actions.

As concerns ICT, such technologies are already playing a central role, in particular considering that new connections are offered by FlixBus and some railroad companies, whose strong point is the easy Internet access to the service. Appropriate information is also a crucial lever for managing change and orientating processes. The urban economy of Trento is changing from a traditional industrial one to a knowledge, culture and outdoor tourism-based one. The use of technological devices is therefore coherent with the attitudes of large part of the population and city users. In this regard, some research institutions and Trento University are active in stimulating the local society and the administrations in innovating practices, and are taking part in the development of smart solutions and the application of new technologies.

8 Conclusions: Open Issues, New Perspectives

A smart territory is expected to make intensive use of knowledge and up-to-date technologies to manage complex issues in order to provide quality services to its citizens and create an innovative environment supporting a sustainable development. A variety of actors must be coordinated, and 'governance' is the keyword to appropriately coordinate actors and actions.

The case study, concerning the city of Trento within its adjoining territories forming the "Tyrol Euroregion", highlights a number of issues concerning both problems to be addressed and innovative practices.

A first issue concerns the spatial scale at the supra-local and regional levels, where consolidated barriers concerning territories and institutional tasks are present and are still difficult to overcome. The reorganisation of the competencies of public administrations in the recent past has been due to European integration, the growing role of the regional administrations, cutbacks in public funding, and the outsourcing of many public services to agencies, companies and private operators. As a result, fewer services are directly provided by the local authorities within the territory of their competence, because new operators are active.

A rapid innovation is taking place, also using the consolidated infrastructure networks, leveraging on ICT to access the service, but mostly within a sectorial, company-oriented, approach. In particular, transport companies (also the public-owned ones) seek to save costs, and a pivotal tool in this effort are ICTs, which can help in optimising organisation and reducing personnel by giving users some tasks, in particular access to information and ticketing.

In the territory considered, some large-scale infrastructural projects are in progress, but the expected effects are not yet fully clear, since they depend on diverse management perspectives. In particular, the interconnection between the local and the national/international scales is still to be defined.

As concerns the urban level, ICTs are being introduced, and the results are promising, but an effective monitoring of the effectiveness of the diverse actions has not yet been activated. New behaviours are taking place, in particular the use of bicycles, together with the spread of technological innovations which are changing mobility, such as hybrid and electric cars, while the prospect of automatic guided vehicles is not so far from becoming a reality. A new ability to analyse phenomena and control processes is therefore required.

A second issue concerns governance, which implies being able to build flexible intervention processes, leveraging on the ability to innovate the way in which opportunities are detected and problems addressed by activating cooperation networks, both formal and informal. While the stress is usually placed on infrastructural projects, the major problem regards coordination of the plurality of actors. The new transport networks can provide a better service only if effective cooperation among operators is activated and more attention is paid to the interaction with customers, both appropriately providing responses and soliciting a change in their behaviours.

Lastly, mobility management requires dealing with some contradictions. In particular, private car mobility generates tax revenues (collected by the state), because of the expensive fuel used, while rail mobility and, in general, public transport implies high costs (mostly covered by the local authorities). The extension across the urbanised territory of the public service gives all citizens equal opportunities, but larger networks are usually costly and perform badly. More flexible mobility systems require connecting different scales, a variety of operators, stimulating an active role of customers. But different rationales confront each other.

In conclusion, the use of ICTs provides opportunities and creates risks, because good and easy access to information and distance interaction with the public offices and economic operators can save journeys, but communication is also used to stimulate the access to specific places, and to boost tourism. An example is the Christmas market—a successful event for Trento—which stimulates unexpected traffic peaks.

Most of the efforts must therefore be concentrated on the potentials provided by ICTs for mobility management, and there is large space for improvement of both private mobility and public transport. Information and control are the key areas of intervention, and they must activate a cooperative environment where all the actors—customers and citizens included—must play an active part. Finally, also the urban space requires due attention, because the use of information methods and technologies affects not only mobility flows but also the allocation of activities and the urban form.

References

Albino V, Berardi U, Dangelico RM (2015) Smart cities: definitions, dimensions, performance, and initiatives. J Urb Technol 22(1):3–21

Banister D (2008) The sustainable mobility paradigm. Transp Policy 15(2):73–80

Banister D (2011) The trilogy of distance, speed and time. J Transp Geogr 19(4):950–959

Banister D, Hickman R (2013) Transport futures: thinking the unthinkable. Transp Policy 29:283–293

Caragliu A, Del Bo C, Nijkamp P (2011) Smart cities in Europe. J Urb Technol 18(2):65–82

CEC (Commission of the European Communities) (1999) ESDP European Spatial Development Perspective. Towards balanced and sustainable development of the Territory of the EU, approved by the Informal Council of the Ministers responsible for Regional/Spatial Planning of the European Union, Luxembourg, European Communities
Delponte I (2012) Trasporti, ICT e la Città. Perché alla città interessano le ICT? TeMA 3:33–45
EUP (2014) European Parliament, Directorate-General for Internal Policies, Policy Department A, Economic and Scientific Policy, Mapping Smart Cities in the EU, European Union. http://www.europarl.europa.eu/studies. Accessed 29 Nov 2017
Fistola R (2013) Smart city. Riflessioni sull'intelligenza umana. TeMA 6(1): 47–60
Giffinger R, Fertner C, Kramar H, Kalasek R, Pichler-Milanović N, Meijers E (2007) Smart cities—ranking of European medium-sized cities, Final report (Oct 2007). http://www.smart-cities.eu/download/smart_cities_final_report.pdf. Accessed 29 Nov 2017
Graham S (2001) The city as sociotechnical process. Networked mobilities and urban inequalities. City 5(3):339–349
Kunzmann K (2014) Smart cities a new paradigm of urban development. Crios (7):9–20
Lyons G (2009) The reshaping of activities and mobility through new technologies. Editorial for special issue on ICT and the shaping of access, mobility and everyday life. J Transp Geogr 17(2):81–82
Mokhtarian PL (1990) A typology of relationships between telecommunications and transportation. Transp Res 24A(3):231–242
Mokhtarian PL, Salomon I (2001) How derived is the demand for travel? Some conceptual and measurement considerations. Transp Res 35A(8):695–719
Monitraf! (2016) Monitraf! Annual Report 2016. Political support for Toll Plus & agenda setting for a new phase, Zurich/Tübingen, 6th Feb 2017. http://www.ferroviabrennero.provincia.tn.it/binary/pat_ferrovia/eventi/iMONITRAF_Annual_Report_2016.1489046087.pdf. Accessed 29 Nov 2017
Nam T, Pardo TA (2011) Conceptualizing smart city with dimensions of technology, people and institutions. In: Proceedings of the 12th annual international digital government research conference: digital government innovation in challenging times, ACM, New York. http://citeseerx.ist.psu.edu/viewdoc/download?doi=10.1.1.432.5479&rep=rep1&type=pdf. Accessed 29 Nov 2017
Papa R, Galderisi A, Vigo Majello MC, Saretta E (2015) Smart and resilient cities. A systemic approach for developing cross-sectoral strategies in the face of climate change. Tema. J Land Use Mob Environ 8(1):19–49
Robinson JB (1990) Futures under glass: a recipe for people who hate to predict. Futures 22(8):820–842
Salomon I (1986) Telecommunications and travel relationships: a review. Transp Res 20A(3):223–238
Territorial Agenda (2011) Territorial Agenda of the European Union 2020. Towards an Inclusive, Smart and Sustainable Europe of Diverse Regions. Agreed at the Informal Ministerial Meeting of Ministers responsible for Spatial Planning and Territorial Development on 19th May 2011 Gödöllő, Hungary. http://www.nweurope.eu/media/1216/territorial_agenda_2020.pdf. Accessed 29 Nov 2017
Wegener M (2012) The future of mobility in cities: challenges for urban modelling. Transp Policy 29(1):275–282
Zanon B (2013) Scaling down and scaling up processes of territorial governance cities and regions facing institutional reform and planning challenges. Urb Res Pract 6(1):19–39
Zanon B (2015) Territorial cooperation and multilevel governance in the Brenner base tunnel project. In: Fabbro S (ed) Mega transport infrastructure planning European corridors in local-regional perspective. Springer, Cham, Heidelberg, New York, Dordrecht, London, pp 99–122

Web Sites

ITU-T—Smart Sustainable Cities: http://www.itu.int/en/ITU-T/ssc/Pages/default.aspx
Osservatorio Nazionale Smart City: http://osservatoriosmartcity.it/
Smart Cities and Communities European Innovation Partnership: http://ec.europa.eu/eip/smartcities/
Trento Smart City: http://www.comune.trento.it/Aree-tematiche/Smart-city/Cos-e-Trento-Smart-City

The Area Trademark for the Launch of Sustainable Processes of Smart Planning in Rural Areas

Mauro Francini, Sara Gaudio, Annunziata Palermo and Maria Francesca Viapiana

Abstract Since the task of urban planning and of the territorial government is also to promote development while maintaining recognition of the identity and the uniqueness of a particular area, this research project seeks to define and implement an Area Trademark (AT) to support the resources and potentialities of a given territorial context. This is achieved by implementing and providing a network of homogeneous, coordinated and complementary services that enable them to promote their identity and specific vocations. In line with the Europe 2020 Strategy and the regional and national programming documents, including the National Strategy of Internal Areas (NSIA), representing the subject of particular interest, our research aims, specifically, to create an Area Trademark in territorial contexts with a rural vocation. In particular, starting precisely from the rural vocation of the area and, therefore, from the rediscovery of agri-food production, from the reinterpretation in an environmentally sustainable key of both the environmental and cultural excellences present, this project seeks to increase the quality of services until they become an integral part of the landscape quality, focusing on the creation of an effective and efficient network of existing accommodation facilities.

1 The Aims of an Area Trademark

The creation of an Area Trademark (AT) is a project with the aim of supporting the resources and the potential of a determined territorial context by implementing and providing for a network of homogenous, coordinated and complimentary services which allow the promotion of their identitary nature and specific vocations.

The creation of an Area Trademark is often connected with critical observations of modern day urban processes which tend to lead to things occurring or located in

M. Francini (✉) · S. Gaudio · A. Palermo · M. F. Viapiana
Laboratory of Environment and Land Use Planning—LabPAT, Department of Civil Engineering, University of Calabria, Cosenza, Italy
e-mail: labpat@unical.it

R. Papa et al. (eds.), *Smart Planning: Sustainability and Mobility in the Age of Change*, Green Energy and Technology,
https://doi.org/10.1007/978-3-319-77682-8_14

different places in the world both similar and comparable. On the one hand, this helps orientation as wherever we are, we can participate in shared ways and costumes, yet on the other hand, this distances us from the hypothesis that there can be different roots and worlds (Leone 2012).

In the need to power the search for authentic and original roots and to exploit the often unique resources of a given place, though synthesizing them in common analysis parameters, the idea of the Area Trademark emerges, also meeting with one of the tasks of urban planning and of the territorial government, namely of promoting development, while maintaining recognizability of the area's identity and uniqueness.

In line with the Europe 2020 Strategy and the regional and national programming documents, including the National Strategy for Internal Areas (NSIA), for which internal areas constitute a subject of particular interest, this research seeks to achieve an Area Trademark in territorial contexts prone to a rural vocation.

Starting from the rural vocation of the area, from the rediscovery of agro-food production, from reinterpreting in the eco-sustainable key of environmental and cultural excellence, the aim of this research project is to increase the quality of services so they become an integral part of landscape quality, with the ultimate aim of creating an effective and efficient network between existing accommodation facilities.

The creation of the Area Trademark generally occurs through the creation of a hospitality network widespread in peri-urban and rural areas. It is an organization that spontaneously becomes shared, a system linked to an offer of beds spread across the territory, which creates a network between the various farm structures, the cellars, the mills whose Trademark will promote productions and will be the guarantor of quality and excellence, realizing thematic routes of naturalistic, cultural and gastronomic valorisation.

The research project pursues a development strategy focused on regeneration, recovery, restoration and maintenance of existing built assets in order to optimize the performance of structures without destroying either the character of the buildings or the identity of the sites. This strategy pursues the principles of sustainability and reduction of soil consumption, the cornerstone of international and national priorities. It is sufficient to consider the recent Draft Law approved in May 2016 by the Chamber of Deputies, which aims precisely at so-called "zero soil consumption" through the reuse and upgrading of the built ground.

Therefore, among the different objectives of creating the Area Trademark, the objective of regulating the restoration of old structures that are and could be used for agritourism-receptive use, emerges, aiming to raise technical-functional performance and elevate the level of building quality.

The adaptation of architecturally significant sites is intended to enhance the typicality of the property and the surrounding environment, which is particularly valuable from an environmental and historical-cultural viewpoint.

Furthermore, the strategic aspect in the creation of an Area Trademark is the ability, through a challenging and highly engaging process, to create a network of communication and collaboration among all the stakeholders involved. It will thus be possible to organize and re-organize the territorial offer while contemplating and complementing both the needs of tourists and the real attractions of the area and the services that the entire territory can offer (Scipioni and Vecchiato 2002).

Like any territorial planning tool, the Area Trademark would thus have the value of coordinating all the activities of the territory, rationalizing and directing the implementation strategies, taking full advantage of the various synergies of possible interventions and planning them in an integrated way.

Moreover, the characterization of a territory through the creation of a Trademark allows better identification and better understanding of tourist expectations: knowledge of territorial typologies allows the offering of new dimensions of space and thus the discovery of new opportunities and experiences.

The Area Trademarks that, to date, have been created focus on the promotion of a brand that guarantees the quality of typical food and wine products. This is the case of the Calabrian Area Tyrrhenian Trademark, which promotes salami, vegetables such as courgettes and pumpkin flowers, olive oil and wine, or the Area Trademark of Alto Casertano, in Campania, mainly dedicated to wines, salami, cheeses, honey, jams and legumes. The Gallo Nero trademark, which is famous on an international level, is present in Tuscany and is a brand that promotes the production of Chianti wine in its motherland.

However, there are but a few examples which have been able to concretize the idea of a brand capable of coordinating services in a broader perspective: among them the Valle Agredo Area Trademark, which promotes rural tourism in central Veneto through the organization of appointments and events, naturalistic, cultural and enogastronomic itineraries, with the possibility of staying in certain hotel structures and B&Bs; the Salento d'Amare Puglia Area trademark promotes mostly seaside tourism, guaranteeing, in addition to the quality of agricultural products, food and handicrafts, as well as a network of services associated with catering and hotel and extra-hotel facilities, such as camping sites.

It would be limiting, however, in this case to think of the creation of the Area Trademark as the simple networking of agritourism structures with certain requirements or as the mere promotion of a brand: in itself it would be a little success for the valorisation of an area that is currently unable to profit from its resources, but would probably not represent a radical and revolutionary revival.

Starting from concrete examples from other Italian regions, the innovative idea lies in linking, according to the criteria of bio-building, that which is produced in the same companies with access to the Area Trademark for architectural adaptation and improvement of structures.

In Sardinia, in the area of Sulcis, the poorest province in Europe, there is an inspirational initiative. An entrepreneur from Cagliari transformed agricultural product surpluses and waste (wool, oil, wine) into excellent materials for building.

Just a few examples are: thermal and acoustic insulators for buildings are produced from the reuse of sheep wool remnants, from the underworking of olive oil a product is obtained that fortifies mortar, from the surplus of viticulture a series of paints for walls and buildings, paint and plaster from the union of lime putty with milk and cheese subproducts, while artichokes, tomato peel and wild vegetables provide floor coverings.

Figure 1 shows a sample external wall which consists, according to the different products, of the Edilana, Edilatte, Editerra and Edimare ranges.

There are multiple benefits to these new methods: from the reduced dependence on non-renewable energy sources to the reduction of serious issues such as waste disposal, from the reuse of materials which would otherwise be lost to the easy traceability of materials, from the relaunch of the local economy to the creation of new employment opportunities.

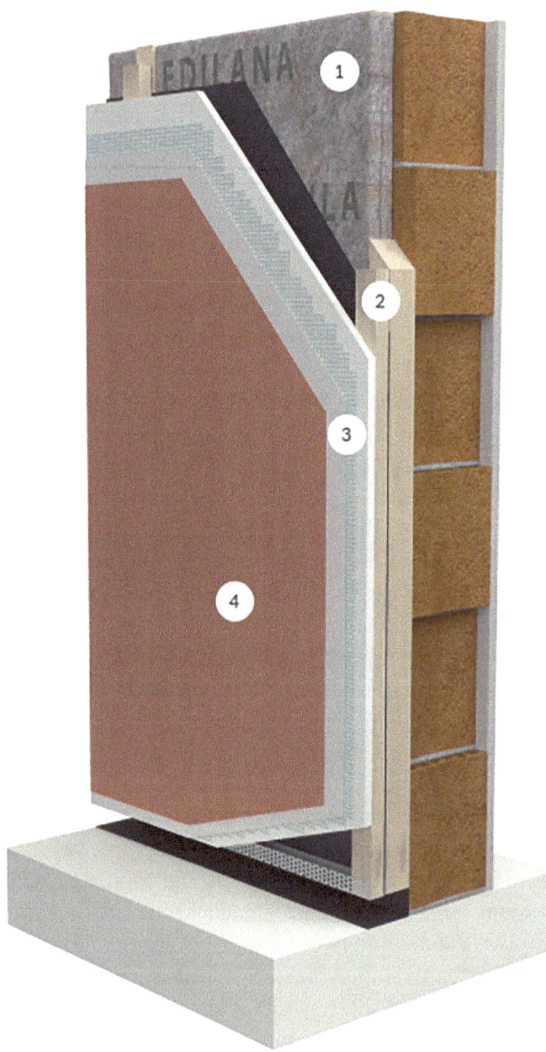

Fig. 1 Example of EDIZERO external wall. (1) Edilana panel made of pure virgin wool from autochthonous Sardinian sheep for thermal and acoustic insulation and hygrometric well-being. (2) Selected clay light brick external walls. (3) Lime putty fine leveling for exteriors made by Edilatte. (4) Edilatte wall paint: Colour: Mother Earth Mix of raw colours with natural ingredients of mineral and vegetable origin (only excess and surplus), a gift from Mother Earth, transformed by Edilatte in mixes of raw colours; Base: Aged lime putty extract by Edilatte. *Source* EDIZERO architecture for peace

The Edilana and Edimare lines constitute some of the services of the "Casa Verde CO2.0" production centre which brings together a group of 78 companies, of which 42 are in Sardinia employing 600 people.

The objectives of an effective territorial government can materialize and bring good results only by activating policies of integrated planning immersed in local realities. The best results can be achieved even by exploiting small opportunities, only if accompanied by a full awareness of them, by detailed studies, by accurate analyses, by collaboration and cooperation.

It is noteworthy that even in the case of the Sulcis province, the collaboration between new companies, the various types of industry that have become research centers over time, and the University of Cagliari have been of fundamental importance and has even led to the realization of some international patents.

2 Technical Requisites of a Performance Specification Between Environmental, Social and Economic Quality

The Area Trademark is placed at the service of accommodation facilities in order to promote and support productions, activities and services within periurban and rural areas relative to the area identified. It simultaneously represents the object of promotion and the promoter of the various companies, becoming a useful tool in guaranteeing both its quality and excellence.

The subjects that can request granting of the Trademark are mainly agricultural businesses, agritourism businesses and hotels. These subjects must conduct their activity within the geographical area identified by the Trademark, they must respect specific requirements of quality and be in compliance with current building and systems regulations.

As stated above, the quality requirements that each structure should possess in order to access the Area Trademark ought to be well-defined in a performance specification which can be divided into three thematic areas: management of structures and open spaces; tourist services; cultivation; food and wine production and tourism.

The management of structures and open spaces mainly regulates the structural and plant properties of the buildings; it is important to guarantee respect for the rural architectural style typical of the place and the use of local materials. Preference is given to measures with the aim of improving environmental sustainability, such as the use of renewable energy, self-sustenance, or natural insulating methods to improve thermal insulation of structures. Particular attention should also be paid to the disposal of waste water and waste and the elimination of architectural barriers to facilitate accessibility of structures to people with reduced mobility.

Tourist services must contribute to guaranteeing an offer that fosters knowledge of the natural and cultural heritage of the place. Depending on the specific vocation of the area, it is necessary to identify the strengths and potential of the territory to promote different tourist itineraries (e.g., gentle walking or cycling paths, historical-cultural itineraries, etc.). Cultivation, production and food and wine tourism should incentivise the enhancement of places and, specifically: company crops must be characteristic of the place and should adopt techniques that respect the environment. Cooking should be typical and based on products grown directly by the company, which can also be sold.

Structures that intend to adhere to the Trademark, if they do not possess the requirements specified by the performance Specifications, must undertake to write up an improvement plan which provides for, with deadlines, the attainment of environmental, social and economic quality objectives. Furthermore, pecuniary and non-pecuniary penalties (temporary suspension of use of the trademark or final revocation) are provided for in case the subject does not comply with the rules imposed.

3 Case Study: The Crati Valley

The experimental research provided for the creation of an Area Trademark in the Crati Valley, a territory in the north of Calabria in a central strip and within the Province of Cosenza. It is characterised by the presence of 30 communes crossed by the Crati River and situated within a variegated landscape context in the vicinity of the Sila and Pollino National Parks, of the Apennine mountainous chain which separates the Ionian and Tyrrhenian coastal areas with an outlet to the north-east towards the Plain of Sibari (Francini 2006).

3.1 The Reference Territorial Context

The first step of the research was characterised by an analysis of the state of the area, focusing on the particularities of its landscape, which is connected to values of nature, history, culture and the economy, and of its rural realities present. The landscape intended as a system of relations: morphological, perceptive, sensorial, ecological, cultural, and is capable of creating a shared imagination, of generating a sense of belonging, of evoking attraction. The landscape synthesises the identity of the territory, an identity that constitutes a resource for its community (Fossa 2012).

The territory covers 19 of the 30 municipalities of the Valley (Altomonte, Bisignano, Cervicati, Cerzeto, Fagnano Castello, Lattarico, Luzzi, Mongrassano, Montalto Uffugo, Roggiano Gravina, Rose, Rota Greca, San Benedetto Ullano, San Marco Argentano, San Martino di Finita, Santa Caterina Albanese, Santa Sofia d'Epiro Tarsia and Torano Castello) and the analysis of the area was conducted following the indications contained in the existing territorial and urban planning instruments: the Province of Cosenza Provincial Coordination Territorial Plan (Province of Cosenza 2009) and the Landscape Valency Territorial Framework of the Region of Calabria (QTR/P) (Region of Calabria 2016).

The morphological properties of the territory highlight a variegated altimetry: the ground, first hilly then mountainous, arises rapidly from the valley areas, in just a few kilometres reaching high quotas. In the immediate vicinity of the Crati River a flat zone extends which, slowly, leads away and transforms into a hilly then mountainous area in correspondence of the areas of the Valle del Crati Nord, Destra Crati and Crati Centrale, identified referring to QTR/P.

Significant altitude variations are responsible for the varied microclimates, vegetation and type of cultivation.

From the geomorphologic viewpoint, there are crystalline and tertiary formations of sandstone sediments, clay and fluvial deposits which, forcing the Crati river to dig a gravel path towards the Sibari Plain, have determined the Tarsia fork, which extends for about 6 km.

The QTR/P of the Region of Calabria, furthermore, as indicated by the Hydrogeological Condition Plan (Basin Authority of Calabria Region 2001), permitted identification of areas at hydraulic risk and landslide risk defining, for each category, four levels:

- R4—very high risk: when there are conditions that determine the possibility of loss of human life or serious injury; serious damage to buildings and infrastructures; serious damage to socio-economic activities;
- R3—high risk: when there is a possibility of damage to people or assets; functional damage to buildings and infrastructures which lead to inaccessibility; interruption of socio-economic activities;
- R2—medium risk: when there are conditions that determine the possibility of minor damage to buildings, to infrastructures and to the environmental heritage without direct prejudice to the safety of people and without compromising the viability and functionality of economic activities;
- R1—low risk: for which social, economic and environmental damage are limited.

The infrastructural system of the Crati Valley has been strongly characterised by the abovementioned morphological formation. The Valley's infrastructure system, today, shows evidence of shortcomings both in terms of infrastructures and of services, also due to the obsolesce of a significant part of the development of networks.

The strengths of the area emerge as those connected to the natural landscape. There are two Reserves (that of Serra Nicolino-Piano d'Albero and the regional one of the Tarsia Basin) and 8 SCI areas (Sites of Community Importance). Furthermore, the Crati Valley is the main director of numerous migratory phenomena and, as such, the place of several archaeological discoveries. The nineteen villages examined, all with ancient origins, have historic centres characterised by local rural architecture and by interesting monuments. The presence of some Albanian communities, which have were established centuries ago in the territory and which have preserved the language and traditions of the country of origin, diversify and enrich the historical landscape and heritage of local culture. The socio-economic landscape of the Valley has also been outlined: the total population of the area, in 2016, was 88,797, approximately 4.5% of the population of Calabria and 12.4% of that of the province of Cosenza. The trend, (Table 1), shows a relative decrease equal to −2.63% compared to 2001 data, in accordance with the population decrease that occurred in the province of Cosenza and the region of Calabria, but greater both than the provincial average (−2.16%) as well as the regional average (1.98%).

From data relating to the occupational situation in the area (Fig. 2), the employment rate in the district is, on average, equal to 35.2%, lower than the regional average of 36.6%.

Regarding the unemployment rate, the average of the area arrives at 22.2%, a value greater than the regional value of 19.5%.

From the data on occupations by professional position and by sections of economic activity related to the last census (ISTAT 2011) (Fig. 3), it results that the tertiary sector absorbs most of those employed in the area, with a percentage of 60%.

The most significant data, in this sense, is represented by the commune of Montalto Uffugo with 75.9% employed people in the tertiary sector, while it is worth highlighting that the data relating to the commune of Cerzeto, in which 28.3% result as being employed in the tertiary sector (lower value compared to the entire Crati Valley), while agriculture dominates with 58.6%.

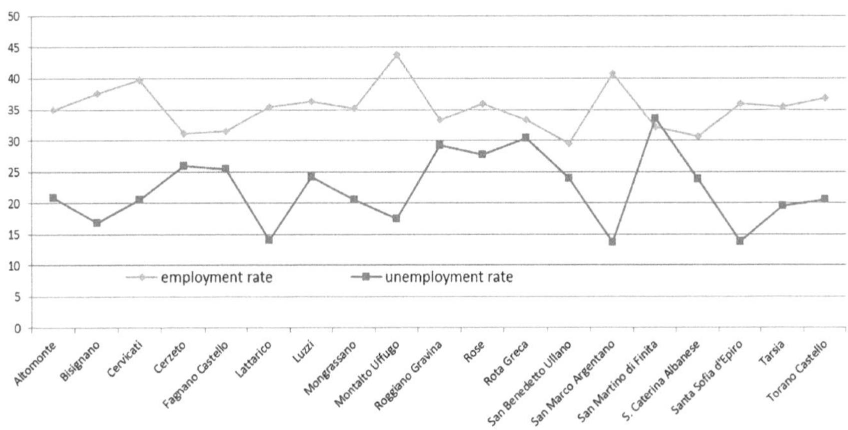

Fig. 2 Employment situation (elaboration ISTAT data)

Fig. 3 Employees by economic sector (elaboration ISTAT data)

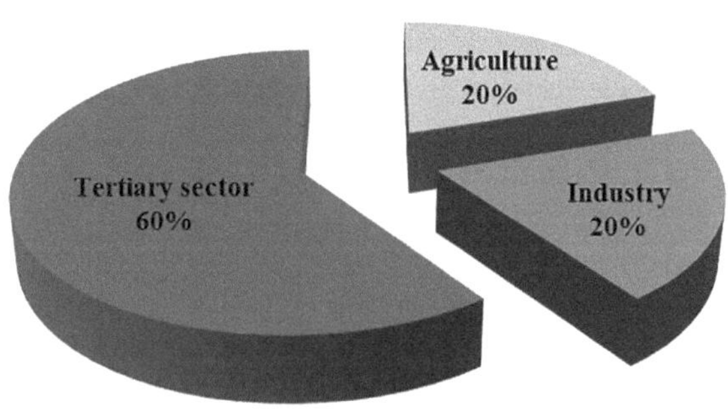

Table 1 Resident population. *Source* ISTAT

Communes	Altitude (m. asl)	Surface area (kms)	1981	1991	2001	2005	2014	2016	Housing density (hs/kms) in 2016
Altomonte	495	65.29	4323	4538	4493	4592	4540	4488	68.7
Bisignano	350	85.28	10400	10760	10877	10472	10219	10203	119.6
Cervicati	485	12.22	1057	1081	1018	980	859	829	67.8
Cerzeto	470	12.21	2418	2245	1467	1381	1354	1373	112.4
Fagnano Castello	516	29	5207	4684	4194	4073	3922	3885	134.0
Lattarico	406	40	3884	4166	4137	4414	4046	4013	100.3
Luzzi	375	77.2	10570	11051	10455	10102	9468	9396	121.7
Mongrassano	540	34.68	1897	1901	1763	1727	1626	1607	46.3
Montalto Uffugo	468	78.43	12181	15016	17258	18272	19517	19669	250.8
Roggiano Gravina	260	44.59	7578	8179	7702	7555	7217	7208	161.7
Rose	435	45	4525	4161	4397	4354	4366	4373	97.2

(continued)

Table 1 (continued)

Communes	Altitude (m. asl)	Surface area (kms)	1981	1991	2001	2005	2014	2016	Housing density (hs/kms) in 2016
Rota Greca	520	12.87	1300	1382	1287	1242	1152	1156	89.8
San Benedetto Ullano	460	19.4	1728	1812	1644	1650	1559	1555	80.2
San Marco Argentano	430	78.28	7682	8294	7646	7535	7417	7424	94.8
San Martino di Finita	550	23.74	1369	1317	1280	1261	1133	1100	46.3
S. Caterina Albanese	472	17.2	1573	1537	1156	1333	1253	1222	71.0
Santa Sofia d'Epiro	550	39.7	2849	2938	3125	3012	2671	2628	66.2
Tarsia	192	39.45	2448	3027	2370	2284	2062	2054	52.1
Torano Castello	370	25.45	4462	4757	4927	4835	4614	4614	181.3
Total		779.99	87451	92846	91196	91074	88995	88797	

From the analysis of the production landscape, the numerous productions of excellence and quality typical of the area emerge. Among the products, the PDO olive oil with a golden colour and fragrant aroma, DOC and IGT wine production, figs (quality assured), often dried and processed in various ways, becoming a Slow Food Presidium in 2002, the production of chestnuts, of forest fruits, of tasty cheeses and vegetables, including the famous Roggiano Gravina pepper, the production of honey and the production of agri-food conserves such as aubergines, peppers, and tomatoes in oil.

3.2 SWOT Analyses

On the basis of the analysis of the state of the Crati Valley and of the natural, historical-cultural, socio-economic and productive aspects examined, it was possible to determine a summary of the data and to identify the fundamental, negative and positive elements on which to focus for the area's development.

In coherence with the interpretative keys identified, the SWOT methodology is used to define an appropriate analytical framework.

Six themes were identified to be used as reading keys:

I infrastructures and accessibility;
II settlement centers;
III environment;
IV culture and society;
V economy and production;
VI public administration and local actors.

The strengths' analysis showed that Crati Valley is characterized by:

I favourable geographical position;
II presence of numerous highly typical historical centers;
III presence of significant natural resources, fundamentally healthy and liveable environment; presence of a highly rural environment;
IV Multiplicity of traditions and local products, strong "sense of welcome" of the local population;
V Existence of typical crops, presence of artisanal activities;
VI Presence of local action groups (lag) in the territory.

The weakness' analysis showed that Crati Valley is characterized by:

I lack of infrastructures and poor territorial accessibility;
II territorial dispersion of settlements with consequent high land consumption;
III hydrogeological instability, seismic risk;
IV poor knowledge of local realities, poor communication of local realities, depopulation due to migratory flows;

V high unemployment rate, absence of integration of chain and of network between productive sectors and poor diffusion of a integration oriented culture, limited effects of agrifood productions with territorial enhancement processes, lack, or poor integration, of a qualified tourism offer and limited tourism support services;
VI lengthy bureaucratic times, difficult implementation of instruments for the planning of a vast area.

The opportunities' analysis showed that Crati Valley is characterized by:

I new forms of eco-sustainable mobility;
II overcoming of current marginality and isolation conditions;
III high biodiversity level;
IV increase of the demand of new forms of tourism linked to the rural vocation of the area and to the environmental and cultural aspects, renewed interest in the history of places;
V potential tourist chain, creation of new jobs in innovative sectors linked to culture and the environment, creation of new forms of economic and productive associations (consortiums, cooperatives, etc.), development potential of new eco-compatible agricultural activities (agritourism and rural tourism);
VI good reference practices for territorial planning, possible collaborations with actors operating in the territory, strengthening of relations with bordering communes to promote associations that include them in shared development projects.

The threats' analysis showed that Crati Valley is characterized by:

I further deterioration of current criticalities in the absence of infrastructural intervention works;
II phenomena of unauthorized buildings with the risk of further pulverization of the settlement;
III Uncontrolled "erosion" of the natural environment;
IV Lack of cooperation between local actors, Loss of cultural identity;
V Promotion of a high land consumption tourist offer, Disappearance of local crafts, Decrease of typical values of local products and difficulties in entering commercial circuits, More aggressive international competition faced with market internationalization;
VI Communal planning as main limit in the absence of instruments for large area planning.

Once the analytical reference framework and highlighting the emerging features of the territory in question has been defined, the reasons for which the definition of an Area Trademark was chosen as a possible intervention strategy become clear.

Resuming the concepts emerging from the SWOT analysis, in fact, it tends to:

- offer new tourism opportunities linked to the rural vocation of the area and to the environmental and cultural aspects;
- create new forms of economic and productive associations;

- aim for the development of new eco-compatible agricultural activities (agritourism and rural tourism);
- create collaborations between the different actors operating in the territory;
- strengthen relations between the neighboring communes to promote associations that include then in shared development projects;

and seeks to offer solutions to problems such as:

- absence of production chain integration and of network between productive sectors;
- poor diffusion of a culture oriented towards collaboration;
- limited integration between food production and territory enhancement processes;
- lack of a qualified tourism offer;
- lack of cooperation between local actors;
- decrease of typical values of local production;
- difficult insertion in commercial circuits;
- limited offer of tourism support services.

Although ISTAT data for the 2011 census shows that the tertiary sector absorbs most of the employed people in the area, careful observation of the landscape makes the true vocation of the Crati Valley clear: namely the rural one. Much of the soil, to date, results as being uncultivated: this almost certainly constitutes one of the greatest opportunities for the Crati Valley, which could focus more on the development of the primary sector for a revival of the entire area.

3.3 *The Crati Valley Area Trademark*

The Crati Valley Area Trademark (Fig. 4) involves farmhouses, B&Bs and hotels that are located in the peri-urban and rural zones of the area, which will promote production and make it a guarantor of quality and excellence, offering a development strategy based on the recovery, restoration and maintenance of the existing built heritage in view of optimizing the performance of structures without destroying the character of the buildings and the identity of the sites, pursuing the principles of sustainability and reducing soil consumption.

Regional Urban Planning Law n. 19/2002 art. 10 para 3 (Region of Calabria 2002), in fact, places as the priority of the Strategic Environmental Assessment[1] "the restoration and requalification of the territory [...] with particular attention to the reduction of soil consumption It is necessary to pursue such objectives in the elaboration and creation of each project, whether it be public or private.

Thus creation of the Trademark focuses on a development strategy imprinted on the recovery, the restoration and the maintenance of the existing built heritage, without destroying the character of buildings or the identity of places.

---

[1]A necessary process during the elaboration of any urban and regional planning instrument, with the aim of preventatively evaluating environmental and territorial sustainability.

It is worth nothing, moreover, that LUR n. 19/2002 neither sets rules nor constraints on the management and requalification of old houses in rural areas, something which occurs, instead, in other territorial areas. One of the objectives of the creation of the Area Trademark is that of regulating maintenance works on ruins and on old structures which could be to be used for agritourism-receptive use, aiming to raise technical-functional performance and increase the level of building quality.

Thanks to the adaptation of the rural sites which are particularly significant from an architectural point of view, it is intended to enhance the typicality of the properties and the surrounding environment, which is particularly valuable from an environmental and historical viewpoint.

The proposal to join the Area Trademark involves the accommodation structures already present in the territory in peri-urban and rural areas. After a detailed analysis of the current tourist-receptive offer, which was also matured thanks to data from the Regional Tourist Information System, an Altomonte agritourism company was identified as an example of "good practice". The property has all the quality requirements to access the Area Trademark: hospitality is offered in a 19th century farmhouse, which was previously in danger of collapsing and with serious structural problems. The farmhouse was subject to reclamation measures designed to secure the farmhouse whose materials had lost, over the years and without maintenance, their original characteristics and properties. Particular attention was paid to the chromatic choices and the use of natural and traditional materials such as wood and stone, respecting the typical rural style of the place. The company is equipped with a photovoltaic system that provides energy self-sufficiency: the annual energy balance is zero, therefore no additional energy is needed.

The walls and floors have been thermally and acoustically insulated and the techniques adopted mainly refer to the use of expanded clay, which is an insulating method which is both natural (it pursues the criteria of bio-building[2]), as well as lasting because there are no materials that deteriorate over time. An Imhoff pit is used for the disposal of waste water. The pit is characterized by a higher sedimentation compartment, where incoming slurries are stationary and slowly disintegrate into light

Fig. 4 "Valle del Crati" area trademark

[2]Bio-building is the set of precautions that lead to the construction of buildings with a low environmental impact which are capable of guaranteeing comfort and well being without equal due to the use of excellent thermal insulation using natural materials which pose no danger to health.

floating substances and heavy ones passing through the lower slurry precipitate in the underlying compartment, and by a lower digestion compartment, where organic substances undergo a process of anaerobic fermentation (absence of oxygen) by bacteria. The effects of this fermentation are the production of methane gas, clarified water uphill and mud depositing on the bottom. A further decantation tank eliminates the rest of the pollutants. Various measures have been adopted for the accessibility of structures to people with reduced motor and sensory capacity: first of all, the provision of sanitation services for the disabled according to the current regulations. The tourist service is guaranteed by the proposal of 8 "type paths": itineraries which can be completed in a day without lengthy car journeys, which turn into opportunities to visit historical villages in the vicinity, but also naturalistic pearls such as the Sila and Pollino National Parks as well as the Tyrrhenian and Ionian coasts.

The business is predominantly olive-growing because it is located in a territory particularly suitable for the production of quality oil: about 3000 olive trees are currently cultivated on 12 hectares. Furthermore, a family orchard, seasonal garden and an artichoke garden are also cultivated.

The oil mill treats olives harvested by hand on the plant when they have not yet fully matured and are worked within 24 h. The transformation is carried out with the traditional system of granite stone mills and one cold pressing. The extra virgin olive oil thus produced, in addition to having a low acidity, has excellent taste and olfactory characteristics thanks to its rich aromas and characteristic fruity flavor.

The sale of oil points to a logic of quality and not of quantity. Generally, production never exceeds what is consumed or sold. The market is mainly non-regional, with sales even in Denmark and Holland, thanks to contacts established with foreign tourists who are guests of the farmhouse.

There are also guided tours to the oil mill and tastings of oil and of traditional local products ranging from specialty preserves to sweet specialties. The cuisine of the farm tourism business is based on typical dishes of the local tradition.

In the farm business, it is possible to participate in a range of agricultural activities (olive harvesting and oil production in the mill), in the preparation of conserves and jams; it is also possible to follow cooking and agriculture courses.

A critical evaluation of the other receptive structures identified in the study was conducted (Table 2).

Each structure was assigned with a class on the basis of its characteristics:

- Class A: the structures possess, for each thematic area, sufficient requirements to adhere to the Area Trademark;
- Class B: the structures must undertake to improve the provision of services belonging to at least one of the three thematic areas;
- Class C: the structures are lacking in at least two of the three thematic areas;
- Class D: the structures do not possess sufficient requirements to become part of the Area Trademark.

Today, 12 out of 42 structures can adhere to the Area Trademark; while other structures require an improvement plan.

Table 2 Evaluation of accommodation facilities structures (A.F.)

		Theme area 1	Theme area 2	Theme area 3	Class
Montalto Uffugo	A.F. 1	✓	✓	✓	A
	A.F. 2	✓	✗	✓	B
	A.F. 3	✓	✗	✓	B
	A.F. 4	✓	✓	✓	A
	A.F. 5	✗	✓	✓	B
	A.F. 6	✗	✓	✗	C
Rose	A.F. 1	✗	✗	✓	C
Luzzi	A.F. 1	✓	✗	✓	B
San Benedetto Ullano	A.F. 1	✗	✗	✓	C
	A.F. 2	✗	✗	✗	D
Lattarico	A.F. 1	✓	✗	✓	B
	A.F. 2	✓	✗	✓	B
	A.F. 3	✗	✗	✓	C
	A.F. 4	✓	✓	✓	A
	A.F. 5	✓	✓	✓	A
S. Martino di Finita	A.F. 1	✓	✓	✓	A
Torano	A.F. 1	✗	✗	✓	C
Santa Caterina Albanese	A.F. 1	✗	✗	✓	C
	A.F. 2	✓	✗	✗	C
Cerzeto	A.F. 1	✗	✗	✓	C
Fagnano Castello	A.F. 1	✗	✗	✓	C
	A.F. 2	✗	✗	✓	C
San Marco Argentano	A.F. 1	✓	✓	✓	A
	A.F. 2	✗	✗	✗	D
	A.F. 3	✓	✓	✓	A
	A.F. 4	✓	✗	✓	B
Bisignano	A.F. 1	✓	✓	✓	A
	A.F. 2	✗	✗	✓	C
	A.F. 3	✓	✗	✓	B
Santa Sofia d'Epiro	A.F. 1	✗	✓	✓	B
	A.F. 2	✓	✓	✓	A

(continued)

Table 2 (continued)

		Theme area 1	Theme area 2	Theme area 3	Class
Tarsia	A.F. 1	✗	✗	✗	D
	A.F. 2	✗	✗	✗	D
	A.F. 3	✗	✗	✗	D
	A.F. 4	✗	✓	✗	C
	A.F. 5	✓	✓	✓	A
Altomonte	A.F. 1	✓	✓	✓	A
	A.F. 2	✓	✗	✗	C
	A.F. 3	✓	✗	✗	C
	A.F. 4	✓	✗	✓	B
	A.F. 5	✗	✗	✗	D
	A.F. 6	✓	✓	✓	A

4 The Area Trademark: Presupposition for the Definition of an Integrated Local System

Starting from the presupposition according to which create a network, or do together, are connotated as an indispensable presupposition on which it is possible to found new visions for local contexts, the constitution of an Area Trademark, even within a wider project such as that referring to the defection of an integral territorial system (Dematteis and Governa 2005), which also includes an Innovation Pole, represents an appropriate means of initiating an effective process pursuing such visions of shared development.

Specifically, referring to the experiment conducted, collaboration between the various receptive structures present proposes a "horizontally" spread hospitality model which is sustainable and representative of a genuine style of life.

The creation of the Area Trademark contributes to implementation of the so-called integrated planning, contributing to building a competitive local system in terms of services, education and cultural systems, as well as values and social cohesion.

A good territorial government requires the convergence of a plurality of interdisciplinary reflections which range from urban engineering to sociology of the economy, to stimulate new approaches and concrete projects, above all in internal areas with a rural vocation, as in the case under examination.

In the idea of integrated planning, the creation of the Trademark can thus represent the first step towards sharing other interventions aimed at, for example, improving and adapting other transversal systems such as mobility (Francini 2011). In this regard, referring to the case study, these interventions underline those aimed at enhancing the so-called "rural roads", to be connected with the main infrastructure of access to the area, as well as the creation of some hiking paths with new accesses and new branches. Underlining that the realization of any objective cannot ignore the current inadequacy of mobility infrastructure, it is necessary first to work to strengthen the relationship system between the various parts of the territory. This has to be done by redefining, primarily, the efficiency of the primary infrastructure network dedicated to vehicular traffic, in order to be able to replenish those traffic units that are inevitably diverted to the secondary network, responding to that new and ever-increasing demand for internal travel within the territory. Therefore, the area trademark can be considered as a prerequisite for making the concept of multifunctionality active in such contexts which can be well inserted within a systemic view of the agricultural sector. In particular, the multifunctional nature of agriculture and rural areas results in the offering, in addition to food and fiber, of services for the community: the contribution to the landscape; the provision of spaces for recreation and tourism; the protection of nature and wildlife; the increase in employment and income in rural areas; the protection of water, soil and air. Of course, all these links are not standard but are highly dependent on the specificities and peculiarities of the socio-cultural context in which they are implemented.

Multifunctional agriculture, which has become a central issue not only in literature (Nazzaro 2009, etc.), but also in political and scientific debates on the role of agriculture in the economy and society, is adopted primarily by the European Union as a guiding principle of rural development.

In this context, multifunctional agriculture acquires a more complete meaning and shows its potential as a driver for integrated development.

As explained by Knickel and Renting (2000), rural development consists of a wide variety of multi-dimensional and integrated activities that fulfill a number of functions not only for the company but also for the region and society. In this perspective the area trademark can be defined as a gear of integrated development and an instrument of valorisation of territorial potential in agricultural activities, determinants of rural sustainability.

References

Basin Authority of Calabria Region (2001) Hydrogeological condition plan

Dematteis G, Governa F (a cura di) (2005) Territorialità, sviluppo locale, sostenibilità: il modello SloT, Franco Angeli, Milano

Fossa G (a cura di) (2012) ITATOUR—Visioni territoriali e nuove mobilità: Progetti integrati per il turismo nella città e nell'ambiente, FRANCOANGELI/Urbanistica, pp 41–49

Francini M (a cura di) (2006) Modelli di sviluppo di aree interne ad alta ruralità, Università della Calabria—Centro Editoriale e Librario

Francini M (a cura di) (2011) Modelli di sviluppo di paesaggi rurali di pregio ambientale, FRANCOANGELI/Urbanistica

ISTAT (2011) 15° Censimento nazionale della popolazione e delle abitazioni

Knickel K, Renting H (2000) Methodological and conceptual issues in the study of multifunctionality and rural development. Socio Rural 40(4):512–528

Leone NG (a cura di) (2012) ITATOUR—Visioni territoriali e nuove mobilità: Progetti integrati per il turismo nella città e nell'ambiente, FRANCOANGELI/Urbanistica

Nazzaro C (2009) Sviluppo rurale, multifunzionalità e diversificazione in agricoltura. Franco Angeli, Milano

Province of Cosenza (2009) Province of cosenza provincial coordination territorial plan
Region of Calabria (2002) Regional urban planning Law n.19/2002, Bur Calabria
Region of Calabria (2016) Landscape valency territorial framework of the region of Calabria (QTR/P)
Scipioni A. e Vecchiato G. (a cura di) (2002) Marchio d'Area: Strumento di competizione turistica, FRANCOANGELI/Urbanistica

Planning and Designing Walkable Cities: A Smart Approach

Elisa Conticelli, Athanasios Maimaris, George Papageorgiou and Simona Tondelli

Abstract Walking may be considered one of the most sustainable and democratic ways of travelling within a city, thus providing benefits not only to pedestrians but also to the urban environment. Besides, walking is also one of the means of transport most likely subjected to factors outside an individual's control, like social or physical abilities to walk and the presence of comfortable and safe street infrastructures and services. Therefore, improving urban conditions provided to pedestrians has positive impacts on walkability. At the same time technological solutions and innovations have the power to encourage and support people to walk by overcoming immaterial barriers due to a lack of information or boring travel and they give to decision makers the possibility to gain data to understand how and where people travel. Merging these two dimensions into a unique approach can drastically improve accessibility, attractiveness, safety, comfort and security of urban spaces. In this context, this paper aims to draw a more multifaceted context for walkability, where new technologies assume a key role for introducing new approaches to pedestrian paths planning and design and thus for enhancing this mode of transport. Indeed, by combining more traditional spatial-based and perceptual analysis of the urban environment with technological applications and social media exploitation there will be room to better support the decision on and to enhance satisfaction of walking as well as to easier plan and design more walkable cities.

E. Conticelli (✉) · S. Tondelli
Department of Architecture, Alma Mater Studiorum, University of Bologna, Bologna, Italy
e-mail: elisa.conticelli@unibo.it

A. Maimaris · G. Papageorgiou (✉)
EUC Research Center, European University Cyprus, Egkomi, Cyprus
e-mail: g.papagerorgiou@euc.ac.cy

R. Papa et al. (eds.), *Smart Planning: Sustainability and Mobility in the Age of Change*, Green Energy and Technology,
https://doi.org/10.1007/978-3-319-77682-8_15

1 Introduction

"Sustainable mobility" is a concept that has been adopted worldwide as a response to the economic, social and environmental issues, such as global GHG emissions and traffic congestion, associated with an extensive use of private cars. At global level, sustainable transport is considered essential to achieve most of the Sustainable Development Goals (SDGs), fixed by the UN 2030 Agenda for Sustainable Development (UN-GA 2015), especially those related to health, energy, infrastructure, cities and human settlements. The UN-Habitat New Urban Agenda (UN-HABITAT 2017) wishes for a significant increase in accessible, safe, efficient, affordable and sustainable infrastructure for public transport, as well as for non-motorized options such as walking and cycling, prioritizing them over private motorized transportation.

In the last decades, also the European Union has developed several initiatives and documents in order to encourage a smarter and more sustainable urban mobility: the Leipzig Charter on Sustainable European Cities (2007), the European "Smart Cities & Communities Initiative" (2009), the Toledo Declaration (2010), the White Paper on Transport (EC, 2011), the Sustainable Urban Mobility Plan concept (COM, 2013), confirmed by the more recent Pact of Amsterdam (2016) hope for a sustainable and efficient urban mobility based on public transport, green vehicles as well as on soft mobility (walking, cycling, public space) that ensures accessibility for all and creates healthy environments.

The overall goal is to encourage active travel, such as cycling and walking, as sustainable travel modes that respond to both environmental and social needs and help people to have healthier and more sustainable lifestyles.

Despite all the efforts made at International and EU level and the improvements in new, clean vehicle technologies and alternative fuels, private motorized transport continues to remain dominant (EEA-TERM 2016), especially within urban areas. In fact, by analysing the modal split in 22 main European cities[1] it emerges that slightly over one third of trips are made by car, another third by public transport, and the last third are made on foot or by bicycle (FIA 2017) and walking covers around one fourth of the overall journeys (Fig. 1).

Although in the period 2009–2011 walking and cycling increased considerably in certain cities due to the effects of the economic crisis, the car still remains dominant in all EU countries (EEA 2015), and reducing air pollution and greenhouse gas emissions as well as traffic congestion generated by transport is therefore a major challenge for any sustainable city.

Furthermore, in dense urban areas and over relatively short distances motorized transport is neither efficient, nor sustainable nor safe and it contributes to a sedentary lifestyle, with dramatic consequences on health (WHO 2016). Conversely, cycling and walking are the most sustainable modes of transport (Kenworthy 2006), and their promotion together with the development of efficient public transport may be the

[1]Paris, London, Madrid, Barcelona, Lisbon, Vienna, Berlin, Amsterdam, Stockholm, Oslo, Ljubljana, Budapest, Bucharest, Brno, Warsaw, Vilnius, Sofia, Athens, Helsinki, Copenhagen, Florence, Bologna, data extracted from Epomm (http://www.epomm.eu) for the year 2011 (source FIA 2017).

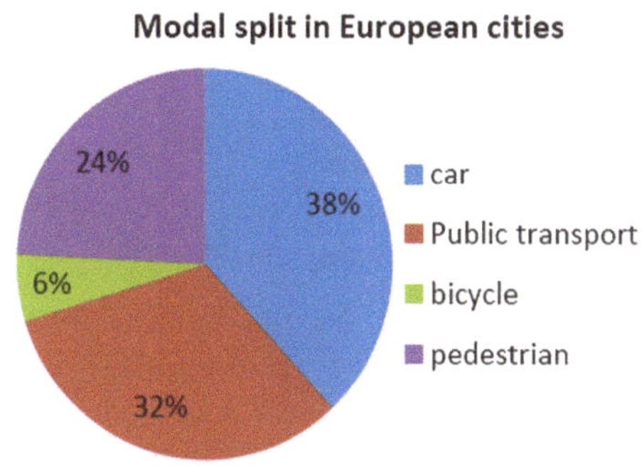

Fig. 1 Modal split in the main European cities (source FIA 2017)

real feasible and convenient alternatives to reduce levels of car use (UN-HABITAT 2017; Banister 2008), to encourage active travels and to make cities healthier and more liveable.

Notably, walking is an important component of almost all trips (Methorst et al. 2010) and may be considered as the most public form of transport (Tolley 2003) as it is accessible to everyone and it still remains an important mode in its own right (Methorst et al. 2010). Walking is also the best option for travelling within a city, experiencing its socio-spatial dimensions (Beyazid 2013), thus providing benefits to pedestrians as well as to the urban environment. Walking is therefore halfway between a transport system and a form of public life (Lamíquiz and López-Domínguez 2015), being a very flexible and socially friendly means of transport. Indeed, pedestrians can easily take decisions on the fly and easier interact with other people or activities, can travel for free, without emitting pollution and enhancing their health conditions, can contribute to make streets more vibrant and safer with their mere presence (Lamíquiz and López-Domínguez 2015). At the same time, walking is the primary form of active travel which has zero emissions, is convenient, and directly addresses lack of physical activity by replacing short, everyday car journeys (Leonard 2014). Finally, positive impacts on crime, foreclosures and housing values seem to be clearly evident in walkable neighbourhoods (Gilderbloom et al. 2015).

Walking has also positive effects for regenerating and for revitalizing the city. Shafray and Kim (2017) highlight that a walkable city concept has been gaining more and more interest as a relevant approach of urban revitalization, since it concurs to reduce congestion and car dependency, to facilitate outdoor walking and exercise thus enhancing healthy life styles, to promote social interaction through "face-to-face collaboration". According to Speck (2012), walkable life generates considerable savings for households and is more appealing for specific segments of population, such as creative communities, who prefer living in cities and who are becoming dominant in urban communities. Arup (2016) identifies relevant regenerative effects due to walking, such as:

- Social benefits affecting: health and wellbeing, safety, placemaking, social cohesion and equality.

- Economic benefits including: city attractiveness, the local economy, urban regeneration, and cost savings.
- Environmental benefits to do with virtuous cycles, ecosystem services, liveability and transport efficiency.
- Political benefits associated with leadership, urban governance, sustainable development and planning opportunities.

Walking is also one of the means of transport most likely affected by factors outside an individual's control, which can positively (or negatively) influence it, therefore a supportive built environment may be ensured to promote walking and cycling and to achieve an enduring increase of human activities. Social or physical abilities to walk (Van Cauwenberg et al. 2012), the presence of elevated residential and employment densities, comfortable, well connected and safe street patterns and amenities, presence of a variety of destinations (Handy 1996; Saelens et al. 2003a, b; Forsyth et al. 2008) are key factors influencing walkability. Since built environment characteristics are of high relevance for walking, and more than for other means of transport (Niemeier and Rutherford 1994), improving urban conditions provided to pedestrians has a very positive impact on walkability.

Despite these advantages and its importance in the everyday trips, walking remains rather poorly considered as a real and effective mode of transport especially over short distances, and few data have been collected concerning pedestrian mobility. Moreover, the importance and the role of walking as means of transport and as generator of aimlessly trips, such as trips for doing physical activity (Handy et al. 2002), is therefore underestimated.

Steering the use of new technologies towards walking, which are traditionally supportive to other means of transport, such as public and shared transports, might increase comfort and quality of pedestrian mobility thus increasing the related modal split. However supporting individual non-motorized mobility with smart solutions is not so frequently adopted as a way for increasing pedestrian mobility, though the smart city concept refers to sustainable cities and sustainable urban mobility (Garau et al. 2016), which are deeply based on walkability.

Indeed, technological solutions and innovations, such as gamification, social media, mobile apps etc., have the power to encourage and support people to walk by overcoming immaterial barriers due to a lack of information or boring travel and to gain data to understand how and where people travel. Furthermore, they can address the lack of data which may be more easily gained through common mobile phones and technological devices. A comprehensive smart city concept also foresees the presence of decisive, independent, and aware citizens (Roche et al. 2012), who become a sort of "human sensors" (Fistola 2013) when they use their personal technologies, such as smartphones, tablets, etc., to gain and to share data and to monitor specific features of an urban phenomenon. Under this viewpoint, pedestrians could be privileged observers and key players for gaining information that public authorities and transport planners hardly manage to achieve.

Moreover, in a smart mobility perspective, where infrastructures and vehicles are closely connected by new technologies and where the introduction of driverless cars and autonomous vehicles is forthcoming, it is essential to equip non-motorized users with proper technological devices and services in order to include them in the Intelligent Transport System, thus providing better accessibility and increasing safety of everyone (Woolsgrove 2016).

Starting from the analysis of spatial based and perceptual factors, which have been commonly identified as strongly influencing walking by previous studies, this paper aims to draw a more multifaceted context for walkability, where new technologies assume a fundamental role for introducing new approaches to pedestrian paths planning and design and thus for enhancing this mode of transport. Indeed, by combining more traditional spatial-based and perceptual analysis of the urban environment with technological applications and social media exploitation there will be room to better support the decision on and to enhance satisfaction of walking as well as to easier plan and design more walkable cities.

2 Issues and Challenges Affecting Walkability

Walking experience is affected by the cumulative impact of multiple factors and interactions (both positive and negative) between people and the urban environment. Spatial-based and perceived factors are frequently recognized as key elements influencing the decision to walk through their interaction with human senses and attitudes. The more recent adoption of new technologies as supporting tools for moving around a city has been opening a new field of study that considers new technologies themselves as an intermediary for processing, supporting and influencing traditional interactions between people and the urban realm.

2.1 Spatial-Based Factors

The influence of the built environment on travel demand, with special focus on walking, has been widespread recognized and agreed by many studies (see: Saelens and Handy 2008; Ewing and Cervero 2010), which have been developed by different disciplines, from transportation planning to public health studies (Saelens and Handy 2008). This "contamination", that conceive walking as a means of transport yielding positive health "side-effects" (Stockton et al. 2016) has presumably made recent transportation research more concerned with human-made environment determinants of non-motorized modes of travel than in the past (Saelens et al. 2003), stressing the importance of identifying which spatial-based factors influence people's attitude of walking and in which ways.

The built environment may be generally defined as "the part of the physical environment that is constructed by human activity" (Saelens and Handy 2008, p. 2) and "a composite of a multitude of characteristics" (Handy 1996, p. 153), but the literature addressing this topic is vast and several are the elements that form and affect it.

By reviewing research on the relationship between built environment and travel demand, Ewing and Cervero (2010) recall first of all Density, Diversity and Design—the first "three Ds" (Cervero and Kockelman 1997), as the most cited factors of the built environment affecting travel patterns. The first one is measured by population, dwelling units, employment, building floor area, per unit area; the second one considers the number of different land uses in a given area, while the third one includes street network characteristics within an area.

These three Ds were followed later on by other two Ds: Destination accessibility, which is defined as the ease of access to trip attractions and Distance to transit, which is usually measured as the average length of the shortest street routes from the residences or workplaces placed in an area to the nearest public transit station/stop.

The above-mentioned factors adhere to a large extent to five main dimensions highlighted later on by Handy et al. (2002) as the elements of the built environment that affect the attitude of walking: (i) density and intensity of development, defined as the amount of activities present in a given area; ii) mixed land uses; (iii) connectivity of the street network, which is the directness and availability of alternative routes through the network; (iv) scale of streets, conceived as the three-dimensional space along a street as bounded by buildings; (v) aesthetic qualities of a place, defined as the attractiveness and appeal of a place. Similar elements have been recognized by other authors, such as Frank and Pivo (1994), Saelens et al. (2003a, b), Forsyth et al. (2008), Ozbil (2009).

An extensive review of previous studies on these topics led by Saelens and Handy (2008) shows that the most cited factors characterizing the built environment that influence the attitude of walking are the following: (i) accessibility based on distance of or proximity to potential destinations, (ii) density, (iii) mixed land use, (iv) sidewalks (and more generally pedestrian infrastructures), (v) connectivity of routes/network. In addition, they highlight neighborhood type, which is generally composite of the above attributes, and aesthetic and safety, which introduce the issue of how the built environment is perceived by pedestrians.

These attributes interlink each other (Saelens and Handy 2008) and should be considered and managed not separately or individually, but interdependently, in order to promote walking (Stockton et al. 2016). For instance, the availability of destinations together with an interconnected street network makes walking more attractive against other mobility options (Saelens et al. 2003; Leslie et al. 2007).

The most relevant factors identified by past research as crucial for encouraging walking are summarized in Table 1.

These factors are frequently mentioned as correlates of the "walkability" concept, which has been investigated by several studies (see Chadwick Spoon 2005). Walkability has been frequently considered more in terms of specific variables that can be used to designate an area as walkable rather than as a concept itself. Indeed, according with Southworth (2005, p. 248), walkability is "the extent to which the built environment supports and encourages walking by providing for pedestrian comfort and safety, connecting people with varied destinations within a reasonable amount of time and effort, and offering visual interest in journeys throughout the network". These variables largely correspond to the above mentioned spatial-based factors.

Table 1 Definition of main factors of the built environment affecting the attitude of walking (authors' elaboration from the analyzed literature)

Factors	Definition
Density and intensity of development	Amount of activities present in a given area
Connectivity of the street network	Directness and availability of alternative routes through the network
Scale of street	Three-dimensional space along a street as bounded by buildings
Accessibility	Based on distance of or proximity to potential destinations
Pedestrian infrastructures	Presence of sidewalks and pedestrian spaces
Mixed land uses	Proximity of different land uses
Aesthetics	Attractiveness and appeal of a place
Safety	Presence of good physical conditions preventing injuries and dangers, such as street lights or pedestrian crossings

2.2 *Perceptual Factors*

Beside analysis concerning spatial based features characterizing the pedestrian environment, other studies have been led under another perspective, letting to identify other important factors influencing pedestrian activity. They refer to pedestrian perception of the built environment, which is derived from subjective values (Lee 2010) and concurs to creating urban environments that support and enhance walking activity as well.

Perceptual factors correlate with the human perception of the built environment, which acts as mediator between the physical features of the environment and walking behaviour (Ewing et al. 2006) since an individual's positive or negative views of the environment may affect his/her attitude of walking (Lee 2010). Decision to walk depends on how difficult or how ease the action to walk is perceived by people (Methorst et al. 2010), therefore it is essential to understand and to assess the influence of perception on walking behaviours.

This 'humanistic perspective' has been firstly underpinned by important planners and urban designers, such as Jacobs (1961), Lynch (1960), Gehl (1987) and Gehl and Gemzoe (2003) who have been studying how the built environment impacts people's city experience and daily activities within an urban area, by using different approaches.

Studies oriented to investigate the relationship between the built environment and the attitude of walking have also taken into account variables related with perception. The above mentioned review performed by Saelens and Handy (2008) identifies attitudes such as "sense of safety" and "aesthetic qualities", as important elements affecting walking behaviors. These elements derive from how the built environment is perceived by people. Other studies have deepened the issue directly by trying to identify specific perceptual factors (Lee 2010).

Notably the study "Measuring Urban Design Qualities" led in the framework of the Active Living Research Program (Ewin et al. 2006) represents an important reference. It has been performed by an Expert Panel who extracted a list of key perceptual qualities of the urban environment based on a review of the most relevant urban design literature. The study focuses on urban design qualities, i.e. qualities of the environment that depend on physical features, but reflect the general way in which people perceive and interact with the environment. Nine main urban design qualities affecting walking behaviors have been identified: imageability, legibility, visual enclosure, human scale, transparency, linkage, complexity, coherence and tidiness (Table 2).

Out of this list another important factor is comfort (Ovstedal and Ryeng 2002) which may influence the decision to walk (Barros et al. 2015). It is affected by several perceived features of the urban environment, such as microclimate, sounds, visibility, smells, etc. Similar to sense of safety, sense of comfort depends on the reaction of an individual to external surroundings therefore it is considered but not as perceptual quality because it is affected by a degree of objectivity by outside observers (Ewing et al. 2006).

These elements are widely recognized as very influent on the decision to walk; at the same time, it is very difficult to map and to measure them, in order to take them into account within urban retrofitting interventions or once mobility changes addressing walking encouragement should be undertaken.

2.3 Real Time Information

People moving around cities are more and more experiencing the need of and the opportunities offered by getting information. It is frequently requested to know about the presence of services and shops, the best way to get to a specific place, how safe it is to travel, and so on. To get such information, individuals increasingly use information and communication technologies (ICT), such as sensor-based networks and geospatial-distributed technologies (Roche et al. 2012) and communication technologies (Papageorgiou and Maimaris 2017). In this way people orient themselves in the physical space and thus effectively navigate themselves (Fang et al. 2015). Mobility can therefore be enhanced with dynamic information for all means of transport, including walking. As a result, a number of Intelligent Transportation Systems (ITS) applications have been developed for all kinds of transport and mobility, showing an interest also for walking (Monterde-i-Bort et al. 2010), even though these applications have been developed mainly for motorized vehicles. Moreover, pedestrian navigation has been one of the most common research topics in the community of spatial behavior of geography, Geographical Information Systems (GIS), as well as outdoor/indoor positioning technology and application (Fang et al. 2015).

Table 2 Definition of the nine Urban design qualities according with the study Measuring Urban Design Qualities (Ewing et al. 2006)

Urban design quality	Description
Imageability	Imageability is the quality of a place that makes it distinct, recognizable, and memorable A place has high imageability when specific physical elements and their arrangement capture attention, evoke feelings, and create a lasting impression
Legibility	Legibility refers to the ease with which the spatial structure of a place can be understood and navigated as a whole. The legibility of a place is improved by a street or pedestrian network that provides travelers with a sense of orientation and relative location and by physical elements that serve as reference points
Enclosure	Enclosure refers to the degree to which streets and other public spaces are visually defined by buildings, walls, trees, and other elements. Spaces where the height of vertical elements is proportionally related to the width of the space between them have a room-like quality
Human Scale	Human scale refers to a size, texture, and articulation of physical elements that match the size and proportions of humans and, equally important, correspond to the speed at which humans walk. Building details, pavement texture, street trees, and street furniture are all physical elements contributing to human scale
Transparency	Transparency refers to the degree to which people can see or perceive what lies beyond the edge of a street or other public space and, more specifically, the degree to which people can see or perceive human activity beyond the edge. Physical elements that influence transparency include walls, windows, doors, fences, landscaping, and openings into midblock spaces
Linkage	Linkage refers to physical and visual connections from building to street, building to building, space to space, or one side of the street to the other which tend to unify disparate elements. Tree lines, building projections, marked crossings all create linkage. Linkage can occur longitudinally along a street or laterally across a street
Complexity	Complexity refers to the visual richness of a place. The complexity of a place depends on the variety of the physical environment, specifically the numbers and kinds of buildings, architectural diversity and ornamentation, landscape elements, street furniture, signage, and human activity
Coherence	Coherence refers to a sense of visual order. The degree of coherence is influenced by consistency and complementarity in the scale, character, and arrangement of buildings, landscaping, street furniture, paving materials, and other physical elements
Tidiness	Tidiness refers to the condition and cleanliness of a place. A place that is untidy has visible signs of decay and disorder; it is in obvious need of cleaning and repair. A place that is tidy is well maintained and shows little sign of wear and tear

The added value of using technological applications is therefore the opportunity that technology provides to automate and integrate spatial and perception factors in order to support pedestrian mobility. Through technology we can integrate components that influence pedestrians' physical and psychological condition by taking into account their individual values, attitudes as well as individual characteristics, such as gender and disability. Further on, we provide an overview of currently available technologies that can be used for developing effective technological applications for pedestrian mobility.

3 ICT for Walkable Cities

3.1 Currently Available Technologies

Technology is advancing very fast nowadays. Urban environments have a plethora of electronic sensor systems, GIS data, mobile communication systems, cameras and big data available via an internet connection. Such technologies can be employed for developing effective smart pedestrian mobility systems in order to promote walkability, reduce GHG emissions and improve the quality of life of citizens. An overview is provided below on the most important currently available technologies to be considered when developing an innovative pedestrian mobility system (Fig. 2) .

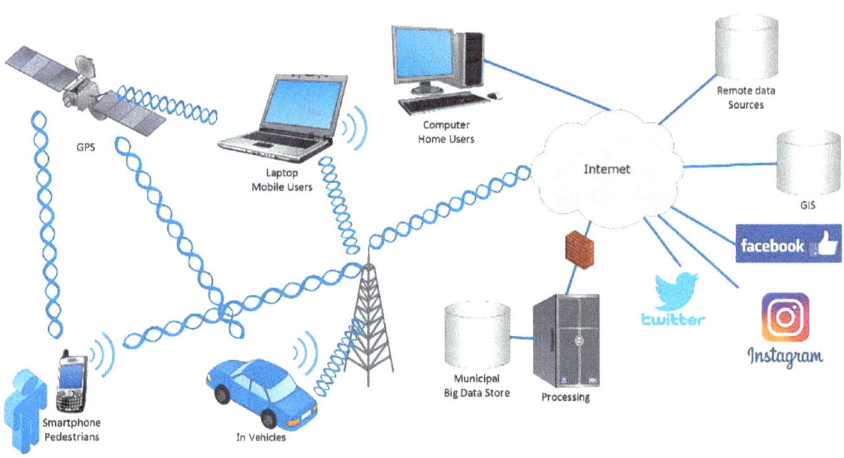

Fig. 2 Concept diagram that depicts various technologies for walkable cities (authors' elaboration)

3.1.1 Geographical Information Systems

Geographical Information Systems (GIS) are an essential component for the development of any smart pedestrian mobility system. GIS data can be collected from various sources, which can either be commercial, free, private or public. GIS data can also be gathered via crowdsourcing, which makes the pedestrians a contributing part of the system, actively identifying problem areas within a traffic network. Also, social media can be an important source of information, which can be converted into useful data via big data machine learning. Publicly available datasets can be incorporated in the GIS engine, such as OpenStreetMap (Haklay 2010). Pedestrian mobility data (commercial and public) can be imported in a central GIS database, which can be further used to generate sophisticated maps for the pedestrian mobility system. GIS can contribute to increasing pedestrian connectivity by providing useful information on plausible walking paths and the use of other transport modes.

3.1.2 Electronic Sensor Systems

Another important technology is electronic sensor systems, which are abundant in our modern cities. Sensors can be used to directly measure the conditions of the pedestrian network. For example, data can be collected via traffic flow inductive loop sensors, air pollution sensors, traffic cameras and so on. Also, smartphones carried by pedestrians, have a whole range of available sensors, including an accelerometer, global positioning system (GPS), gyroscope, sound, picture, video, magnetic field strength, and barometer that can collect geotagged environmental data.

3.1.3 Radio Frequency Identification

An interesting communication technology that can facilitate navigation is radio frequency identification (RFID). RFID can provide infrastructure to pedestrian communication (Cheng et al. 2012). Pedestrians can be provided with information on traffic signage which can be retrieved from the Internet. Further, RFID tags can monitor pedestrian flows, density, movement and speed of pedestrians. Before implementation of RFID, any privacy issues should be resolved.

3.1.4 Bluetooth

Bluetooth (Mamdouhi et al. 2009) can provide important local information to pedestrians without a need for an internet connection. It can also be used to monitor pedestrian mobility by collecting the MAC address of pedestrians' smartphones and can be used to build origin-destination matrices. Bluetooth can also be very useful for a number of associated devices or sensors such as wearable technology for physical exercise, health monitoring and so on.

3.1.5 Wifi Connection

As internet connectivity is essential for any smart pedestrian mobility system, Wifi (Shlayan et al. 2016) can be an important provider of internet access. Since Wifi is a duplex communication technology and supports TCP/IP it can be used by pedestrians for information exchange and services based on location. Further, the various generations of mobile communication—2G (Ma et al. 2008), 3G (Zhao et al. 2013), 4G (Zhioua et al. 2015) and the upcoming 5G (Mumtaz et al. 2015)—are the main technologies for connection with the internet. As a result, pedestrians have available abundant information which can be fed to them in real time.

3.1.6 GPS

Obviously, GPS would be the most important component of any smart mobility system regarding navigation—GPS (Ansari et al. 2014) (USA), GLONASS (Cozzetti et al. 2011) (Russia), Galileo (Margaria and Falletti 2014) (EU) and BeiDou (Jan and Tao 2016) (China). Also, a number of services based on location are facilitated with the use of GPS, such a visiting sites and tourist attractions.

3.1.7 ZigBee Protocol

As we are living in the era of the Internet of Things (IoT), the ZigBee protocol (Chen et al. 2016) can further enhance the pedestrian mobility experience. ZigBee can facilitate wireless sensor networks (WSN) which can connect a number of devices in the pedestrian network. A less restrictive wireless communication technology is ISM RF (Sichitiu and Kihl 2008) which provides a more flexible way of connecting devices on a smart pedestrian mobility system.

3.2 Areas of Application

By considering the opportunities offered by the available technologies described above, and relating them to issues ad challenges discussed in par. 2, it is possible to identify three main integration areas where ICT can sensibly improve cities walkability:

- measuring and assessing walkability,
- changing behavior,
- gaining data by monitoring pedestrian mobility and providing real time information in order to meet specific and diverse mobility-related needs.

3.2.1 Walkability Assessment and Measurement

As already mentioned, the spatial-based dimension is highly correlated to the walkability of an urban area. Walkability is frequently considered in terms of specific factors that can be used to designate an area as walkable and there is a number of the GIS-based applications that process different factors allowing to measure and assess the spatial walkability conditions.

Such applications have shown important advantages for assessing walkability in a certain neighborhood in comparison with traditional tools, such as surveys, audits or observational data (Gilderbloom et al. 2015). Furthermore, there is a potential to facilitate the development of indices of walkability that could be used to evaluate new environmental and policy initiatives (Sallis et al. 1998; Bauman et al. 2002; Leslie et al. 2007).

One such walkability index is Walk ScoreTM which automatically calculates walkability for NYC and for all major cities of USA, Canada, Australia, and New Zealand. It is based on a set of two different types of data: Walking routes and distances to amenities and Road connectivity metrics. Walk Score rates the walkability of an address by determining the distance to educational, retail, food, recreational and entertainment destinations (Pivo and Xudong 2016). Its main objective is to aid the home-buying process (Agampatian 2014) but it has also become a reference for ranking walkable urban developments in the US (Leinberger and Rodriguez 2016). Therefore, merging different dimensions made by spatial based factors into a unique tool can drastically improve accessibility, attractiveness, safety, comfort and security of urban spaces.

3.2.2 Supporting Walking Behaviors

Perceptual factors of the built environment as well as design of urban space are widely recognized as very influent on the decision to walk thus affecting people's behaviors, especially for transportation purposes (Hoehner et al. 2005). At the same time, it is very hard to map and to clearly identify them.

Currently available technologies can be employed in the walking domain for providing a series of different information to pedestrians. As a matter of fact, pedestrian navigation has received increasing attention by the GIS discipline (Fang et al. 2015), and more and more frequently people use GIS-based applications and technological devices for navigating and for assisting pedestrian wayfinding, thus increasing confidence in an unknown environment, or for being alerted on unsafe conditions due to traffic or on insecure environments. GIS based applications can also increase pedestrian comfort navigation using multi-modal environmental sensors, and particularly helping people with special needs (Monterde-i-Bort et al. 2010; Fang et al. 2015). Processed information may be related both to physical and to perceptual factors and are mainly addressed to citizens and to city users in order to support their walking travels and thus to influence their mobility behaviors.

Beyond new technologies supporting pedestrian navigation in pleasant environment, persuasive strategies based on the adoption of new technologies can be useful in facilitating a behavioural change (Wunsch et al. 2015), thus encouraging walking mobility. Many approaches based on challenges and goal setting, self-monitoring, personalized messages, social comparison, gamification and rewards have been developed so far (Anagnostopoulou et al. 2016).

Self-monitoring is the most frequently used approach and takes the form of visual feedbacks on e.g. CO_2 emissions caused by the users' trips, cost savings and burned calories, etc. Gamification and reward-based competitions have a role in boosting behavioural change as well, especially among youngsters and new generations. Gamification is now incorporated into many running and cycling mobile phone applications where users collect points for the miles they monitor or for reaching a specific target (Coombes and Jones 2016). In order to achieve good results in this field in terms of engaging even the most reluctant people it is important to personalize persuasion profiles which will be used to tailor interventions to individual user characteristics (Anagnostopoulou et al. 2016).

3.2.3 Data Gaining and Real-Time Information

Lack of data about pedestrian trips is a twofold issue. On one side, it affects the knowledge and therefore the decision making that frequently does not consider walking as a real means of transport, but only a necessary part of a trip represented by other—more important—means; on the other side, real time information to pedestrians could improve their awareness about the different trips possibilities (both in terms of alternatives paths and of trips length). As a matter of fact, National travel surveys often register neither the shorter trips nor the walking parts of trips made by public transport (Wittink 2001) or more generally by motorized vehicles (ITF 2012).

Walking is also considered as a derived demand and consequently walking trips not associated with a specific destination, e.g. trips related to leisure and physical activity, are not usually taken into account in the total amount of urban travels (Handy et al. 2002).

Therefore, there is a tendency to underestimate walking flows especially if the walking trips are short (ITF 2012) both by people who generate them and by transport planners. Even if citizens are all pedestrians at one point in time, pedestrians are not sufficiently represented in urban policies; consequently, urban and transport planners are not adequately influenced by pedestrian needs when they fix strategies and interventions on urban mobility and infrastructures.

Despite this poor attention on walking as a real means of transport, some studies have been performed for reducing the gaps on data collection and for standardizing walking data, in order to raise awareness among urban and transport planners on the importance of walking as a means of transport and on how the traffic measurements can include walking more appropriately. Indeed, walking should be considered as an important allay of public transport because, without walking, buses, trams and trains would have no passengers. A valuable example is the International walking data standards report (Sauter et al. 2016), which aims at providing a standard way of defining and measuring walking. The objective is to establish comprehensive data on walking trips, according with the same degree of accuracy and diligence as other modes, in order to facilitate policy makers and planners to face the extent of walking and to take effective decision for its promotion. London have already adopted this approach for analyzing and then for better planning a walkable city and cities and mobility agencies from Europe and abroad have been strongly invited to do the same, but a strong commitment to invest in this direction is still needed.

According with this issue the European Parliament resolution of 23 April 2009 (COM 2009) on an Action Plan on Urban Mobility (APUM) gives special attention to potentially useful data on urban transport and mobility, which could be collected through information technology. Crowdsourcing for data collection may be financially effective, especially in cases where the user base is small but enthusiastic and motivated; in such cases crowdsourcing has a huge potential in augmenting the standard data collection procedures by including the opinions of otherwise marginalized groups of users (Misra et al. 2014).

If we consider gaining data systems already in place, both 'Google Better Cities' and 'COWI City Sense–Signal Re-identification' appear to be promising for data collection on active transportation modes use in the future (Steenberghen et al. 2017). COWI has undertaken projects on pedestrians over the past two years, based on increasingly accurate GPS based systems. These systems are used to analyse human mobility around sporting events or community gatherings. However, this application has only been applied in Denmark so far.

4 Conclusions

The above described technologies and applications can greatly contribute to the development of smart walkable cities by strengthening and by supporting all the factors related with perception and features of the built environment in a positive way. Notably, connectivity can be enhanced with GIS, Bluetooth, and mobile communications; safety can be improved by crowdsourcing and image processing from traffic cameras; the choice of comfortable routes can also be facilitated in the pedestrian network through smart sensor technologies and GIS mapping; navigation systems may encourage interactions between pedestrians´ behaviours and the urban environment.

Technological applications may be very supportive also for the decision-making process oriented to develop more walkable cities. By modelling physical and perceptual factors that mostly affect walkability and people's mobility behaviors it is possible to develop innovative decision support systems able first of all to systematically determine and map minimal existing conditions for walkability and consequently to determine new requirements and interventions to be planned and designed for each street in order to develop more walkable cities. Assessing the walkable conditions provided by the streets is essential to detect if cities have a network of suitable pedestrian streets or not. Identifying necessary interventions and design requirements may also help transport and urban planners and policy makers in better programming infrastructure investments and adopting differentiated and targeted actions for encouraging walkability, such as: improving street lighting, improving footpaths and sidewalks, prioritizing some streets to pedestrians in order to regenerate public spaces and neighborhoods.

Clearly, currently available technologies have a lot to offer when it comes to developing an effective smart pedestrian mobility system. Technology can be a source for innovative solutions to many walkability issues.

Acknowledgements The outcomes presented in this paper have been developed within the SPN ongoing research project funded in the framework of the ERA–NET Cofund Smart Urban Futures (ENSUF) Joint Programming Initiative (JPI) Urban Europe, under the HORIZON 2020 ERA–NET Cofund scheme.

References

Agampatian R (2014) Using GIS to measure walkability: a case study in New York City. Master's of Science Thesis in Geoinformatics TRITA-GIT EX 14-002, School of Architecture and the Built Environment, Royal Institute of Technology (KTH), Stockholm, Sweden

Anagnostopoulou E, Bothos E, Magoutas B, Schrammel J, Mentzas G (2016) Persuasive technologies for sustainable urban mobility. arXiv Preprint arXiv:1604.05957

Ansari K, Feng Y, Tang M (2014) A runtime integrity monitoring framework for real-time relative positioning systems based on GPS and DSRC. IEEE Trans Intell Transp Syst 16:980–992. https://doi.org/10.1109/TITS.2014.2349011

ARUP (2016) Cities alive. Towards a walking world. London

Banister D (2008) The sustainable mobility paradigm. Transp Policy 15:73–80. https://doi.org/10.1016/j.tranpol.2007.10.005

Barros AP, Martínez LM, Viegas JM (2015) A new approach to understand modal and pedestrians route in Portugal. Transp Res Procedia 10:860–869. https://doi.org/10.1016/j.trpro.2015.09.039

Bauman A, Sallis JF, Owen N (2002) Environmental and policy measurement in physical activity research. In: Welk G, Dale D (eds) Physical activity assessments for health-related research. Human Kinetics, Champaign, pp 241–251

Beyazid E (2013) Mobility cultures. In: Givoni M, Banister D (eds) Moving towards low carbon mobility. Edward Elgar, Chelternham

Cervero R, Kockelman K (1997) Travel demand and the 3Ds: density, diversity, and design. Transp Res D 2(3):199–219

Chadwick Spoon S (2005) What defines walkability: walking behavior correlates. Masters Project submitted to the faculty of the University of North Carolina at Chapel Hill in partial fulfillment of the requirements for the degree of Master of Regional Planning in the Department of City and Regional Planning. Chapel Hill, North Carolina

Chen C, Pei Q, Li X (2016) A GTS allocation scheme to improve multiple-access performance in vehicular sensor networks. IEEE Trans Veh Technol 65:1549–1563. https://doi.org/10.1109/TVT.2015.2412613

Cheng W, Cheng X, Song M et al (2012) On the design and deployment of RFID assisted navigation systems for VANETs. IEEE Trans Parallel Distrib Syst 23:1267–1274. https://doi.org/10.1109/TPDS.2011.259

Commission of the European Communities (COM) (2009) Communication from the Commission to the European Parliament, the Council, the European Economic and Social Committee and the Committee of the Regions—action plan on urban mobility, COM (2009), 490 final

Commission of the European Communities (COM) (2013). Communication from the Commission to the European Parliament, the Council, the European Economic and Social Committee and the Committee of the Regions—together towards competitive and resource-efficient urban mobility, COM (2013), 913 final
Coombes E, Jones A (2016) Gamification of active travel to school: a pilot evaluation of the beat the street physical activity intervention. Health Place 39:62–69
Cozzetti A, Scopigno R, Lo Presti L (2011) Tight coupling benefits of GNSS with VANETs. IEEE Aerosp Electron Syst Mag 26:15–23. https://doi.org/10.1109/MAES.2011.5763339
European Environment Agency (EEA) (2013) A closer look at urban transport. TERM 2013: transport indicators tracking progress towards environmental targets in Europe. EEA Report No. 11/2013, European Environment Agency
European Environment Agency (EEA) (2015) The European environment—state and outlook 2015: synthesis report, European Environment Agency, Copenhagen, https://www.eea.europa.eu/soer. Accessed 29 Sept 2017
Ewing R, Cervero R (2010) Travel and the built environment. J Am Plan Assoc 76(3):265–294. https://doi.org/10.1080/01944361003766766
Ewing R, Handy S, Brownson RC, Clemente O, Winston E (2006) Identifying and measuring urban design qualities related to walkability. J Phys Act Health 3(Suppl 1):S223–S240
Fang Z, Li Q, Shaw S (2015) What about people in pedestrian navigation? Geo-Spat Inf Sci 18(4):135–150. https://doi.org/10.1080/10095020.2015.1126071
Federation internationale de l'automobile, Region I – Europe the Middle East and Africa (FIA 2017) Briefing on urban mobility
Fistola R (2013) Smart City: riflessioni sull'intelligenza urbana [Smart city: thinking about urban intelligence]. TeMA 1:47–60. https://doi.org/10.6092/1970-9870/1460
Forsyth A, Hearst M, Oakes JM, Schmitz KH (2008) Design and destinations: factors influencing walking and total physical activity. Urban Stud 45(9):1973–1996
Frank LD, Pivo G (1994) Impacts of mixed use and density utilization of three modes of travel: single-occupant vehicle, transit, and walking. Transp Res Rec 1466:44–52
Garau C, Masala F, Pinna F (2016) Cagliari and smart urban mobility: analysis and comparison. Cities 56:35–46. https://doi.org/10.1016/j.cities.2016.02.012
Gehl J (1987) Life between buildings: using public space. Van Nostrand Reinhold, New York
Gehl J, Gemzoe L (2003) New city spaces, 3rd edn. The Danish Architectural Press, Copenhagen
Gilderbloom JI, Riggs WW, Meares WL (2015) Does walkability matter? An examination of walkability's impact on housing values, foreclosures and crime. Cities 42:13–24. https://doi.org/10.1016/j.cities.2014.08.001
Haklay M (2010) How good is volunteered geographical information? A comparative study of OpenStreetMap and Ordnance Survey datasets. Environ Plan 37:682–703
Handy S (1996) Methodologies for exploring the link between urban form and travel behavior. Transp Res Rec 1(2):151–165
Handy SL, Boarnet MG, Ewing R, Killingsworth RE (2002) How the built environment affects physical activity. Views Urban Plan Am J Prev Med 23(2S):64–73
Hoehner CM, Brennan Ramirez LK, Elliott MB, Handy SL, Brownson RC (2005) Perceived and objective environmental measures and physical activity among urban adults. Am J Prev Med 28(2S2):105–116
ITF (2012) Pedestrian safety, urban space and health. OECD Publishing. https://doi.org/10.1787/9789282103654-en
Jacobs J (1961) The death and life of great American cities, Random House, New York
Jan S-S, Tao A-L (2016) Comprehensive comparisons of satellite data, signals, and measurements between the BeiDou Navigation satellite system and the global positioning system. Sensors 14248220(16):1–24
Kenworthy JR (2006) The eco-city: ten key transport and planning dimensions for sustainable city development. Environ Urbanization 18(1):67–85. https://doi.org/10.1177/0956247806063947

Lamíquiz PJ, López-Domínguez J (2015) Effects of built environment on walking at the neighbourhood scale. A new role for street networks by modelling their configurational accessibility? Transp Res Part A 74(2015):148–163. https://doi.org/10.1016/j.tra.2015.02.003
Lee EH (2010) Perceptions and evaluation of an urban environment for pedestrian friendliness: a case study. A thesis presented to the Faculty of California Polytechnic State University, San Luis Obispo, in partial fulfillment of the requirements for the Degree Master of City and Regional Planning in College of Architecture & Environmental Design
Leinberger CB, Rodriguez M (eds) (2016) Foot traffic ahead: ranking walkable urbanism in America's Largest Metros, George Washington University
Leonard C (ed) (2014) Civitas, innovative urban transport solutions. Civitas makes the difference. How 25 cities learned to make urban transport cleaner and better, ICLEI—Local governments for sustainability, Freiburg
Leslie E, Coffee N, Frank L, Owen N, Bauman A, Hugo G (2007) Walkability of local communities: using geographic information systems to objectively assess relevant environmental attributes. Health Place 13:111–122. https://doi.org/10.1016/j.healthplace.2005.11.001
Lynch K (1960) The image of the city, The MIT press, Cambridge MA
Ma Z, Chen D, Cui L (2008) Wireless monitoring system of vehicle violation of running red led based on GPRS. In: 3rd International Conference on Innovative Computing Information and Control, ICICIC'08. IEEE Computer Society
Mamdouhi H, Khatun S, Zarrin J (2009) Bluetooth wireless monitoring, managing and control for inter vehicle in vehicular Ad-Hoc networks. J Comput Sci 5:922–929. https://doi.org/10.3844/jcssp.2009.922.929
Margaria D, Falletti E (2014) A novel local integrity concept for GNSS receivers in urban vehicular contexts. In: 2014 IEEE/ION Position, Location and Navigation Symposium-PLANS 2014. IEEE, pp 413–425
Methorst R, Monterde i Bort H, Risser R, Sauter D, Tight M, Walker J (2010) PQN final report, WALK21, Cheltenham
Misra A, Gooze A, Watkins K, Asad M, Le Dantec CA (2014) Crowdsourcing and its application to transportation data collection and management. Transp Res Rec J Transp Res Board 2414:1–8
Monterde i Bort H, Johanssen C, Leden L, Basbas S (2010) ITS and on-trip tasks while walking. In: Johannessen S, van Wee B (eds) PQN final report—Part B: documentation
Mumtaz S, Saidul Huq KM, Ashraf MI et al (2015) Cognitive vehicular communication for 5G. IEEE Commun Mag 53:109–116. https://doi.org/10.1109/MCOM.2015.7158273
Niemeier DA, Rutherford GS (1994) Non-motorized transportation, NPTS: Travel Mode Special Reports, No. 3, U.S. Department of Transportation, Federal Highway Administration, Office of Highway Information Management, Washington, D.C
Ovstedal L, Ryeng EO (2002) Understanding pedestrian comfort in Europena cities: how to improve walking conditions? Association for European Transport
Ozbil A (2009) Walking to the station: the effects of urban form on walkability and transit ridership. Dissertation Topic Proposal. College of Architecture, Georgia Institute of Technology
Papageorgiou G, Maimaris A (2017) Towards the Development of Intelligent Pedestrian Mobility Systems (IPMS), 2017 International Conference on Electrical Engineering and Informatics (ICELTICs), At Banda Aceh, Aceh, Indonesia, 268–273, DOI: 10.1109/ICELTICS.2017.8253267
Pivo G, Xudong A (2016) Sustainable development and commercial real estate financing: evidence from CMBS loans. http://capla.arizona.edu/sites/default/files/faculty_papers/Sustainable%20Development%20and%20Real%20Estate%20Financing%20-%20Evidence%20from%20CMBS%20Loans.pdf
Roche S, Nabian N, Kloeckl K, Ratti C (2012) Are 'smart cities' smart enough? In: Global geospatial conference 2012, spatially enabling government, industry and citizens, Québec City, Canada, 14–17 May 2012
Saelens BE, Handy SL (2008) Built environment correlates of walking: a review. Med Sci Sports Exerc 40(7 Suppl):S550–S566. https://doi.org/10.1249/MSS.0b013e31817c67a

Saelens BE, Sallis JF, Frank LD (2003a) Environmental correlates of walking and cycling: findings from the transportation. Urban Des Plan Literatures Ann Behav Med 25(2):80–91
Saelens BE, Sallis JF, Black JB, Chen D (2003b) Neighborhood-based differences in physical activity: an environment scale evaluation. Am J Publ Health 93(9):1552–1558
Sallis J, Bauman A, Pratt M (1998) Environmental and policy interventions to promote physical activity. Am J Prev Med 15(4):379–397
Sauter D, Tight M, Pharoah T, Martinson R, Wedderburn M (2016) International walking data standard. Treatment of walking in travel surveys. Internationally standardized monitoring methods of walking and public space. www.measuring-walking.org
Shafray E, Kim S (2017) A study of walkable spaces with natural elements for urban regeneration: a focus on cases in Seoul, South Korea. Sustainability 9(587):1–20
Shlayan N, Kurkcu A, Ozbay K (2016) Exploring pedestrian Bluetooth and WiFi detection at public transportation terminals. In: 2016 IEEE 19th International Conference on Intelligent Transportation Systems (ITSC), pp 229–234
Sichitiu M, Kihl M (2008) Inter-vehicle communication systems: a survey. IEEE Commun Surv Tutorials 10:88–105. https://doi.org/10.1109/COMST.2008.4564481
Southworth M (2005) Designing the walkable city. J Urban Plan Dev 131(4):246–257
Speck J (2012) Walkable city: how downtown can save America, one step at a time. North Point Press, New York
Steenberghen T, Tavares T, Richardson J, Himpe W, Crabbé A (2017) Support study on data collection and analysis of active modes use and infrastructure in Europe. Final Report. European Union, Luxembourg
Stockton JC, Duke-Williams O, Stamatakis E, Mindell JS, Brunner EJ, Shelton NJ (2016) Development of a novel walkability index for London, United Kingdom: cross-sectional application to the Whitehall II Study. BMC Public Health 16:416. https://doi.org/10.1186/s12889-016-3012-2
Tolley R (ed) (2003) Sustainable transport: planning for walking and cycling in urban environments. Woodhead Publishing Ltd, Cambridge
United Nations, General Assembly (UN-GA) (2015) Transforming our world: the 2030 agenda for sustainable development. A/RES/70/1
United Nations Habitat III (UN-HABITAT) (2017). New urban agenda, United Nations
Van Cauwenberg J, Van Holle V, Simons D, Deridder R, Clarys P, Goubert L, Nasar J, Salmon J, De Bourdeaudhuij I, Deforche B (2012) Environmental factors influencing older adults' walking for transportation: a study using walk-along interviews. Int J Behav Nutr Phys Act 9:85. https://doi.org/10.1186/1479-5868-9-85
Wittink R (2001) Promotion of mobility and safety of vulnerable road users: final report of the European research project PROMISING (Promotion of Measures for Vulnerable Road Users). D-2001-3. SWOV Institute for Road Safety Research, Leidschendam
Woolsgrove C (2016) Advanced vehicle technologies, autonomous vehicles and cycling. European Cyclists' Federation
World Health Organization (WHO) (2016) Global report on urban health. Equitable, healthtier cities for sustainable development. WHO Press, Geneva
Wunsch M, Stibe A, Millonig A, Seer S, Dai C, Schechtner K, Chin RCC (2015) What makes you bike? Exploring persuasive strategies to encourage low-energy mobility. In: MacTavish T, Basapur S (eds) Persuasive technology. 10th international conference, persuasive 2015, Chicago, Proceedings. Springer, New York–London, June 3–5, 2015
Zhao Q, Zhu Y, Chen C et al (2013) When 3G meets VANET: 3G-assisted data delivery in VANETs. IEEE Sens J 13:3575–3584. https://doi.org/10.1109/JSEN.2013.2265304
Zhioua GEM, Tabbane N, Labiod H, Tabbane S (2015) A fuzzy multi-metric QoS-balancing gateway selection algorithm in a clustered VANET to LTE advanced hybrid cellular network. IEEE Trans Veh Technol 64:804–817. https://doi.org/10.1109/TVT.2014.2323693

A Spatial Multi-criteria Analysis Method for the Assessment of Walkability to Guide the Design of Interventions in Urban Crossings

Dario Canu, Tanja Congiu and Giovanna Fancello

Abstract In last decades, we assist to a constant, growing attention on urban walkability. Despite copious studies analyse the components of pedestrian spaces, little attention is given to road intersections, elements of interruption and obstacle of the walking paths. This study proposes a multi-criteria spatial analysis method of road intersections aimed to classify and prioritise actions for the implementation of pedestrian-oriented mobility policies. The method reveals the spatial elements to improve for enhancing walkability. An application to the city of Alghero revealed safety and comfort as the main factors that limit the propensity to walk and therefore need to be tackled through specific policies and interventions.

1 Introduction

Urban and transport planning give increasing attention to the influence of the urban environment on the propensity to walk (i.e. the choice of walking as transportation). Pedestrian accessibility analysis models mainly study road segments with little attention to the influence of intersections. Such road components are generally seen as a positive factor in increasing urban connectivity. However, if we consider the topology of the network, intersections represent a break in the route that can significantly limit the perception and the use of urban space by people.

This study aims to investigate and overcome this limitation in the analysis of the pedestrian urban network. For this purpose, a multi-criteria analysis method

D. Canu · T. Congiu
DADU, University of Sassari, Alghero, Italy
e-mail: dacanu@uniss.it

T. Congiu
e-mail: tancon@uniss.it

G. Fancello (✉)
LAMSADE, University of Paris-Dauphine, Paris, France
e-mail: giovanna.fancello@dauphine.fr

R. Papa et al. (eds.), *Smart Planning: Sustainability and Mobility in the Age of Change*, Green Energy and Technology,
https://doi.org/10.1007/978-3-319-77682-8_16

is proposed for measuring the hindering effect of road intersections for pedestrians crossing the road. The study goes beyond the conventional measure of urban connectivity and focuses on spatial and functional configuration of intersections (geometry, signage, control devices, behaviours of different road users, etc.), investigating how they influence the quality of the walk in terms of practicability, safety, and comfort.

The developed multi-criteria spatial analysis method of urban intersections is conceived as a planning a design tool that support planners and policymakers in the definition of integrated transport policies as well as in the evaluation and classification of alternative actions aimed at improving the walkability of crossings.

In fact, administrations have often limited budgets and need aiding tools in order to define priorities.

In addition, the proposed instrument highlights the elements of the urban environment that positively and negatively affect pedestrian accessibility, providing useful suggestions for the design of spatial measures and public policies.

The article is structured as follows: in the first section we illustrate the state of the art of urban walkability analysis and evaluation methods with particular attention to intersections; the second section contains the description of the model and its application in the city of Alghero; in the last section we discuss results and draw conclusions.

2 State of the Art

Studies on urban walkability claim the idea that origin-destination distance is a critical characteristic in the decision of people to move by foot, together with directness and absence of interruptions along the way. The usability of a space is influenced by its connectivity, which in turn depends on the urban fabric and the spatial distribution of functions and activities. Pedestrian accessibility measures analyse intersections mainly with respect to their density (number of intersections per area) and typology (number of converging network branches/arches) (Cervero and Kockelman 1997; Dill 2004; Forsyth et al. 2008; Schlossberg et al. 2000) according to a principle of increasing possibilities of movement. These measures are based on large spatial units (km^2, block, neighbourhood) and neglect analysis of built environment at a finer-grain scale (Berrigan et al. 2010; Stangl and Guinn 2011). As a result, the integration of spatial connectivity indicators into composite indicators of walkability can lead to unreliable aggregate results. Such a methods combine detailed data of street segments, with rougher and less detailed description of intersections, thus ignoring the negative effect of intersections on the walking experience.

More detailed analysis of intersection characteristics is provided by studies of the pedestrian level of service (PLOS) (TRB 2000), aimed at measuring road performance from the pedestrian point of view. The PLOS classifies roads and intersections into classes, each of which describes a different level of the conduciveness to walk of road space. Factors mainly considered are traffic volume, vehicle speed, visibility and the presence of physical barriers (TRB 2000). According to the PLOS literature, there are different qualities of intersections related to walkability. Among them, traffic safety, comfort requirements and pedestrian delay are some of the key performance indicators of walking impedance of crossings (Landis et al. 2005). Researchers broadly employ the PLOS method to enrich the set of factors affecting pedestrian behaviour, both along sidewalks and at crossings. Recent studies combine the physical and operational variables with the pedestrian's perception of safety and comfort, measured by collecting data on pedestrians through direct observation or by means of surveys. Results show that the first (and perhaps the most obvious) factors influencing pedestrian safety are traffic volume and vehicle speed. High-speed traffic makes pedestrian intersections difficult and dangerous to cross: wide roads with high car volumes and high speeds cause safety risks and long waiting times for pedestrian crossings (Muraleetharan et al. 2005; Jacobsen et al. 2009). According to Litman (2000), the barrier effect of vehicles pushes pedestrians to change route or method of travel. The barrier effect (caused by both motorised and non-motorised traffic) and the crossing distance represent two main obstacles perceived by pedestrians at intersections (Petritsch et al. 2005; Muraleetharan et al. 2005; Bian et al. 2013; Hubbard et al. 2009). Confirming this, Sisiopiku and Akin (2003) studied the pedestrian perceptions of various signalised and un-signalised intersections and found that the location and distance of the crosswalk, relative to the destination of the trip, is the most influential factor for pedestrian deciding to cross. Therefore, the physical distance of the crossing, the number of crossings, their geometry and control system influence pedestrian behaviour (Bian et al. 2013), reducing their freedom of choosing where to go by foot, with this limitation affecting particularly people with limited abilities.

Exploring the difficulties in crossing of people with motor and visual problems (disabled, elderly, etc.) Langlois et al. (1997) pointed out that long distances between sidewalks and poor visibility lengthen crossing times making pedestrian green intervals often insufficient. Balfour and Kaplan (2002) observed that excessive noise, insufficient lighting and heavy traffic negatively influence the capacities and independence of elderly people. With regard to children, studies on parents' propensity to allow their kids to walk alone to school identified traffic close to the residences, risk of accidents and long distances as main dangerous and discouraging factors (Kadali and Vedagiri 2015; Nasar et al. 2015). Since comfort is often concerned as pedestrian influencing factors at intersections, ramps for disable at opposite sides of the crossing are considered to be a particularly important element. The same role is played by the crossing islands, better if raised and wide enough to provide adequate protection for pedestrians during the crossing. Other important attributes of pedestrian-friendly space are: good lighting, land use mix, enclosure, benches, trees and signs (Ewing and Handy 2009). In the case of intersections, the presence of lamps, signs, trees, bus shelters, furniture, is at once both useful and dangerous as on the one hand these elements support and regulate road user behaviours, and on the other hand, if wrongly positioned, they can compromise the view and therefore the safety of pedestrians crossing the road. A limit to visibility is also given by the presence of parked cars in close proximity to the intersection (Martin 2012; Peprah et al. 2014). Finally, the waiting time at intersections and the absence of traffic lights have resulted in several studies important risk factors (TRB 2000; Li 2013).

3 Method

The goal of this study is to evaluate the walkability of the street environment, with particular attention to crossings. Indeed, we measured intersection features which positively and negatively affect pedestrian accessibility. We propose an evaluation model to guide decision-makers in the definition of new public policies and in the selection of actions to prioritise.

In this section we provide a short description of the study area (3.1), a brief technical explication of the multi-criteria assessment method (3.2), the list of the criteria analysed (3.2.1), the method and results the evaluation of intersections (3.2.2) and the method implemented for their prioritisation (3.2.3).

3.1 The Study Area

We tested the method in the city of Alghero, on nine intersections of "La Pietraia" neighbourhood with about 10,000 inhabitants (Fig. 1). "La Pietraia" is a medium urban density district characterised mostly by residential land use and some retail activities and urban services located along the main roads. Medium-high volumes of motorized and non-motorized traffic move along the road network, depending on the time of day and of the season. Two major roads cross the neighbourhood. They are at once collectors and local roads as they connect different parts of the city and distribute local flows as well.

The first one, via Don Minzoni, divides the study area into two parts: the east side mainly residential with permanent inhabitants and neighbourhood level services; the west side, with a touristic vocation, characterised by a great variability in the number of inhabitants during the year and consequently by different intensity of flows (motorized and non-motorized) and the related accessibility problems. The majority of activities and services of this sector of Alghero are located along via Don Minzoni, and include local retail activities and urban scale services (hospital, weekly street market, secondary schools, railway station, etc.). The second important road, via Lido, extends along the seaside and accommodates amenities and touristic facilities (the waterfront promenade, hotels, restaurants, camping) together with other urban services.

The nine intersections analysed in the study (Fig. 1) are representative of the whole set of intersections of the neighbourhood and include the most critical features of walkability (Table 1). In addition, intersections have been selected according to their strategic position with respect to local and urban functions (Fig. 2) of the neighbourhood such as the street markets, the civil hospital, basic commercial activ-

Fig. 1 Geographical distribution of the nine intersections analysed ("La Pietraia" neighbourhood, Alghero)

ities, etc. With regard to the roads that approach the intersections they have different functionalities (from local roads to inter-district collectors) and dimensions (diverse lanes numbers and width ranging from 6 to 20 meters).

3.2 The Multi-Criteria Evaluation Method

We proposed a multi-criteria evaluation method based on the ELECTRE TRI family methods (Bouyssou et al. 2006; Roy 1990; Roy and Bouyssou 1993) in order to assess the limiting factors of pedestrian accessibility at intersections. Such a method is widely used in literature for selecting public policies. Some examples of mobility are given by Fancello et al. (2014) and Sousa et al. (2017) which use the ELECTRE TRI for the evaluation of street segments and footpaths.

The ELECTRE TRI method compares a set of alternatives through a binary outranking relation of concordance (at least as good as) or discordance (not better than) generating the so-called "hierarchical ranking" (Roy 1990). Given a set of criteria $h_1 \ldots h_m$, it evaluates n actions by assigning a rating class with respect to a set of ordered classes $C^1 \ldots C^m$. Each C^k class is defined by means of a limit profile π^k given a m set of criteria.

This method has some desirable properties for this study: (i) it allows obtaining an exhaustive classification of elements in categories with ordered priorities; (ii) the aggregation of criteria is flexible and allows to include weights, coalitions in clusters (majority rule) and possible veto rules: (iii) it allows for a non-compensatory

Table 1 Attribute of analysis of intersections

Attribute		Type/units
Number of approaches at each intersection[a]		Number
Number of lanes of each road		
Dedicated bike lanes (B_{lanes})[a]		Present (1) Absent (0)
Traffic lights (T_{light})		
Crossing markings (Z)		
Couples of curb cut (C_{cutx2})		
Sidewalk extensions (S_{ext})		
Elevated sidewalk (S_{elev})		
Physical obstacles to visibility (V_{ObEl})	Parked car (C_{parked})	Present (1) Absent (0)
	Sticking out buildings (B_{protr})	
	Signs (S)	
Physical elements directing pedestrian movements (D_{PEl})	Fencings (F)	Present (1) Absent (0)
	Expedients to direct pedestrians (D_{exp})	
Corner area ($A_P = H_P + W_P$)	Holding space (H_P)	m^2
	Space for walking (W_P)	
Carriageway area (A_v)		m^2
Road width from curb to curb (R_{width})		m
Number of pedestrians in all directions (P_T)[ab]		n_{ped}/h
Number of motorized vehicles (in vehicle equivalency units) (M_T)		VEU[a]/h
Speed limit (V_S)[a]		km/h

[a]Attributes used for the selection of intersections
[b]Attributes used to assess the importance of crossing

aggregation of data, with a reduced loss of information during consecutive evaluation phases; (iv) it is a procedure that resembles individual thinking. These features make the ELECTRE TRI an appropriate procedure for our evaluation aim.

3.2.1 Evaluation Criteria

As mentioned above, the final aim of our multi-criteria analysis and evaluation method is to provide a tool for aiding policymakers to prioritise actions to improve the walkability of urban intersections. In addition, the proposed method supports the design of walkable-oriented urban policies, by addressing interventions toward the enhancement of specific spatial requirements (i.e. safety, comfort, convenience,

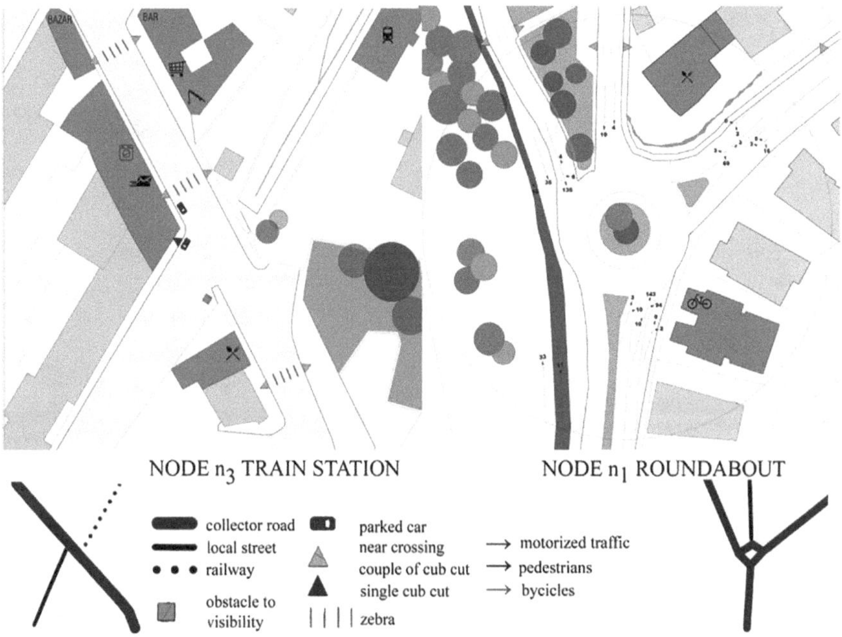

Fig. 2 Nodes n_1, n_3—Spatial and practical characteristics measured in the field survey

attractiveness, of pedestrian walkways) that result critical for people deciding to walk.

With this purpose, we selected a set of criteria for the walkability analysis aimed to: (i) detect critical factors, such as the negative state of some characteristics of the intersection; (ii) evaluate their effect on individuals' ability to use and live the city by foot; (iii) provide suggestions to planners and designers involved in the formulation of policies and projects; and (iv) address decision-makers to define action priorities.

The resulting five evaluation criteria $H = \{H_1,\ldots,H_5\}$ depict the performance of the crossing with respect to its geometry, design and possibilities of use by pedestrians: H_1—Intersection control devices; H_2—Continuity and ease of movement; H_3—Comfort; H_4—Safety (from traffic); H_5—Barrier effect (Table 2).

In order to consider all possible route alternatives, these criteria have been analysed with respect to each individual crossing that compose an intersection. For example, the intersections n_1 and n_3 in Fig. 2 are composed of four crossings.

With regard to the attributes considered in the analysis, priority was given to those characteristics of the built environment that are directly observable and objectively measurable. We also incorporated some attributes used in the CAWS walkability assessment model (implemented in *Walkability Explorer*) (Blečić et al. 2015) to make the two systems interoperable and to lay the foundations for further research developments. Table 1 summarizes the elements considered for the evaluation of intersections.

Table 2 Evaluation criteria and class profiles (see Table 1 for variables)

Criteria	Attribute	Class profiles (π)		
		C^1	C^2	C^3
H_1—Crossing control	T_{light} (Traffic lights)	T_{light} and Z	T_{light} xor Z	Neither
	Z (Crossing markings)			
H_2—Continuity and ease of movement	C_{cutx2} (Couples of curb cut)	(C_{Cut} and S_{ext}) or S_{elev}	C_{Cut} xor S_{ext} if no S_{elev}	Neither
	S_{ext} (Sidewalk extensions)			
	S_{elev} (Elevated sidewalk)			
H_3—Comfort	$P_{ratio} = A_P/A_v$ (Area side-walk/carriageway)	$P_{ratio} >> 1$	$P_{ratio} \cong 1$	$0 << P_{ratio} << 1$
H_4—Safety	V_{ObEl} (Physical obstacles to visibility)	no V_{ObEl} and D_{PEl}	no V_{ObEl} xor D_{PEl}	V_{ObEl} and no D_{PEl}
	D_{PEl} (Physical elements direction pedestrian movements)			
H_5—Barrier effect	$(R_{width} + M_T + V_S)/3$ (Road width; Number of vehicles; Speed limit)	$0 \leq H_5 \leq 1.4$	$1.5 \leq H_5 \leq 2.4$	$2.5 \leq H_5 \leq 3$

The collection of data, related to the physical characteristics of crossings, has been carried out with the help of some students, that surveyed every single component of the crossing. Figure 2 shows two examples of crossings analysed.

Traffic data has been collected through video recordings during peak hours of weekly days (from 10:00 to 12:00 and from 17:30 to 19:30) and in good weather. The videos quantified the traffic volumes (motor vehicles, bicycles, pedestrians) for each intersection in different directions.

3.2.2 Evaluation of Crossings

The analysis and evaluation of intersections are carried out by assigning a *rating* class with the ELECTRE TRI method. The model employs three classes of walkability: C^1 "Very good to walk", C^2 "Support walking", C^3 "Obstacle to walk". Such classes are defined according to the variables measured at intersections and are summarised in Table 2.

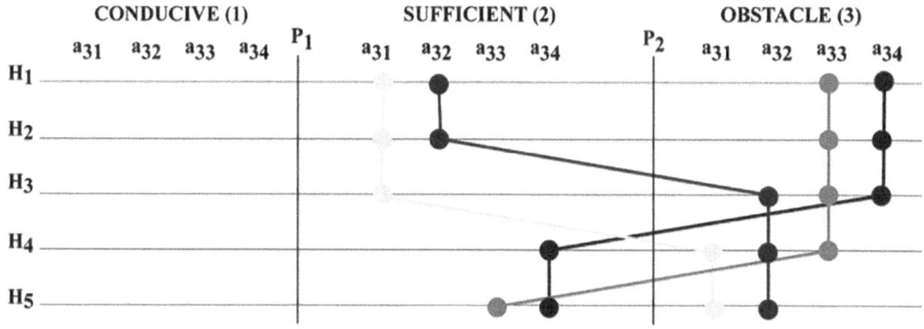

Fig. 3 Evaluation of the intersection n_3

According to the decision-maker's goal of considering coalitions of criteria composed of three components (e.g. H_1, H_2, H_3; *or* H_2, H_4, H_5), we adopted equal weights for the five criteria ($w = 1/5$) and we established two boundary conditions (thresholds) for assigning the walkability class to each crossing.

The two thresholds, a more conservative one ($\lambda = 75\%$) and a less conservative one ($\lambda = 50\%$), were defined considering the possible coalitions of the criteria as stated by the decision-maker; upper or lower thresholds would not allow the declared preference to be respected. To be considered in a class, an alternative must be at least as good as the attributes that define the three walkability classes (C^1, C^2, C^3), in three out of five criteria ($\lambda = 75\%$) or in two out of five criteria ($\lambda = 50\%$).

To better understand the method, in Fig. 3 is represented the diagram of ELECTRE TRI for the four crossings (a_{31}, a_{32}, a_{33}, a_{34}) the intersection "Train Station" is composed of (n_3).

Based on this classification, the crossing a_{31} is classified in class C^3 "Obstacle to walk" according to the conservative threshold ($\lambda = 75\%$) and in class C^2 "Supports walking" by adopting a less strict rule ($\lambda = 50\%$). Similarly, a_{32}, a_{33} e a_{34} are classified in C^3 under both conditions.

Following this procedure, each crossing is assigned to a walkability class for each of the two majority thresholds.

3.2.3 Evaluation of Intersections

The classification of intersections highlights the physical and operational characteristics that act as impeding factors for walking and which require possible corrective actions. Once each constituent crossing has been classified, the overall assessment of the intersection is carried out to establish a prioritization of crossroads in terms of their need for improvement. This sorting is made using two criteria:

Class of walkability: according to the previous classification for each crossing, a walkability class (C^1; C^2; C^3) is assigned with respect to the two majority thresholds;

Table 3 Evaluation of intersection n_3—"Train station" starting from the classification of constituent crossings

Intersection	Crossing	Weight (relative importance)	Class		Score	
			$\lambda = 75\%$	$\lambda = 50\%$	$\lambda = 75\%$	$\lambda = 50\%$
n_3	a_{31}	0.07	C^3	C^2	3	2
	a_{32}	0.03	C^3	C^3	3	3
	a_{33}	0.23	C^3	C^3	3	3
	a_{34}	0.67	C^3	C^3	3	3
	Final class (weighted average)		$\mathbf{C^3}$	$\mathbf{C^3}$	**3**	**2.93**

Table 4 Evaluation of intersections

Intersection	n_1		n_2		n_3		n_4		n_5		n_6		n_7		n_8		n_9	
λ	(a)	(b)	(a)	(b)	(a)	(b)	(a)	(b)	(a)	(b)	(a)	(b)	(a)	(b)	(a)	(b)	(a)	(b)
Score	2	2.2	2.5	3	2.9	3	2	2.7	2	3	2.5	2.6	2.9	3	2	2	3	3
Class	C^3	C^2	C^3	C^3	C^3	C^3	C^3	C^3	C^3	C^2	C^3	C^3	C^3	C^3	C^2	C^2	C^3	C^3
W	W_2		W_1		W_1		W_2		W_2		W_1		W_1		W_3		W_1	

with $\lambda(a) \rightarrow \lambda = 50\%$ *e* $\lambda(b) \rightarrow \lambda = 75\%$

Importance of crossing: the class score is weighted according to the relative importance (%) of each crossing as defined by the total number of pedestrians at the intersection (Table 3).

After the walkability class of intersection is defined, it is possible to set the priority of the intervention by considering the importance of the same intersection:

Walkability class of the intersection (Table 4);

Class W_1, if the intersection is classified as C^3 for both majority thresholds ($\lambda = 75\%$ e $\lambda = 50\%$);

Class W_2, if the intersection is classified in C^3 by applying the majority threshold $\lambda = 50\%$, but does not belong to C^3 for the threshold $\lambda = 75\%$;

Class W_3, if the intersection is classified as C^2 for both majority thresholds ($\lambda = 75\%$ e $\lambda = 50\%$);

Class W_4, in all other cases.

Intersection Importance: intersections are classified into four importance classes ($I_1,\ldots,I_4$) according to the percentage of pedestrians at each intersection in respect of the total number of pedestrians at all intersections detected, according to data observed during the same monitoring period.

The evaluations of the nine intersections of Alghero, according with the two criteria, are shown in Table 5.

We defined the order scale of intersections by assigning numerical values to the walkability classes ($W_1,\ldots,W_4$) and to the levels of importance of the intersection

Table 5 Walkability class and importance of intersections

Class of walkability	Importance			
	I_1	I_2	I_3	I_4
W_1	n_3	n_2, n_7	n_6	n_9
W_2	n_1	n_4		n_5
W_3	n_8			
W_4				

(I_1,…,I_4), thus defining the intersections order scale (Borda rule). By calculating the weighted sum (with weight 60% for "walkability class" and 40% for "importance of crossing") it is possible to have the final order of priority between intersections. For example, intersection n_3 has the highest priority among all nine intersections surveyed as it is in the first priority class for walkability assessment (W_1) as well as for importance (I_1). In this way it assumes the final score 1 (defined as $0.6*1 + 0.4*1 = 1$). This is followed by intersections n_2 and n_7 (score 1,4), n_6 (1,8), n_1 (1,6), n_4 (2), n_8 (2,2), n_9 (2,2) and n_5 (2,8).

4 Discussion of Results and Conclusions

Studies on urban walkability are generally based on the spatial characteristics of continuous street segments and are focused on safety, comfort, convenience, and pleasantness as conditions that influence the quality of pedestrian paths. Whereas, attention to intersections as obstacles to walking is limited to a few studies. In this paper, we presented the application of a spatial multi-criteria analysis and evaluation model based on the ELECTRE TRI method, aimed at classifying intersections with respect to their conduciveness to favour or hinder walking and to help policymakers to define priorities of interventions. The method we propose provides significant information about spatial conditions that mainly influence pedestrian behaviour at intersections. Therefore, we believe that it represents a useful tool for planners and decision-makers, to address choices and actions directed to enhance the liveability of the city.

With regard to the analysed intersections in Alghero, our method revealed a general necessity for improvement of walkability factors. According to the most conservative procedure (majority threshold $\lambda = 75\%$), 8 out of 9 crossings were classified as C^3 "Obstacle to walk". The scenario improves when using a less rigid majority threshold ($\lambda = 50\%$); in this case, five out of nine crossings are classified in C^3, while the remaining four are classified as C^2 "Supports walking".

Going backwards in the evaluation procedure, it is possible to identify the characteristics of geometry and functionality of intersections which influence most pedestrian accessibility. In the case of Alghero, the main impedance factor is the barrier effect (criterion H_5), determined by the distance of the crossing (roadway width) and the traffic intensity (frequency and speed of vehicles). This result suggests the use of traffic calming measures such as the reconfiguration of road space by lane narrowing to reduce vehicular speeds and to promote more careful and respectful driving behaviour, even for non-motorized users. Interventions such as carriageway narrowing, raised crosswalks and intersections, median islands, use of special pavements and markings, have proved to be effective in different contexts and have become part of urban policies for improving the safety and the quality of urban environment (Gonzalo-Orden et al. 2016).

Often, crossings are not favourable for walking even with respect to criterion H_3, which expresses the ratio between the space available for pedestrians and vehicles: in the case of Alghero, 77% of crossings were classified as "Obstacle to walk", highlighting the prevalence of an auto-oriented conception of the urban structure and its land use.

On the other hand, a conception of the road space, including intersections, which combines the dimension of connection with the one of social interaction can guide future public policies aimed at enhancing the liveability of cities. The Dutch woonerf model, the recent superillas in Barcelona, or the implementation of the shared space approach to spatial contexts larger than the neighbourhood street scale (Karndacharuk et al. 2014) are concrete examples of plans and projects that deal with road space as a constituent of the system of public spaces in cities. Such policies improve the quality of walkability through spatial solutions and regulations based on the coexistence in the same environment of various users with special care for non-motorized ones.

The implementation of the proposed method to the case study of Alghero, made possible to define priorities among actions needed to encourage walkability. By taking into account the level of walkability and the importance of places expressed in terms of intensity of use (frequency by pedestrians), intersection n_3 reached the highest priority for intervention, followed by intersections n_2, n_7, and n_6 that are characterised by a lower level of walkability and high pedestrian frequency, and intersections n_1, n_4 and n_8 have a higher priority with respect to intersections n_5 and n_9 characterised by low levels of walkability and frequentation. The definition of priority classes is useful for decision-makers, both for the identification of the most critical situations, as well as for the management of economic resources, often limited, and for the organization of the interventions.

In conclusion, the evaluation model developed represents an operational tool to support public administrations in planning interventions for achieving greater pedestrian accessibility. It also gave the research group the opportunity to deepen the study on urban walkability (Blečić et al. 2014), by integrating the original CAWS model (Blečić et al. 2015, 2016), that concerns street segments, with the considerations on the effect of intersections (Blečić et al. 2017). Indeed, the identification, through the method proposed here, of the characteristics of intersections that affect pedestrian behaviour, represents a first step towards the integration of intersections in the measure of urban walkability. In such a way the effects (positive or negative) on walkability of the combination of streets and intersections characteristics, will be explored more thoroughly.

Acknowledgements Thanks for the thesis: "A special thanks goes to Andrea Cappai who collected the data of this research during his thesis".

References

Balfour JL, Kaplan GA (2002) Neighborhood environment and loss of physical function in older adults: evidence from the Alameda County study. Am J Epidemiol 155(6):507–515

Berrigan D, Pickle LZ, Dill J (2010) Associations between street connectivity and active transportation. Int J Health Geogr 9:20

Bian Y, Lu J, Zhao L (2013) Method to determine pedestrians level of service for unsignalized intersections. Appl Mech Mater 253–255:1936–1943

Blečić I, Cecchini A, Congiu T, Fancello, G, Trunfio GA (2014) Walkability explorer: An evaluation and design support tool for walkability. Comput Sci Appl 8582:511–521

Blečić I, Cecchini A, Congiu T, Fancello G, Trunfio GA (2015) Evaluating walkability: a capability-wise planning and design support system. Int J Geogr Inf Sci 29(8):1350–1374

Blečić I, Canu D, Cecchini A, Congiu T, Fancello G (2016) Factors of perceived walkability: a pilot empirical study. Comput Sci Appl 9789:125–137

Blečić I, Canu D, Cecchini A, Congiu T, Fancello G (2017) Walking across rural-urban fringe: the hindering effect of intersections. Sustainability 9(6):883

Bouyssou D, Marchant T, Pirlot M, Tsoukiàs A, Vincke P (2006) Evaluation and decision models with multiple criteria: stepping stones for the analyst. Springer Science & Business Media Inc., New York, USA

Cervero R, Kockelman K (1997) Travel demand and the 3Ds: density, diversity, and design. Transp Res D 2:199–219

Dill J (2004) Measuring network connectivity for bicycling and walking. In: 83rd annual meeting of the Transportation Research Board. Washington DC, USA, pp 11–15

Ewing R, Handy S (2009) Measuring the unmeasurable: urban design qualities related to walkability. J Urban Des 14(1):65–84

Fancello G, Carta M, Fadda P (2014) A decision support system based on electre III for safety analysis in a suburban road network. Transp Res Procedia 3:175–184

Forsyth A, Hearst M, Oakes JM, Schmitz KH (2008) Design and destinations: factors influencing walking and total physical activity. Urban Stud 45(9):1973–1996

Gonzalo-Orden H, Rojo M, Pérez-Acebo H, Linares A (2016) Traffic calming measures and their effect on the variation of speed. Transp Res Procedia 18:349–356

Hubbard SML, Bullock DM, Mannering FL (2009) Right turns on green and pedestrian level of service. Stat Assess J Transp Eng 135:153 159

Jacobsen PL, Racioppi F, Rutter H (2009) Who owns the roads? How motorised traffic discourages walking and bicycling. Inj Prev 15(6):369–373

Kadali BR, Vedagiri P (2015) Evaluation of pedestrian crosswalk level of service (LOS) in perspective of type of land-use. Transp Res A 73:113–124

Karndacharuk A, Wilson D, Dunn R (2014) A review of the evolution of shared (street) space concepts in urban environments. Transp Rev 34(2):190–220

Landis BW, Petritsch TA, McLeod PS, Huang HF, Guttenplan M (2005) Video simulation of pedestrian crossings at signalized intersections. Transp Res Rec 1920:49–56

Langlois JA, Keyl PM, Guralnik JM, Foley DJ, Marottoli RA, Wallace RB (1997) Characteristics of older pedestrians who have difficulty crossing the street. Am J Public Health 87(3):393–397

Li B (2013) A model of pedestrians' intended waiting times for street crossings at signalized intersections. Transp Res B 51:17–28

Litman T (2000) Transportation cost analysis; techniques, estimates and implications. Victoria Transport Policy Institute, Victoria. www.vtpi.org

Martin A (2012) Factors influencing pedestrian safety: a literature review. TRL Limited, London, UK
Muraleetharan T, Adachi T, Hagiwara T, Kagaya S (2005) Method to determine pedestrian level-of-service for crosswalks at urban intersections. J East Asia Soc Transp Stud 6:127–136
Nasar JL, Holloman C, Abdulkarim D (2015) Street characteristics to encourage children to walk. Transp Res A 72:62–70
Peprah C, Oduro CY, Afi Ocloo K (2014) On-street parking and pedestrian safety in the Kumasi metropolis: issues of culture and attitude. Dev Country Stud 4(20):85–94
Petritsch TA, Landis BW, Mcleod HF, Huang HF, Challa HS, Guttenplan M (2005) Level of service model for signalized intersections for pedestrians. Transp Res Rec 1939:55–62
Roy B (1990) Decision aid and decision making. Eur J Oper Res 45:324–331
Roy B, Bouyssou D (1993) Aide Multicritère à la Décision: Méthodes et Cas. Economica, Paris, FR
Schlossberg M, Johnson-Shelton D, Evers C, Moreno G (2000) Refining the grain: using resident-based walkability audits to better understand walkable urban form. J urbanism 8(3):260–278
Sisiopiku VP, Akin D (2003) Pedestrian behaviors at and perceptions towards various pedestrian facilities: An examination based on observation and survey data. Transp Res F 6:249–274
Sousa N, Coutinho-Rodrigues J, Natividade-Jesus E (2017) Sidewalk infrastructure assessment using a multicriteria methodology for maintenance planning. J Infrastruct Syst 23(4):05017002
Stangl P, Guinn JM (2011) Neighborhood design, connectivity assessment and obstruction. Urban Des Int 16:285–296
Transportation Research Board (2000) Highway capacity manual. TRB National Research Council, Washington D.C., USA

A TOD Classification of Metro Stations: An Application in Naples

Enrica Papa, Gerardo Carpentieri and Gennaro Angiello

Abstract Due to extensive use of private cars, cities are facing negative social and environmental impacts, including poor air quality, noise, health issues. Responding to these urgent challenges, some metropolitan areas are building new metro lines and stations. Certain studies demonstrate that a number of factors of the station's catchment areas may influence public transport use. Starting from this premise, this chapter applies a cluster analysis with the aim of understanding and classifying station areas. In details, the contribution proposes (i) a literature review on indicators used for station areas analysis; and (ii) a method to identify station areas typologies. The study proposes an application at the sixty-two rail stations of the city of Naples, where several interventions have been recently carried out on the urban rail network and in station areas, within an integrated land-use and transport planning approach.

1 Introduction

Urban rail systems have had the role of shaping metropolitan areas and providing a sustainable alternative to private car use. A considerable body of professional and academic research has analysed the factors influencing rail public transport use. Evidence suggests a compelling conclusion: while transit fares, safety and service quality have a significant impact on transit ridership (Redman et al. 2013), in the long run, public transport ridership is also influenced by attributes of the station catchment areas (Ryan and Frank 2009; Ewing and Cervero 2010; Lindsey et al. 2010; Sung and Oh 2011). These factors can indeed provide the necessary (albeit insufficient) conditions for shaping transit use, without which, urban rail policies would have

E. Papa
Department of Planning and Transport, University of Westminster, London, UK

G. Carpentieri (✉) · G. Angiello
Department of Civil, Architectural and Environmental Engineering, University of Naples Federico II, Naples, Italy
e-mail: gerardo.carpentieri@unina.it

R. Papa et al. (eds.), *Smart Planning: Sustainability and Mobility in the Age of Change*, Green Energy and Technology,
https://doi.org/10.1007/978-3-319-77682-8_17

limited effects (Suzuki et al. 2013). An assessment of such attributes around railway nodes could provide interesting insights for policy formulation aimed at changing public transport use through land use policies.

The integration of land use and transport policy around station areas has not just the potential to influence public transport use, but also to generate development returns that could help fund infrastructure improvements and enable sustainable high-density development.

Those are the principles of a Transit Oriented Development (TOD) approach (Bertolini et al. 2012; Cervero 2004; Dittmar and Poticha 2004), which are currently on the agenda of many metropolitan areas.

Studies show how each city and various different stations, could apply a TOD approach in diverse ways (Lyu et al. 2016). Context-based TOD typologies are thus essential to identify a more targeted set of strategies.

In the light of these issues, the primary objective of this study is to provide a simple and replicable GIS-based methodology for analysing station areas and their TOD classification. By classification, we mean a particular set of land-use and transport attributes of station areas. The identification of station areas classification is the first step towards the identification of integrated land-use and transport policy.

Using a similar methodology already applied in other studies (Singh et al. 2014, 2017; Atkinson-Palombo and Kuby 2011; Masoumi and Shaygan 2016; Papa et al. 2016), we propose a set of indicators describing TOD attributes and an application of Cluster Analysis that identify TOD classification of station areas. This methodology is applied in the city of Naples where several interventions have been carried out on the urban rail network with the opening of new stations and the re-development of associated urban areas, both in the central and peripheral areas of the city. Since 1994, as better explained in Sect. 3, Naples has been implementing an integrated approach to land use and public transport planning based on the integration of new lines and stations with urban development interventions or the renewal of station areas. For this reason, Naples has been selected as a study case for this research.

The remainder of this chapter is organised as follows: Sect. 2 proposes a literature review of TOD indicators, Sect. 3 describes the study area, the indicators used in the application and data sources, Sect. 4 discusses the methodology and the spatial analysis tools used in this study, while Sect. 5 demonstrates the results of the cluster analysis. The final section summarises the conclusions and proposes future research directions.

2 Literature Review: TOD Indicators

Quantifying the TOD degree of urban environments around railway stations requires measuring the attributes that might influence public transport use and the identification of significant indicators. Cervero and Kockelman (1997) propose a well-known model, called the 3Ds—land use density, diversity and urban design—as factors affecting public transport use. The relevance of these 3Ds factors is also emphasised in other studies (Calthorpe 1993; Curtis 2009; Lund et al. 2004; Evans and Pratt 2007; Ewing and Dumbaugh 2009; Renne 2009a, b) and can be summarised as follows:

- *Density* reflects the intensity of opportunities for interaction within a station area and its effects on travel demand. High densities are fundamental to supporting high-frequency public transport services. High density development around stations leads to high demand and customers patronage. Studies demonstrate however, that compact developments results in negative consequences for public health (Naess 2013);
- *Diversity* relates to the presence of different land uses (e.g. housing, offices, public services). Diversity produces a balanced demand for public transport over time (reducing differences between peak and off-peak periods) and space (in terms of direction of flow) while contributing to the vitality of an area;
- *(Pedestrian-oriented) Design* includes a range of elements that promote walking. Station areas characterised by highly connected street networks, small block sizes, and a continuous footpath system promoting access to stations on foot, thus increasing public transport ridership.

Besides the 3Ds factors, another important aspect that has been considered in recent literature relates to the intensity of economic activities in the catchment areas of stations. The reason for this is that higher economic activities lead to higher levels of travel activity and a higher public transport modal split.

Each of these factors can be measured in various ways, resulting in different indicators. To identify meaningful measures, we reviewed indicators proposed in the literature, as described in Table 1.

The articles reviewed have been selected from a broad assortment of literature according to the following criteria: (i) articles published in the last seven years in (ii) highly ranked academic journals and (iii) only focusing on factors affecting transit use in station areas.

The literature review suggests the recurrence of some indicators that could be clustered in four main categories: density, geographical proximity, functional mix and property values. The indicators used in this study reflect this categorisation.

Table 1 TOD indicators used in literature

	Residents density	Jobs density	Economic Activity density	Entropy	Job-housing balance	Land use mix	Street density	Intersection density	Average block length	Distance to the city center	Proximity of activities at the station	Average travel time	Number of business	Property values	Employment levels
Atkinson-Palombo and Kuby (2011)						•					•			•	
Bhattacharjee and Goetz (2016)	•	•	•			•							•		•
Cao and Porter-Nelson (2016)						•								•	
Cervero and Kang (2011)	•	•		•					•		•			•	
Hess and Almeida (2007)										•	•			•	
Duffhuers et al. (2014)	•		•	•		•							•		
Lyu et al. (2016)	•	•	•	•				•	•		•	•			
Hasibuan et al. (2014)	•				•	•									

(continued)

Table 1 (continued)

Higgins and Kanaroglou (2016)	•	•		•	•		•				•			•	
Jun et al. (2015)	•	•		•		•	•	•							•
Kamruzzaman et al. (2014)	•					•			•			•			
Kerkman et al. (2017)	•	•				•				•					
Monajem and Nosratian (2015)	•	•			•						•	•			•
Mulley and Tsai (2016)										•			•	•	
Nasri and Zhang (2014)	•	•		•					•	•	•				
Pagliara and Papa (2011)														•	
Reusser et al. (2008)	•					•									•
Singh et al. (2014)	•	•	•	•		•		•							•

(continued)

Table 1 (continued)

Singh et al. (2017)	•		•	•				•			•	•	•		•
Sohn and Shim (2010)	•	•				•	•	•		•					•
Sun et al. (2016)	•	•	•				•	•	•			•	•		
Sung and Oh (2011)	•		•	•		•	•		•	•		•			
Vale (2015)	•	•			•										
Zemp et al. (2011)	•	•											•		
Zhao et al. (2011)	•				•						•				

3 The Study Area

The study area corresponds to the administrative boundary of the city of Naples (see Fig. 1), the third largest city in Italy, with an urban population of 956,919 inhabitants, and one of the highest residential densities in Europe of 8,273,34 inhabitants per square km (Istat 2011). The urban structure of the city is characterised by a predominantly historical urban fabric.

The socio-economic structure of Naples is in a structural crisis, and influenced by the surrounding urban hinterland. The city has an extensive urban rail network, with low levels of service. The rail transport system has 62 stations and 79.4 km of track. There are six different lines.

Since 1994, the Naples Municipality has implemented an integrated approach to land use and transport planning based on the integration of new lines and stations with associated urban development and renewal (Cascetta and Pagliara 2008). *Land Use Strategies*, approved in 1994, was the first planning document that guided urban development towards an integration of new transport infrastructure with urban transformational processes.

Following this, the Naples Council developed further plans, such as the *Municipal Transport Plan* (1997), the *Primary Road Network Plan* (2000), the *100 Stations Plan* (2003) and the new *Urban Spatial Plan* (2004). At the same time the Regional and Provincial authorities approved some normative and programmatic tools, includ-

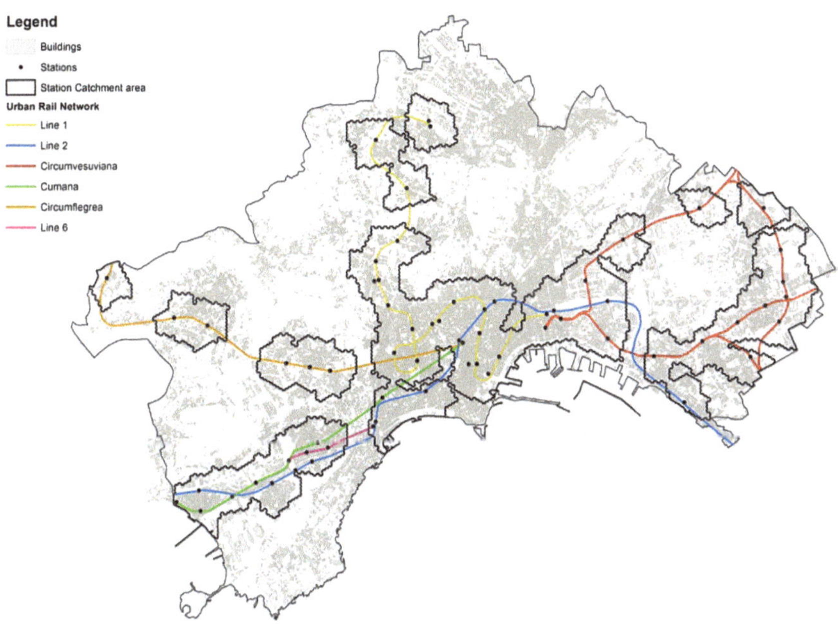

Fig. 1 Study area: the administrative boundaries and the urban rail network of the city of Naples

Table 2 Naples's Metro Network

Metro line name	Number of stations	Line length [km]
Linea 1	18	18.0
Linea 2	10	17.9
Circumvesuviana	16	22.2
Cumana	8	8.4
Circumflegrea	7	10.7
Linea 6	4	2.2
Total	**62**	**79.4**

ing the *Regional Law of Reform for the Mobility System* (2002) and the draft of the *Provincial Coordination Territorial Plan* (2007), which further encouraged the implementation of land use transport integrated policy. In these plans, the authorities underline the urgency of fostering the interconnection between transport and urban components. One of the innovative elements of these policies is the role of the urban railway transport network in supporting the renewal and development of central and peripheral areas (Cascetta and Pagliara 2008).

With this integrated approach, the metropolitan area of Naples benefitted from substantial investments that involved the construction of new metro lines and stations (Table 2). This generated an increase of public transport accessibility levels supporting redevelopment processes. The new *Urban Spatial Plan* for Naples identifies a specific strategy (Scope N° 30) for stations and interchange nodes. For these areas, the plan identifies interventions which seeks to "maximize accessibility, the redevelopment of buildings and roads and the introduction of new services that can support the localisation of new economic activities" (Comune di Napoli 2004).

4 Methodology

The methodology used in this study consists of five steps aimed at identifying a set of TOD indicators and TOD typologies of station areas. A desktop GIS package (ArcGIS 10.5) was used to develop this GIS-based procedure.

Step 1. Identifying the transport network. Open Street Map (OSM) was used as the primary data source to identify the pedestrian and public transport network. This platform provides open access spatial dataset using contributions from Internet users. It has been extensively used as a source of spatial data in many applications. However, the accuracy and the completeness of OSM data depends, among other things, on the number of contributors (Haklay 2010). To measure walking routes to transit stations, some topographic corrections were made to the original data and some missing street links where manually added.

Step 2. Defining the spatial unit of analysis. The diverse data sources that make use of different spatial units of analysis could potentially be problematic when measuring socio-economic data within a walking distance of a transit stations (Papa

Table 3 TOD indicators used in this study

Category	Indicators
Density	Population density
	Job density
	Economic Activities density
Proximity	Population proximity
	Job proximity
	Economic Activities proximity
Functional Mix	Population-job mix
	Entropy
	Land use mix
Property value	Housing property value
	Commercial property value
	Services property value
	Industrial property value

and Bertolini 2015). Consequently, we have selected a hexagonal grid element (with a sidelength of 50m) as the spatial unit of our analysis (Papa et al. 2017). We then allocate data to grid cells using the area-ratio method (Gutiérrez and García-Palomares 2008);

Step 3. Defining station catchment areas. Station catchment areas are broadly based on an understanding of how far people are willing to walk to ride public transport. One critical aspect to consider when defining station catchment areas is the threshold which determines the station's radius of influence. Various radii has been used in the literature. As already applied in other studies focusing on Naples, in this research we used 500 m radius that corresponds to an eight-minutes walking time (Pagliara and Papa 2011). Using the ArcGIS ERSI Network Analysis tool, catchment areas around each station were identified;

Step 4. Calculating indicators. The final set of indicators measured for each station is illustrated by the following Table 3. This selection of indicators reflects the literature review results. These indicators are those that can most easily be measured in the Italian context. In fact, all the data used to estimate these indicators are freely open accessible in the Italian context (see Table 4). The set of indicators include three different diversity measures: the entropy index, the mixed-use index and the jobs-housing balance. The entropy index measures the location of the different types of jobs, the mix use index measures the presence of different types of public services jobs, the jobs-housing balance measure the presence of both jobs and residents.

Using the ArcGIS ESRI Spatial Join Analysis tool, the following data was associated with each station: population and jobs (by sector); buildings footprints; street lengths; Euclidean and network distances from centroids (of the spatial unit) to the associated station. Following these steps, built environment indicators were calcu-

Table 4 Data sources

Data	Format	Source
Census unit area	Geometric (Polygon)	National Institute for Statistics ISTAT
Population	Alfa numeric	National Institute for Statistics ISTAT
Employees	Alfa numeric	National Institute for Statistics ISTAT
Economic Activity	Alfa numeric	National Institute for Statistics ISTAT
Property values	Alfa numeric	Tax Revenue Agency
Stations	Geometric (Point)	Open Street Map OSM
Walking street network	Geometric (Line)	Open Street Map OSM
Built environment	Geometric (Polygon)	Open Street Map OSM
OMI unit area	Geometric (Polygon)	Tax Revenue Agency

Table 5 Summary of the Cluster Analysis

Group	N° stations	Standard distance
I	17	0.256
II	8	0.584
III	7	0.684
IV	18	0.350
V	12	0.5038

lated for each station. Additionally, the value of the indicators was normalised to a 0 to 1 range.

Step 5. Classifying the station's catchment area. To identify station area typologies, we used the Arc GIS ESRI Grouping Analysis tool. The optimal number of groups was automatically determined by using the Calinski-Harabasz pseudo F-statistic (ESRI 2015). The final dataset consisted of 62 rows, one for each station area, and 13 columns, one for each indicator.

5 TOD Cluster Analysis Results

The final results of the cluster analysis were the identification of TOD categories of station areas. In the following tables, the main summary (Table 5) and the statistical results (Table 6) of the cluster analysis are reported.

The cluster analysis provided the identification of five TOD station areas clusters (Fig. 2): (1) Underdeveloped stations; (2) Jobs oriented stations; (3) Mixed central stations; (4) Medium Residential Stations; (5) Highly Residential Stations.

Table 6 Statistic result of the Cluster Analysis

Indicator	Average	Standard deviation	R^2
Activities density	0.3030	0.2771	0.8752
Activities proximity	0.2871	0.2581	0.8435
Land use mix	0.3648	0.2242	0.8382
Services property value	0.4175	0.2438	0.7939
Commercial property value	0.3935	0.2520	0.7828
Industrial property value	0.3190	0.2652	0.7593
Jobs proximity	0.2601	0.2609	0.7383
Jobs density	0.2554	0.2667	0.7310
Housing property value	0.3559	0.2463	0.7214
Entropy	0.2877	0.2125	0.7000
Population proximity	0.5081	0.2692	0.6650
Population-job mix	0.1509	0.1886	0.6591
Population density	0.4962	0.2731	0.6398

1. Seventeen stations classified in "Group I—Undeveloped Stations" are located in the peripheral areas of the study area, in east and north-west neighbourhoods. All indicators show values significantly lower than average values. Real estate values and densities are particularly low.
2. "Group II—employment oriented stations" is composed of eight stations. These are located in the hospital area (the line 1) and the east financial district. The positive indicators of the group show a density of employees situated close to the station above the average.
3. The seven stations of "Group III—mixed use central stations" are located in the central and historic neighbourhoods of the city. In these station areas, the density of economic activities is particularly high compared to the average of other stations. Real estate values for all land uses and the mixed use values are higher than the average.
4. "Group IV—medium value residential stations" station areas are located in neighbourhoods with a predominantly residential land use. The indicators of population density and proximity are higher than the average elsewhere, while the functional mix is low.
5. The indicators for the twelve station areas included in the "Group V—high value residential stations" have very high land values compared to the metropolitan average. These results indicate settlement densities and functional mixes higher near stations. Furthermore, in these areas, there are higher real estate values than other areas of the city.

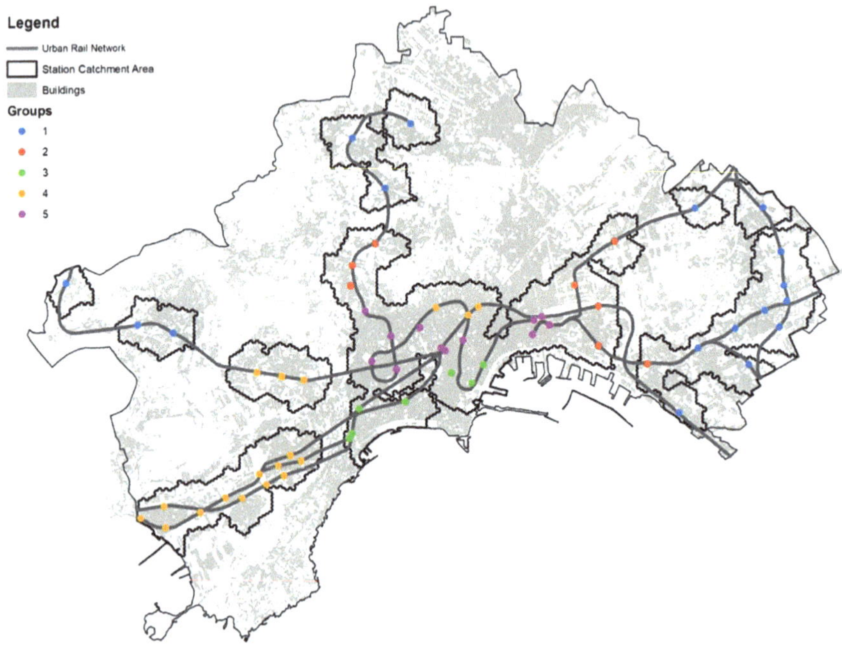

Fig. 2 The TOD station areas clusters

6 Conclusions

In this research, we developed and applied a GIS-based methodology for the classification of station catchment areas. By reviewing of international literature, we selected a set of indicators to analyse urban environments around urban railway stations to generate inputs for the future land uses and transport planning policies.

Furthermore, this methodology utilises the potentiality of spatial analysis and grouping tools of ESRI ArcGIS 10.5 (Fig. 3).

We believe that this GIS-based methodology represents a useful tool that could also be applied in other urban contexts to initiate a discussion for future uses.

Also, we applied this GIS-based methodology at the sixty-two urban rail stations in the city of Naples. The results could be useful for stakeholders involved in station areas transformation (local authorities, rail companies, private developers) to implement integrated land use and transport development strategies and projects. The similar characteristics of station catchment areas for the different groups could potentially allow local urban planners, designers and policymakers to develop a set of new strategies for each group. For the case of Naples, the results show an untapped potential for peripheral stations where the low values of density and functional mix indicators are important indications as input into decision-making processes.

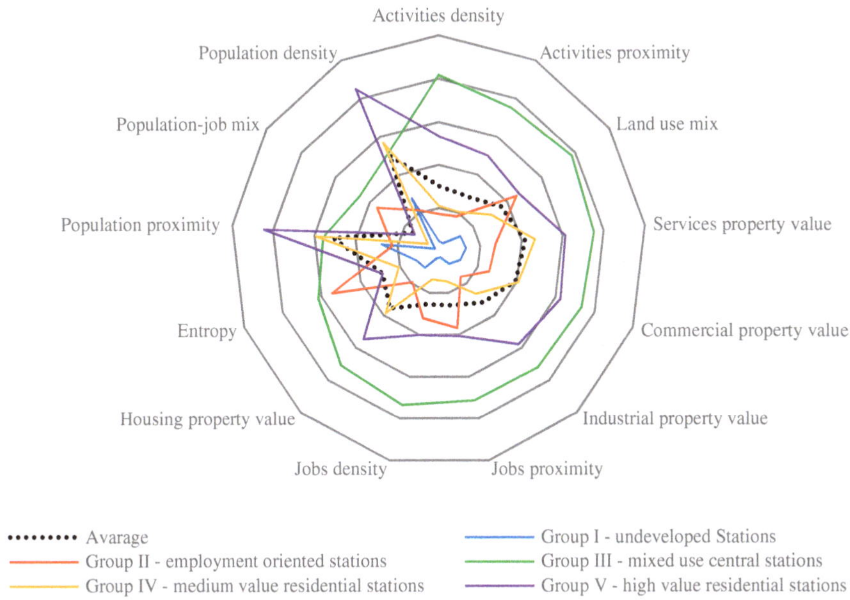

Fig. 3 The TOD station areas clusters, values of the indicators

As for the future research on these topics, to test the accuracy of GIS-based methodology, one future possibility might involve applying the model to different urban areas.

References

Atkinson-Palombo C, Kuby MJ (2011) The geography of advance transit-oriented development in metropolitan Phoenix, Arizona, 2000–2007. J Transp Geogr 19(2):189–199. https://doi.org/10.1016/j.jtrangeo.2010.03.014

Bertolini L, Curtis C, Renne J (2012) Station area projects in Europe and beyond: towards transit oriented development? Built Environ 38(1):31–50. https://doi.org/10.2148/benv.38.1.31

Bhattacharjee S, Goetz AR (2016) The rail transit system and land use change in the Denver metro region. J Transp Geogr 54:440–450. https://doi.org/10.1016/j.jtrangeo.2016.02.004

Calthorpe P (1993) The next American metropolis: ecology, community, and the American dream. Princeton architectural press, New York. ISBN 1-878271-68-7

Cascetta E, Pagliara F (2008) Integrated railways-based policies: the Regional Metro System (RMS) project of Naples and Campania. Trans Policy 15(2):81–93

Cao XJ, Porter-Nelson D (2016) Real estate development in anticipation of the Green Line light rail transit in St. Paul. Transp Policy 51:24–32. https://doi.org/10.1016/j.tranpol.2016.01.007

Cervero R (2004) Transit-Oriented Development in the United States: Experiences, Challenges, and Prospects. Transportation Research Board of the National Academies, Washington, D.C. Available at: https://www.valleymetro.org/sites/default/files/legacy-images/uploads/general_publications/TCRP-Report-102_TOD-in-the-US-Experiences-Challenges-and-Prospects_10-04.pdf

Cervero R, Kang CD (2011) Bus rapid transit impacts on land uses and land values in Seoul, Korea. Transport Policy 18(1):102–116. https://doi.org/10.1016/j.tranpol.2010.06.005
Cervero R, Kockelman K (1997) Travel demand and the 3Ds: density, diversity, and design. Transp Res Part D: Transp Environ 2(3):199–219. https://doi.org/10.1016/S1361-9209(97)00009-6
Comune di Napoli (2004) Variante al piano regolatore generale centro storico, zona orientale, zona nord-occidentale. Naples.
Curtis C (2009) Implementing transit oriented development through regional plans: a case study of Western Australia. ISBN: 9780754673156. Available at: http://worldcat.org/isbn/9780754673156
Dittmar H, Poticha S (2004) Defining transit-oriented development: the new regional building block. In: Dittmar H, Ohland G (eds) The new transit town—best practices in transit oriented development. Island Press, Washington D.C., pp 19–39. ISBN: 1559631171
Duffhues J, Mayer IS, Nefs M, van der Vliet M (2014) Breaking barriers to transit-oriented development: insights from the serious game SPRINTCITY. Environ Plan B: Plan Des 41(5):770–791. https://doi.org/10.1068/b39130
ESRI (2015) ArcGIS 10.5 Software and user manual. Environmental Systems Research Institute, Redlands, CA. Available at: http://www.esri.com/
Evans JE, Pratt RH (2007) Transit Cooperative Research Program (TCRP) Report 95: traveller response to transportation system changes. Chapter 14—road value pricing. Transportation Research Board, Washington, D.C. Available at: http://onlinepubs.trb.org/onlinepubs/tcrp/tcrp_rpt_95c14.pdf
Ewing R, Cervero R (2010) Travel and the built environment: a meta-analysis. J Am plan Assoc 76(3):265–294. https://doi.org/10.1080/01944361003766766
Ewing R, Dumbaugh E (2009) The built environment and traffic safety: a review of empirical evidence. J Plan Lit 23(4):347–367. https://doi.org/10.1177/0885412209335553
Gutiérrez J, García-Palomares JC (2008) Distance-measure impacts on the calculation of transport service areas using GIS. Environ Plan 35(3):480–503. https://doi.org/10.1068/b33043
Haklay M (2010) How good is volunteered geographical information? A comparative study of OpenStreetMap and Ordnance Survey datasets. Environ Plan 37(4):682–703. https://doi.org/10.1068/b35097
Hasibuan HS, Soemardi TP, Koestoer R, Moersidik S (2014) The role of transit oriented development in constructing urban environment sustainability, the case of Jabodetabek, Indonesia. Procedia Environ Sci 20:622–631. https://doi.org/10.1016/j.proenv.2014.03.075
Hess DB, Almeida TM (2007) Impact of proximity to light rail rapid transit on station-area property values in Buffalo, New York. Urban stud 44(5–6):1041–1068. https://doi.org/10.1080/00420980701256005
Higgins CD, Kanaroglou PS (2016) A latent class method for classifying and evaluating the performance of station area transit-oriented development in the Toronto region. J Transp Geogr 52:61–72. https://doi.org/10.1016/j.jtrangeo.2016.02.012
Jun MJ, Choi K, Jeong JE, Kwon KH, Kim HJ (2015) Land use characteristics of subway catchment areas and their influence on subway ridership in Seoul. J Transp Geogr 48:30–40. https://doi.org/10.1016/j.jtrangeo.2015.08.002
Kamruzzaman M, Baker D, Washington S, Turrell G (2014) Advance transit oriented development typology: case study in Brisbane, Australia. J Transp Geogr 34:54–70. https://doi.org/10.1016/j.jtrangeo.2013.11.002
Kerkman K, Martens K, Meurs H (2017) A multilevel spatial interaction model of transit flows incorporating spatial and network autocorrelation. J Transp Geogr 60:155–166. https://doi.org/10.1016/j.jtrangeo.2017.02.016
Lindsey M, Schofer JL, Durango-Cohen P, Gray KA (2010) Relationship between proximity to transit and ridership for journey-to-work trips in Chicago. Transp Res Part A: Policy Pract 44(9):697–709. https://doi.org/10.1016/j.tra.2010.07.003
Lund HM, Cervero R, Willson R (2004) Travel characteristics of transit-oriented development in California. California Department of Transportation, Sacramento, CA. Available at: http://

staging.community-wealth.org/sites/clone.community-wealth.org/files/downloads/report-lund-cerv-wil.pdf
Lyu G, Bertolini L, Pfeffer K (2016) Developing a TOD typology for Beijing metro station areas. J Transp Geogr 55:40–50. https://doi.org/10.1016/j.jtrangeo.2016.07.002
Masoumi HE, Shaygan M (2016) A longitudinal analysis of densities within the pedestrian sheds around metro stations. The case of Tehran. Tema J Land Use Mobility Environ 0:5–20. https://doi.org/10.6092/1970-9870/3908
Monajem S, Nosratian FE (2015) The evaluation of the spatial integration of station areas via the node place model; an application to subway station areas in Tehran. Transp Res Part D: Trans Environ 40:14–27. https://doi.org/10.1016/j.trd.2015.07.009
Mulley C, Tsai CHP (2016) When and how much does new transport infrastructure add to property values? Evidence from the bus rapid transit system in Sydney, Australia. Transp Policy 51:15–23. https://doi.org/10.1016/j.tranpol.2016.01.011
Naess P (2013) Urban form, sustainability and health: the case of greater oslo. Europ Plan Stud 22(7):1524–1543
Nasri A, Zhang L (2014) The analysis of transit-oriented development (TOD) in Washington, DC and Baltimore metropolitan areas. Transp Policy 32:172–179. https://doi.org/10.1016/j.tranpol.2013.12.009
Pagliara F, Papa E (2011) Urban rail systems investments: an analysis of the impacts on property values and residents' location. J Transp Geogr 19(2):200–211. https://doi.org/10.1016/j.jtrangeo.2010.02.006
Papa E, Bertolini L (2015) Accessibility and Transit-Oriented Development in European metropolitan areas. J Transp Geogr 47:70–83. https://doi.org/10.1016/j.jtrangeo.2015.07.003
Papa R, Angiello G, Carpentieri G (2016) Accessibility and built environment surrounding metro stations: a GISbased comparison of Naples line 1, Milan line 3 and London Jubilee line. In: Proceedings of 9th international conference on innovation in urban and regional planning (INPUT), pp 269–274. ISBN 978-88-9052-964-1
Papa R, Angiello G, Carpentieri G (2017) Il Governo del Sistema Integrato Città-Trasporti-Energia. FedOA Press (Federico II Open Access University Press), Naples. http://dx.doi.org/10.6093/978-88-6887-013-3. ISBN: 978-88-6887-013-3
Redman L, Friman M, Gärling T, Hartig T (2013) Quality attributes of public transport that attract car users: a research review. Transp Policy 25:119–127. https://doi.org/10.1016/j.tranpol.2012.11.005
Renne JL (2009a) Evaluating transit-oriented development using a sustainability framework: lessons from Perth's network city. In: Tsenkova s (ed) Planning sustainable communities: diversity of approaches and implementation challenges. University of Calgary, Calgary, pp 115–148
Renne JL (2009b) From transit-adjacent to transit-oriented development. Local Environ 14(1):1–15. https://doi.org/10.1080/13549830802522376
Reusser DE, Loukopoulos P, Stauffacher M, Scholz RW (2008) Classifying railway stations for sustainable transitions—balancing node and place functions. J Transp Geogr 16(3):191–202. https://doi.org/10.1016/j.jtrangeo.2007.05.004
Ryan S, Frank LF (2009) Pedestrian environments and transit ridership. J Public Transp 12(1):3. https://doi.org/10.5038/2375-0901.12.1.3
Singh YJ, Fard P, Zuidgeest M, Brussel M, van Maarseveen M (2014) Measuring transit oriented development: a spatial multi criteria assessment approach for the City Region Arnhem and Nijmegen. J Transp Geogr 35:130–143. https://doi.org/10.1016/j.jtrangeo.2014.01.014
Singh YJ, Lukman A, Flacke J, Zuidgeest M, van Maarseveen MFAM (2017) Measuring TOD around transit nodes—towards TOD policy. Transp Policy 5:96–111. https://doi.org/10.1016/j.tranpol.2017.03.013
Sohn K, Shim H (2010) Factors generating boardings at metro stations in the Seoul metropolitan area. Cities 27(5):358–368. https://doi.org/10.1016/j.cities.2010.05.001

Sun G, Zacharias J, Ma B, Oreskovic NM (2016) How do metro stations integrate with walking environments? Results from walking access within three types of built environment in Beijing. Cities 56:91–98. https://doi.org/10.1016/j.cities.2016.03.001

Sung H, Oh JT (2011) Transit-oriented development in a high-density city: identifying its association with transit ridership in Seoul, Korea. Cities 28(1):70–82. https://doi.org/10.1016/j.cities.2010.09.004

Suzuki H, Cervero R, Iuchi K (2013) Transforming cities with transit: transit and land-use integration for sustainable urban development. World Bank Publications, Washington, D.C. ISBN: 978-0-8213-9745-9 e-ISBN: 978-0-8213-9750-3

Vale DS (2015) Transit-oriented development, integration of land use and transport, and pedestrian accessibility: combining node-place model with pedestrian shed ratio to evaluate and classify station areas in Lisbon. J Transp Geogr 45:70–80. https://doi.org/10.1016/j.jtrangeo.2015.04.009

Zemp S, Stauffacher M, Lang DJ, Scholz RW (2011) Classifying railway stations for strategic transport and land use planning: context matters! J Transp Geogr 19(4):670–679. https://doi.org/10.1016/j.jtrangeo.2010.08.008

Zhao P, Lü B, De Roo G (2011) Impact of the jobs-housing balance on urban commuting in Beijing in the transformation era. J Transp Geogr 19(1):59–69. https://doi.org/10.1016/j.jtrangeo.2009.09.008

Slow Mobility and Cultural Tourism. Walking on Historical Paths

Romano Fistola and Rosa Anna La Rocca

Abstract Slow mobility could stand as an occasion to foster new sustainable forms of territorial fruition. In this sense, the design of methods and technical tools, able to support the decision-makers, amounts to fundamental exigence of a form of town planning oriented both towards safeguarding and promoting territorial resources. The pursuit of this aim requires an accurate political and administrative strategy based on integration among actors involved in territorial development, as well as being oriented towards attaining improved tourist attractiveness. Tourism, in fact, can be a facilitator of territorial development if it is embedded in the general process of territorial governance. Cultural and historical paths represent physical infrastructures for supporting this sustainable and slow form of tourism involving walking across territories. Using these premises as a starting point, this paper aims to provide a methodology for designing or recovering historical paths suitable for slow mobility. The paper, thus, is articulated in three parts. The first part focuses on the characteristics of slow mobility. The second part highlights the potentialities connected with the revitalization of cultural paths, considered physical infrastructures able to promote sustainable tourism. The third part proposes a methodology for the recovery of a historical path linked to the Via Francigena.

1 Slow Mobility as a New Way of Exploring Territories

In recent years, the concept of *soft mobility* as an opportunity to enhance territorial resources has been widely accepted in urban practices and territorial policies oriented towards sustainability. This concept, indeed, refers to a conscious approach to mobility, rather than to a generic mode of movement (La Rocca 2010). Creating a system

R. Fistola (✉)
DING, University of Sannio, Benevento, Italy
e-mail: rfistola@unisannio.it

R. A. La Rocca
DICEA, University of Napoli, Naples, Italy

R. Papa et al. (eds.), *Smart Planning: Sustainability and Mobility in the Age of Change*, Green Energy and Technology,
https://doi.org/10.1007/978-3-319-77682-8_18

for slow mobility, thus, can be planned as a tool for promoting alternative behaviors and sustainable lifestyles involving visiting and travelling across territories. In this sense, the use of slow mobility, as a form of sustainable mobility, could also be a factor in improving the development of sustainable tourism in internal regions, generating positive effects for the territorial economies.

The rising attention of urban policies towards sustainable mobility is strongly linked to the need to affirm a model of development based on the correct use of resources (including water, energy and soil) to face environmental exigencies and to reduce energy consumption and the emission of pollutants produced by vehicles. Even though a unique definition of sustainable mobility does not yet exist, its key role in recent urban policies is widely acknowledged.

Among the remarkable number of statements produced at the European level, and the consequent strategies adopted at the local level to improve sustainable mobility, the use of *soft mobility* as a practice able to improve both urban livability and safety for residents and urban visitors is, in fact, on the rise. In this sense, it can be defined as a type of mobility that includes all forms of non-motorized transport (NMT) that exclusively uses "human energy" (Human Powered Mobility).[1] According to such a definition, *soft mobility* (pedestrianism, cycling and other non-motorized displacements) can be meant as a zero impact mobility. In a broader sense, soft mobility should be an alternative way of thinking about mobility, improving cooperation among users and policymakers for the achievement of common objectives of livability, both in the urban context and in the larger territorial environment.

Against the backdrop of these premises, the present study considers the reuse of cultural and historical paths as a possible practice for improving slow mobility oriented towards visiting and exploring territories. This form of mobility could also be particularly adapted to reducing impact derived from the tourism sector[2] and integrating slow mobility opportunities with territorial valorization (La Rocca 2014).

2 Historical Paths and Slow Tourism as Tools for Territorial Fruition

The 2016 Directive of the Italian Ministry of Cultural Heritage and Activities and Tourism in proclaiming the national "Year of Paths" has highlighted the importance of historical paths in improving territorial economies. This can be considered a legal recognition of the need to both preserve and reuse historical paths to promote a slow form of mobility for tourism and other cultural activities connected to territorial

[1]Definition by the Swiss Department of the Environment, Transport, Energy and Communications (DETEC 2007) in the Guidelines for Slow Traffic, Berna 2007, available at www.astra.admin.ch, accessed in March 2010.

[2]It has been estimated that tourism is responsible for around 5% of CO_2 emissions (UNWTO–UNEP–WMO 2008), a significant fact if related to the trend of growth in tourism over the last twenty years and its forecast to 2030 (UNWTO 2016).

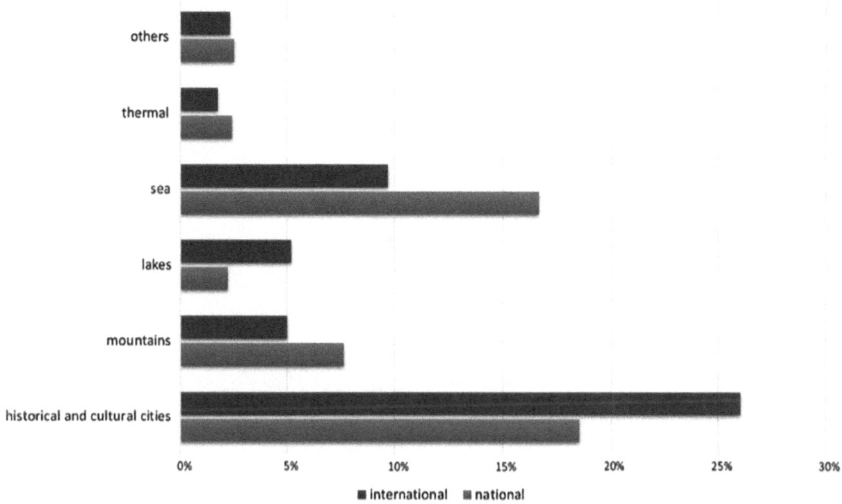

Fig. 1 Percentage of national and international arrivals in accommodations for destinations (*source* Elaboration on ONT and ISTAT data 2015)

fruition (enjoyment). The directive, in particular, points out two important aspects. On the one hand, it outlines the need for a census of paths in order to create an "Atlas of the Paths." On the other hand, it outlines the exigence in promoting "slow tourism" as a form of sustainable tourism able to reduce impacts on the environment and natural resources. As stated in the Directive, the achievement of this last objective, in particular, requires coordination among different administrative levels and cooperation among territorial institutions (Regional Administrations, Universities, Associations, Companies Scientific Institutions, etc.) to valorize historical and cultural heritage. These conditions could improve the development of regions that are not included inside the traditional tourism circuits but have strong potentiality to be attractive for cultural tourism (Figs. 1 and 2).

This type of tourism is considered strategic for the enhancement of sustainability (Richards 2011). While slow tourism can be considered a recent notion within a sustainable type of tourism, the project to activate a cooperative network through the revaluation of cultural routes is not new. It dates back to the Sixties, when the European Council assigned official certification to the first two historical routes: the Way of Saint James and the Rural Habitat.[3] The first one is probably the best known among European routes. Set up in the ninth century, as a route for pilgrims to reach the grave of Saint James, at present, the Way of Saint James can be considered as an "icon" for cultural and slow tourism, one that people may choose to walk for reasons beyond religious motivations (Table 1).

[3]The path of Rural Habit refers to a system of itineraries characterized by the presence of rural ecosystems and local peculiarities (particular handicraft and/or gastronomic productions, specifics of the natural landscape, etc.).

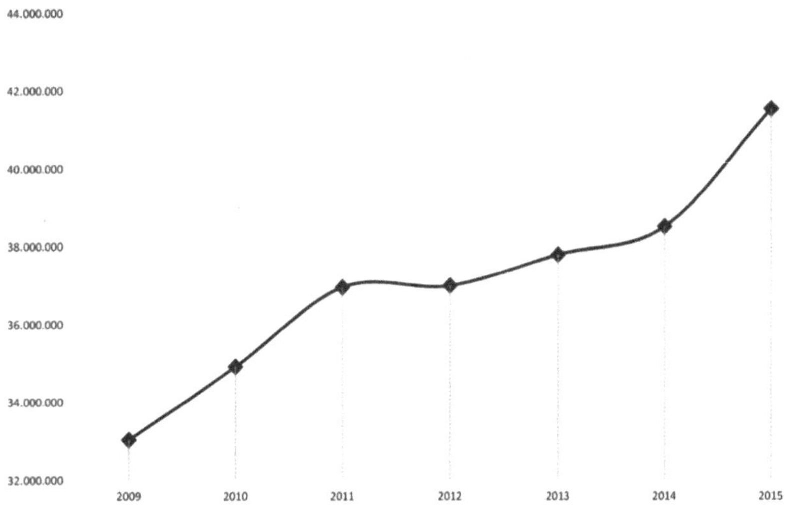

Fig. 2 The growth of cultural tourism in 2009–2015. National and international arrivals in cities of historical and cultural interest in Italy (*source* Elaboration on ONT and ISTAT)

Table 1 Motivations for traveling the Saint James Way 2012–2015 (*source* www.pergrinossantiago.es accessed on August 2017)

Motivations	2012	2013	2014	2015
Cultural and religious	101.171	117.715	120.412	141.969
Only religious	79.490	86.291	101.013	99.680
Only cultural	11.827	11.804	16.461	20.809
Total	192.488	215.810	237.886	262.458

The use of historical paths, in fact, is no longer exclusively connected to the pilgrimage, as has been shown in a recent survey developed by the Italian Tourism Club (TCI) about the stated motivations for walking the Via Francigena[4] (Fig. 3). The results of the survey allow us to observe that a new tourist demand is rising that is more sensitive to the "experiential aspects" connected with the culture and the history of the destinations they visit.

The recovery of these ancient paths could become a mainstream town planning tool in which the rules for their maintenance, valorization and design should be defined. In this sense, slow mobility could contribute to the definition of a sustainable design of urban and regional planning.

[4]The "Via Francigena" is composed of the set of paths that connected Rome with central-western Europe. The name underlined that the route originated in France. It was also called "Romea," because it led to Rome, which was the center of Christianity. It has always played a strategic role, being the main link between north and south of Europe, along which merchants, armies and pilgrims passed.

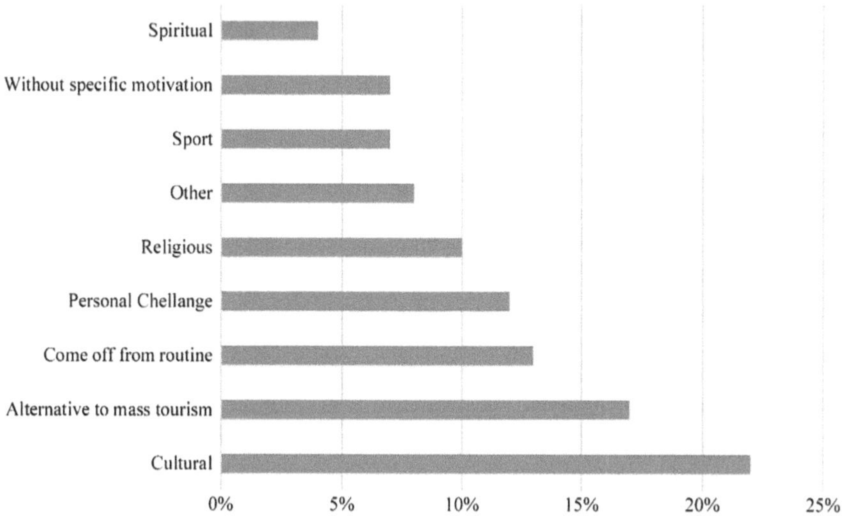

Fig. 3 Percentage of types of motivation for walking the Via Francigena in Italy (Elaboration on TCI data 2015)

3 Cultural Paths as a Distributed Museum

The Cultural Routes Program launched by the Council of Europe in 1987 is based on the idea that the valorization of leisure through culture can activate cooperation among different countries. In this sense, the wandering journey can represent a means to rediscovering heterogeneous heritages still interconnected within common historical and cultural roots.

The route, thus, becomes a shared "*common cultural heritage*," able to link different places and resources that characterize the regions across which it runs. The path itself constitutes an element of tourist attraction, as it can be, simultaneously, the object and the means for carrying out the "tourist experience." The real fruition of the paths, thus, is the fundamental condition of activating positive processes of territorial transformation. Cultural tourism, in spite of its continued lack of a truly unique definition, is considered as a positive form of tourism that complies with the characteristics of this particular element (the historical or cultural path) of the territorial supply. Richards (2003) defines cultural tourism as "*The movement of persons to cultural attractions away from their normal place of residence, with the intention to gather new information and experiences to satisfy their cultural needs*" (Richards, p. 6). The definition by the UNWTO is more focused on the need expressed through cultural tourism to be in contact with aspects connected to local identity and history as an expression of the culture of the tourist destination chosen (ONT 2013). In spite of these differences, general opinion refers to cultural tourism as a "good tourism" that is able to attract "sensible visitors[5]" who contribute both to improving local economies and to limiting environmental impacts.

In this sense, cultural tourism would be a form of sustainable tourism that could be defined as the movement of people inspired by the aptitude to discover new destinations for the purpose of increasing their own cultural background, being respectful of local characteristics (social, physical and functional[6]), and adopting sensible behaviors that limit the impacts on the hosting system. Such a definition seems to synthesize the peculiarities of both cultural tourism and sustainable tourism, maintaining a vision of "good tourism." This vision, indeed, also seems to prevail in the official realm of word development. 2017 has, in fact, been defined as the international "Year of Sustainable Tourism"[7] by ONU with the aim of finding and proposing concrete solutions for sustainable development by considering tourism as an activity driven by the promotion of a change in lifestyles and consumption. In particular, tourism has been considered as a catalyst for positive change in policies, business practices and consumer behavior, leading towards a more sustainable tourism sector that can contribute to the Sustainable Development Goals (SDGs) based initially on cooperation among governments, the private sector and the common people.[8] Within the SDGs, connected to the urban and territorial contexts, efforts to protect and safeguard the world's cultural and natural heritage play a key role in the pursuit of sustainable development.

From this perspective, and considering this political framework, cultural heritage assumes a key role, both as an economic factor (Trimarchi 2002; Santagata 2002; Greffe 2003; Santagata 2009), which generates direct positive externalities (i.e., all services directly connected to tourism), and also as a factor that can produce indirect positive externalities (e.g., an increase in the citizen's sense of place). This second effect must be considered in the planning of the rediscovery of historical and cultural

[5]We refer to a type of tourism that takes care of the local population, culture and history and that is characterized by visitors who are willing to spend.

[6]We refer to the systemic approach to the study of regional and urban planning that considers the territorial system as being composed of three main subsystems: social (people), physical (material resources), functional (activities) (for more details, see von Bertalanffy 1972; Mc Loughlin 1969; Fistola and La Rocca 2016).

[7]The international year 2017 has promoted the role of tourism in five areas:

- Inclusive and sustainable economic growth.
- Social inclusiveness, employment and poverty reduction.
- Efficient use of resources, environmental protection and climate change.
- Cultural values, diversity and heritage.
- Mutual understanding, peace and security.

The promotion of cultural roads can be brought into the mainstream in the fourth area, as the roads represent the physical link between different territories able to join diverse cultures.

[8]The seventeen goals have been embedded in the 2030 Agenda for Sustainable Development and were endorsed in 2015. Even though they are not legally binding, many governments are involved in realizing their achievement. The SDGs mainly concern the reduction of poverty at every level, the protection of the planet and the spread of equality for all the earth's inhabitants.

paths as physical infrastructure to support alternative forms of moving and travelling across regions.

They represent a "dynamic dimension" of the memory of the places they pass through. In this, they are radically different from museums, archeological sites and historic urban centers, which represent a "static dimension" of the culture of a place. In this perspective, historical and cultural paths can be considered as a "scattered museum," linking destinations that normally would not be included in the tourist circuits (Al-hagla 2010). At the same time, historical paths are a possible solution to the dichotomy between "heritage preservation" and "territorial development." In view of this, they can play a strategic role in improving approaches to the valorization of historical heritage in its double nature as resource and tourist product.

We must also consider the key role that the rediscovery of cultural and historic paths can play as promoters of co-operation between the public component (administrations and politics) and the private component (investors, financiers, stakeholders), because of the fact that they pass across different territorial contexts (physical, social, administrative).

4 Historical Path Recovery: New Technologies for the Fruition of Cultural Heritage

Over the last ten years, a radical change in tourist behavior and habits has taken place: personal use of tablets and smartphones has, in fact, exceeded those purposes exclusively connected to business needs, with that use spreading to recreation and entertainment as well. Within this change, tourism can play an active role in promoting sustainable lifestyles, contributing to the opening up of new perspectives related to the recent concept of urban smartness (Papa and Fistola 2016). In truth, our entire interpersonal system of communication has changed quite rapidly. Referring to tourism, in particular, the whole information supply system is modifying itself, and a multiple "virtual dimension" (Fistola and La Rocca 2001) is being established beside the traditional supply of tourists.

The number of so-called "mTravellers[9]" is increasing, and it has been estimated that about 42% of world travelers use mobile devices to plan their holidays. Even though the "traditional users" and the "digital users" coexist, there is a need to renew the management process of the supply of services and facilities related to those actors (public authorities, as well as private providers) who have responsibilities in the planning of territorial development. This supply cannot be accomplished exclusively by the increasing number of apps for tourism.[10]

[9]Tourists who use their personal devices (smartphone, tablet, pc, etc.) to organize their travels (booking and planning). In Italy, this kind of user numbers about 1.3 million, with a correspondent market value of about 835 million euros (Amadeus 2016).

[10]In Europe, the number of tour operators engaged in the production of apps for mobile devices is about 82% of the entire sector (Amadeus 2016).

Up to now, it has been possible to indicate three main types of use of new technologies applied to the fruition of cultural heritage by tourists (Orlando 2013; Garau 2014).

The first is a *communicative* type, in which the use of technologies refers to the diffusion of information to a wide audience (audio-guides, QR codes, smart plaques, etc.) using narration.

The second is an *educational* type, in which the use of technologies refers to the transmission of complex information to a specialized audience (students, scholars, researchers, teachers, etc.), using augmented reality, visual ambience, interactive reconstructions of virtual environments, immersive virtual environments, etc.

The third type can be defined as *device-centered*, as it refers to the use of mobile devices (smartphones, tablets, cameras, etc.) capable of combining geographical data with cultural information to be used on site (via an app, for example).

The development of these technologies has been widely recognized in the cultural tourism sector and is increasingly linked to the composition of the supply-side, oriented toward this type of tourist demand.

The most frequent applications involve integration with geolocation tools inserted into mobile devices. These applications allow users to select and personalize their paths according to their own composition (individual; family with children, older people, etc.) Normally, these applications merge the use of a Global Position System (GPS) with that of a Web GIS (Geographic Information System). Through the GPS, thus, users can visualize their localization on a geo-referenced map, and then they can also plan their route according to their own exigencies (disposition of time, difficulty, interests, curiosity, etc.).

If, then, the use of technologies allows users to form their own active participation and the customization of their experience according to their exigencies, on the supply-side, it is necessary to set out adequate support services, mainly aimed at:

- indicating safety standards for the fruition of the paths, both those existing and those to be planned;
- promoting the territorial potentialities according to sustainable modalities;
- enhancing tourist typologies that generate low environmental impacts (eco-tourism, slow tourism).

These objectives, in turn, need the definition of tools and rules aimed at:

- ensuring the implementation of the project aimed at valorizing the existing paths or creating new paths;
- defining criteria for the recognizability of the paths at the local, regional and national levels;
- regulating the use and walkability of the paths;
- delineating the rules for the management, maintenance and usability of the paths;
- supervising the relationship between the private and public sectors;
- coordinating the different levels of town planning;
- verifying the compatibility of the plan with the existence of areas that may be subject to environmental restrictions;

- providing the alignment of interests in territorial development strategies and technical tools.

Cooperation, in particular, is an essential condition if we consider that the paths run across different counties, beyond regional and administrative borders.

Partnership among public and private actors and among countries through which the paths pass, thus, represents the condition for achieving objectives of development based on sustainability and on the possibilities that tourism offers (Briedenhann and Wickens 2004). In this view, the paths are the element of a network connecting actors and levels involved in the town and regional planning processes, and cooperation among different levels (administrative, economic sector, local population) can improve conditions for the fulfillment of the objective of development (Meyer 2004; Mitchell and Hall 2005). The positive effect derived from cooperation can be indicated in five main points:

- the establishment of a network among poles that might not hold the same attraction individually;
- the enhancement of lesser-known destinations;
- the increasing of employment and the training of workers in new professional skills;
- the diffusion of such new tourist typologies as slow tourism;
- the decreasing of negative impacts on natural environments.

These conditions should improve the distribution of benefits derived from the development of tourism, including for those work categories not directly engaged in the tourism sector.

5 The Historical Paths as a Complex Tourist Cultural Product: The Recovery of via Francigena in Campania

The project of a tourist route represents, first of all, a territorial project, where aspects of material and immaterial heritage should both be considered.[11] In the framing of the project, this heritage must necessarily be interpreted according to a holistic and systemic approach that allows for the valorization and, at the same time, the preservation of the local identities of the territories crossed (Papa and La Rocca 2017). This approach, moreover, allows for construction of a varied supply based on the creation of connections between different nodes and poles within a regional or trans-regional territory. The connections system is not only physical, but also perceptive, because crossing realities, scenarios, and different landscapes allows the tourist to build their travel experience.

[11] Fioravanti (2016) titled "I cammini storici per la valorizzazione del territorio: la via Francigena nel comune di Montecalvo Irpino" supervisioned by R. Fistola and R. A. La Rocca at University of Sannio.

Furthermore, in the systemic approach, the focus is on both the whole set of physical and immaterial relationships between the path and the territory crossed and among the elements that make up the itinerary itself. It is possible to distinguish at least two "dimensions" in the path project:

- the path represents the link between points of interest (POIs), which may also be physically far one from another and belong to geographically different territories;
- the path is itself an attraction based on its own elements that make it somewhat unique.

In the first case, relationships are external and affect the "path system" in relation to the territorial context that this crosses. Such relationships may affect the design and physical structure of the itinerary, which places itself as a material link among different poles of attraction located inside the territory. Because of different relationships, paths may have various geometries and typologies, despite being structurally dependent or deriving from aggregations of simple geometric kinds (Papa and La Rocca 2017).

In the second case, relationships are developed among the elements that are part of the path and which alone constitute the potential of attraction; the attraction elements, that is, are within the same path and are not necessarily specific places.

To the previous two dimensions, in the case of historical cultural itineraries, it is necessary to join a perceptual dimension to which the nature and the very existence of the itinerary are closely connected. Choosing to take a path, in fact, is strongly linked to a contemporary user's emotional quest, in which it is perceived as an opportunity to develop an inner knowledge before even meeting a need for curiosity. In this dimension, the traveler and the journey merge, and the cultural itinerary can be understood as a conceptual evolution of what the Grand Tour was for the bourgeois young people in the 18th century.

The concept of integration is also emphasized in the same definition of cultural itinerary provided by UNESCO,[12] which considers both the merely material aspect of the itinerary as a means of promoting trade between different regions and countries, as well as the intangible aspect associated with its function of linking cultures and stories that, in some cases, may be radically different.

Since 1997, the European Institute of Cultural Itineraries has been the Council of Europe's body, which, also through the activation of partnerships and funding, supports both the re-evaluation and monitoring of existing itineraries and the creation of new projects aimed at promotion of the cultural aspects of the territories crossed. At present, there are twenty-nine certified itineraries that cover different geographic areas with reference to different themes inspired by Europe's culture and history (Table 2). The routes certified by the Council of Europe are recognized as a common heritage and are an expression of a project of cultural cooperation involving all political and administrative levels (from local to ministerial and European),

[12]"A path is made up of tangible elements, whose cultural significance derives from exchanges and multidimensional dialogues between countries and regions and illustrating the interactive and continuous movement of people along the itinerary, in space and time" (UNESCO 1994).

also based on the participation of the populations involved in the valorization program (Council of Europe 2005). These principles are referred to in the resolution of Cultural Itineraries (CM Res. 2010), which defines them as projects representing common European values. Italy is among the European countries most concerned by the itinerary program of cultural, educational, patrimonial and tourist cooperation, respectively. Over the last decade, the significance of the themes represented by this program has been recognized at the institutional level and is present in the program of the current government with the aim of networking a widespread heritage in compliance with European directives.

The focus on issues of sustainability and the promotion of slow tourism towards the creation of transregional and transnational networks for promoting new forms of use of widespread cultural heritage is also highlighted in the recent Ministerial directive when defining pathways as "cultural itineraries of particular European and/or national importance, passable on foot or with other forms of sustainable soft mobility, and which represent a way of enjoying the natural heritage and Cultural heritage, as well as an opportunity to enhance the natural, cultural and territorial attractors involved. In coherence with the vision of the Council of Europe, the paths cross one or more regions, can be part of European paths, [and] are organized around themes of historical, cultural, artistic, religious or social interest" (Mibact Decree 567 of 16/12/2015).

If, therefore, there emerges an awareness of the existence of a widespread heritage that can be valued through the involvement of the local populations supported by European incentives, it is also necessary to highlight the need to make these resources available through the vision of appropriate planning and design actions.

5.1 The Recovery of the via Francigena in the South

The project concerning the Via Francigena of southern Italy is part of a wider scheme aimed at the reconstruction of the original path used by pilgrims from northern Europe in order to reach Jerusalem. The project, in addition to being an opportunity to re-evaluate the Christian origins of European culture, also has strong value for the development, at the national level, of the territories crossed. The design, in fact, foresees the identification of the main routes of pilgrimage in the areas of southern Italy, with the aim of defining some first and effective actions of valorization of the Territory. In the European context, the recovery of these paths testifies to the commitment and interest in promoting the defense of cultural heritage.

The Via Francigena that brought visitors from Canterbury to Rome on their way to arriving in the Holy Land, in fact, is a historical itinerary, traveled in the past by thousands of pilgrims making their way to Rome and Jerusalem. This road also attests to the importance of the precise rules that attended the pilgrimage in the Middle Ages: it had to be mainly on foot (for penitential reasons), and demanded that a minimum of between 20 and 25 km be traversed per day, to reach the points

Table 2 Typologies of certified paths (*source* Council of Europe)

Type	Route	Year
Pilgrimage	The Way of St. James	1987
	Via Francigena	1994
	The Route of Saint Olav Ways	2010
Rural habitat	The Rural Habitat of the Pyrenees.	2002
European personalities	The European Mozart Ways	2004
	The Heinrich Schickhardt European Cultural Route	1992
	The Saint Martin of Tours Route	2005
	Via Habsburg	2014
	The European Route of Emperor Charles IV	2015
	Destination Napoleon	2015
	The Huguenot and Waldensian trail	2013
	In the Footsteps of Robert Louis Stevenson	2015
Abbeys and monasteries	The European Route of Cistercian abbeys	2010
	The Cluniac Sites in Europe	2005
Historical and architectonical heritage	The Routes of El legado andalusí	1997
	TRANSROMANICA-The Romanesque Routes	2007
	The European Cemeteries Route	2010
	ATRIUM-Architecture of Totalitarian Regimes	2014
	Fortified towns of the Grande Region	2016
	Réseau Art Nouveau Network	2014
	Prehistoric Rock Art Trails	2010
Agricultural landscape	The Routes of the Olive Tree	2005
Populations	The Phoenicians' Route	2003
	The European Route of Jewish Heritage	2004
	The Roman Emperors and Danube Wine Route	2015
	The Viking Routes	1993
Historical treaties and city networks	The Hansa	1991
Industrial and artisanal cultures	The Pyrenean Iron Route	2003
	The European Route of Ceramics	2012
	The European Route of Megalithic Culture	2013
Thermal towns	European Route of Historic Thermal Towns	2010

associated with the Saints of the Christian religion. During the journey, the Pilgrim had to spend time travelling in meditation and prayer.

The decision to undertake the journey along these routes, therefore, in itself had a high devotional value.

Although these motives are no longer the sole reason for making the trip, the Via Francigena preserves a highly symbolic cultural value and represents an element of valorization of the territory that is highly significant up till now.

In Italy, the region that first understood the importance of recovering these values within a coordinated design of territorial development has been the Tuscany region. On the Tuscan territory, the Via Francigena represents a spine that crosses the entire regional territory, including UNESCO World Heritage sites, artistically-identified cities, small villages at risk of depopulation and rural areas traditionally excluded from the most significant tourist flows. The route extends for 375 km, from the Passo della Cisa to the Lazio border, crossing 37 communes and 5 provinces with very different territorial features. Since 2006, the Tuscany Region has acted as the interregional leader of the "Via Francigena" project, along with six other Italian regions crossed by the route.

During 2009, the "Master Plan of Via Francigena[13]" was launched, oriented towards the pursuit of the development of sustainable tourism (Council of Europe 2010). The Master Plan has been shared with local authorities and regional associations, under an institutional co-ordination that has allowed the implementation of the planned interventions and the pursuit of results according to a logic based on cooperation in decision-making and sharing of burdens and benefits.[14]

The Tuscan segment was inaugurated in 2014, with an expenditure of about 24 million euros. The recovery of the Francigena route represents 13% of the total movement of regional tourism and affects 6% of GDP and 10% of active employment (Conti et al. 2015).

In the overall frame of actions aimed towards the valorization of the landscaping, cultural, economic and environmental contexts of the territories crossed by the Via Francigena, tourism is a strategic resource, especially for small communes or territorial spheres that are not yet able to express autonomous territorial attractiveness that is capable of attracting interest, in regards to tourists, resources, and investments. For these territorial realities, the recovery of the Via Francigena is also an opportunity to revive their identities. The acknowledgment of the "Great European Route" by the Council of Europe, in fact, has strengthened the strategic role of this historic artery for territorial enhancement. In addition, being a long-running route in Italy (from the Great St. Bernard to Rome and Brindisi), it represents an opportunity to unite

[13]This refers to tourist typologies with reduced impact on the environment: naturalistic and sports tourism, historical, cultural, religious, food and wine-based tours, and so on.

[14]The actions were primarily aimed at the infrastructure and the safety of the route, through the setting up of "stage points," the design of facilities for accommodation, the forecast of plans for the recovery of monuments, the adaptation of roads and paths to the criteria of practicability in regard to safety conditions, the preparation of plans for local signage, the definition of programs to guarantee the routine maintenance of the route. The plan of action for implementation has led to, among other interventions, the realization of a dedicated Wi-Fi network.

factors of local diversity linked to the monumental, artistic, historical and landscape patrimony.

In order for these potentials to become effective, it is necessary to define a timely and articulated supply system along the territories crossed, which is, at the same time, a guarantee of efficiency and safety, as well as an additional element of attraction.

The regulation of the path and the verification of its validity, from a technical point of view, are carried out by the "European Association of the Vie Francigene" (EAVF), which verifies the characteristics that the path must have in order to be considered a tourist product.[15] From an administrative point of view, the local authority involved in the recovery project (usually the municipal administration) has to include the path within the urban plan in order to define costs and benefits, considering both the properties involved in the project and ordinary maintenance operations, including via conventions with private individuals or associations.

6 Recovering Historical Paths: A Methodological Proposal

By adopting the principles of both the Vademecum and the Manual for the realization of an itinerary along Via Francigena, this part of the work deals with the definition of a project hypothesis for the recovery of a part of the historical route that crosses the municipality of Montecalvo Irpino in the province of Avellino in Campania. Montecalvo's medieval origins are borne witness to by the presence of a feudal hamlet located on the slopes of the Campania Apennines on the border between the two provinces of Avellino and Benevento. With a population of 3663 inhabitants and a predominantly agricultural vocation, Montecalvo is one of the numerous historic cities within the region that has high potential for the development of "slow tourism" that appreciates the historical, cultural and environmental resources. According to the considerations outlined in the first part of this paper, the design proposal is based on the awareness that a development plan for sustainable tourism must necessarily involve building a "network of actions" that interprets the territory and its resources through a systemic approach. The Via Francigena of Southern Italy can be considered a system of routes that developed alongside the main route (the Via Francigena) connecting Rome with Brindisi, the original location from which pilgrims left to

[15]The requirements for the realization of the route are given in the Handbook of European Standards (2016), which indicates the criteria for the selection of itineraries within the territories crossed by the Via Francigena. The criteria for selection refer to:

- the safety of pilgrims (wayfarers);
- historical importance;
- cultural significance;
- proximity to points of hospitality;
- safety conditions necessary for the monitoring of solitary and shady traits;
- the presence of water and refreshment points.

These criteria have been considered for the development of the project described in this paper.

reach Jerusalem. More appropriately, therefore, is the southern part of the route is best known as Via “Romea”, so as to emphasize the role that Rome played as the main pole on the path to the Holy Land.

Nowadays, the route is designed and divided into 32 stages that make up the network of potentially viable itineraries, awaiting recognition as a Cultural Route of the Council of Europe. The network of paths so far consists of five routes crossing the regional territories of the lower Lazio, Campania, Molise, Basilicata and Puglia. The primary paths are surrounded by a network of alternative routes that the local territorial authorities have proposed or can propose.

In this context, the proposal is framed with the aim of identifying such opportunities within the territory of Campania that are coherent and compatible for development with local vocations. Those local vocations themselves will be activated through recovery and territorial upgrading; the proposed path is built along the route of the Appia Traiana, conceived and built by Emperor Trajan (108 BC) to speed up the connections with the ports of Apulia (Fig. 4).

The structure of the project proposed in this paper, by acquiring the criteria set out in the guidelines, can be articulated in the following main phases:

- identification of the historical track within the municipal territory of Montecalvo Irpino (this choice involves a narrowing of the path to 7 km, compared to the 25 km standard indicated in the Guidelines for a complete stage);
- definition of the type of route (pedestrian, cycling, equestrian);
- tracing out of the route through geolocation to allow for applicability to any GPS device;
- designing of the elements of the signage, including safety signs for travelers, based on a single and coordinated image to give the route a sense of identity;
- locations selected for signposting;
- design of lighting equipment for the sake of the safety of the route;
- proposal to reclaim and refinance a dismantled building, called the “Tavern of the Duke” to be used as accommodation in accordance with the standards indicated in the Guidelines.

6.1 Identification of the Historical Path Within the Municipal Territory

The identification of the pre-existence of the historical path has taken place through the consultation of historical sources and direct investigations aimed at recovering the historical memory of the places crossed by the path. The phase of the gathering of knowledge enabled the development of an exhaustive overview of the critical issues on the site that needed to be taken into account in the design phase. The surveys carried out revealed that the route goes through heterogeneous contexts, as well as the serious lack of security conditions necessary for its fruition.

Fig. 4 The Via Francigena in the South of Italy can be regarded as a system of routes that develops next to the main line (the Via Francigena), which connects Rome with Brindisi, the original location from which pilgrims sailed to reach Jerusalem. (rielaboration from www.viefrancigene.org)

6.2 *Definition of the Type of Path (Pedestrian, Cycle, Equestrian, Carriage)*

The presence of a heterogeneous condition along the path has led to the decision to maintain the heterogeneity as a fundamental feature of the route. Therefore, all four typologies outlined in the Guidelines have been ignored, taking care to differentiate the path's traits anyway and to place the necessary security measures in the less developed areas where the path will be cultivated. In these specific cases, it will be necessary to introduce mitigation measures regarding traveling speed through the introduction of a suitable signaling system.

6.3 Rediscovering the Path: Tracing by Mean the Geolocation to Allow for Applicability on Any GPS Device

The planned path follows almost the entirety of the route of the Via Traiana after the connection with the Regional Tratturo Pescasseroli-Candela (the tratturo was the route through several Italian regions that the shepherds followed to transfer their herds from the mountain pastures to those of the plains, and vice versa; this migration is known by the term "transhumance"). The path, after reaching the confluence of the Ginestra stream with the Miscano River, goes across the territory of Montecalvo, through the bridge of S. Spirito, or "of the Devil," and heads to the site of Malvizza di Montecalvo Irpino. Veering away from the right, the Regio Tratturo continues for another kilometer, where, due to the intrusion of private land, it changes 6.1 course towards the Bubbles of Malvizza and the Tavern of the Duke.

Tracing was done using an Android application with GPS connection: Locus Map (free software), which lets you create a virtual path and take photographs of the most significant points. In using this app to track the path, you need to provide it with information on the type of route you want to create (pedestrian, cycling, equestrian, etc.). The application encodes the geolocation parameters (the minimum distance in meters between the points to be recorded, the interval in seconds between two subsequent recordings, the required precision, etc.). Once the path is recorded, a .KMZ file is created; a KMZ is a file in the Keyhole Markup Language format (in a compressed version) that can be downloaded inside a web environment, and specifically with the geographic web software Google Earth (Fig. 5).

6.4 Designing the Signage and Positioning of the Path Signs

The design of the signage must be considered another important activity for the general process of marking out the path. The signage system, in addition to its purely informative function, must allow the area to be used in full compliance with the environmental values that make it unique. In this sense, it is necessary to adopt solutions that can communicate an immediate message about the orientation and "sacredness" of the place through which you are passing. For this project, it was decided that the signage should be standardized, adopting the recognized European Standards on cultural historical itineraries with an environmental value and using symbols and materials corresponding to the guidelines. Furthermore, original notices have been designed and located along the path in order to underline the importance

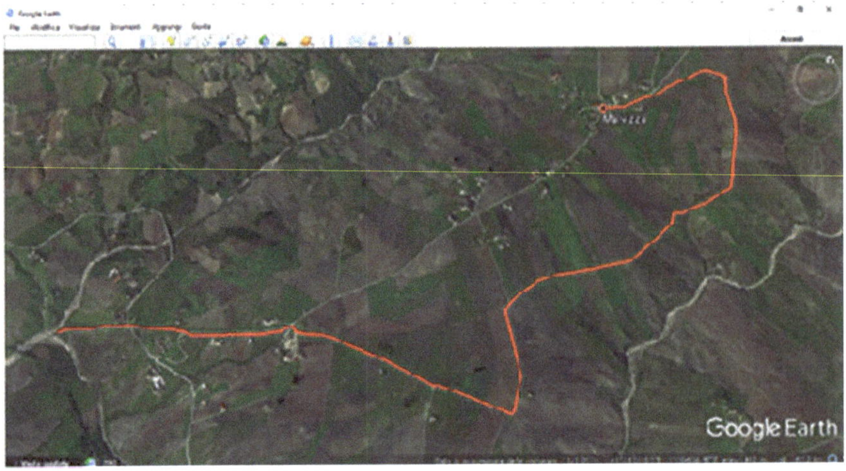

Fig. 5 View in Google Earth of the geolocalized path

of specific sites. Another important activity to take into account is the design of lightweight structures to ensure the safety of the path. The path could pass through potentially hazardous sections because of the great reach of the flow of wayfarers, so, where it is needed, special structures (e.g., docks) can be constructed out of wood and natural materials to ensure safe crossing.

6.5 *Recovery and Reuse for Accommodation of a Disused Building Called the "Tavern of the Duke"*

Along the path, there is an old tavern called the "Tavern of the Duke" an important 17th century house that offered hospitality and refreshment to people and livestock in transit along the *tratturo* (which originally joined the regions of Molise and Puglia). The tavern is currently a house in serious disrepair (Fig. 6), but could be particularly suitable for a careful recovery operation aimed at providing hospitality services along the path.

Taking this possibility into account, the project proposes renovation of the tavern in order to create a place along the new path useful for the hospitality and accommodation of the wayfarers. By using a type of cad software in its free release (SketchUp by Trimble), a 3D mock-up of the new Tavern of the Duke has been designed, restoring the original architectural elements of the eighteenth century. A 3d rendering has subsequently been placed on Google Earth to complete the project of the entire path (Fig. 7).

Fig. 6 The Tavern of the Duke at the present

Fig. 7 Rendering (on Google Earth) of the project of the reuse of the Tavern of the Duke

7 Conclusions and Possible Research Developments

The recovery and revaluation of historical paths can be an opportunity to promote territories that are barely known and appreciated on the tourism market. Such revaluation can promote a virtuous territorial regeneration based on logic geared towards land consumption control and environmental sustainability. The rediscovery of cultural paths within a territory thus becomes an opportunity for development, even in regard to tourism, compatible with the requirements of sustainability and the preservation of values of environment and landscape. Furthermore, it becomes an opportunity to promote alternative forms of mobility, as opposed to invasive and polluting modes of use.

This article attempts to highlight how Italy can play a key role in such an endeavor due to the significant historical assets located inside the country and on the basis of a renewed interest in conscious tourism and slow mobility from the administrative policy arena.

An indispensable condition for the occurrence of positives is the construction of transversal cooperative environments involving public and private actors, as well as the retraining of the populations concerned for development of the program/project.

On the other hand, the redesign of a network of paths, according to a holistic and systemic approach, must consider the relationships that arise among the different components. In this vision, one must not neglect the potential role of integration that the path can play in regard to territories that may be profoundly different from one another. It is therefore important to emphasize the action of social mending that the path can facilitate, particularly within rural areas where boundaries and old conflicts among the owners have marked the territorial contexts. Defining a common development project can be a process that overcomes cultural distances that are sometimes greater than physical barriers.

The paper also attempted to demonstrate how new technologies can offer important support in the process of design and recovery of the path, also allowing for real time updates of the offer of services and equipment that the wayfarers may need.

It has also been emphasized that the recovery of these paths can support the spread of new forms of enjoyment and use of the most sustainable territory. In this vision, cultural paths are configured as physical infrastructures, supporting a non-motorized gentle mobility, able to connect diversified users with the resources of the territory (Massa and Campagna 2014). Further development of the research can also be oriented towards the improvement of the design methodology through the definition of specific planning tools, supplementary to the general urban planning tools, capable of supporting decision-making aimed at the recovery of such territorial infrastructures.

Acknowledgements The authors have made a joint contribution to the paper's conception and design, background and concluding remarks. Rosa Anna La Rocca designed the methodological approach to the problem and the knowledge framework (§§ 1, 2, 3 and 4). Romano Fistola developed the systemic methodology for the path identification (§§ 5 and 6).

References

Al-hagla KS (2010) Sustainable urban development in historical areas using the tourist trail approach: a case study of the Cultural Heritage and Urban Development (CHUD) project in Saida, Lebanon. Cities 27:234–248

Amadeus (2016) Global Report 2016. A business, financial, sustainability overview. Available at http://www.amadeus.com/globalreport2016

Briedenhann J, Wickens E (2004) Tourism routes as a tool for the economic development of rural areas—vibrant hope or impossible dream? Tourism Manag 25(I):71–79

Conti E, Ommi S, Piccini L, Rosignoli S (2015) Itinerari culturali europei e sviluppo sostenibile: il caso della via Francigena. EyesReg 5(4). Available at http://www.eyesreg.it

Council of Europe (2005) Promotion of cultural tourism as a factor of development of the regions. Council of Europe Publishing, Strasbourg

Council of Europe (2010) Resolution CM/Res (2010)52 on the rules for the award of the "Cultural Route of the Council of Europe" certification. Strasbourg

DETEC (2007) Guidelines for slow traffic. Berna 2007. Available at www.astra.admin.ch. Accessed Mar 2010

Fioravanti G (2016) I cammini storici per la valorizzazione del territorio: la via Francigena nel comune di Montecalvo Irpino. Tesi di laurea aa 2016–2017 Università degli Studi del Sannio

Fistola R, La Rocca RA (2001) The virtualization of urban functions. In: Bakis H, Huh W (eds) Geocyberspace: building territories on the geographical space of the 21st century, NETCOM, vol. 15, nn. 1–2 2001. Montpellier, France, pp 39–48

Fistola R, La Rocca RA (2016) Driving functions for urban sustainability: the double-edged nature of urban tourism. Int J Sustain Dev Plann 12(3):425–434 (2017). WIT Press Special Issue, Southampton, Boston. e-ISSN 1743-761X, print ISSN 1743-7601

Garau C (2014) From territory to smartphone: smart fruition of cultural heritage for dynamic tourism development. Plann Pract Res 29(3):238–255. https://doi.org/10.1080/02697459.2014.929837

Greffe X (2003) La valorisation économique du patrimoine. La Documentation Française. Ministère de la Culture. Département des Etudes de la Prospective et des Statistiques (DEPS)

La Rocca RA (2010) Soft mobility and urban transformation. TeMA. J Land Use Mobility Environ 3(SP):85–90. University of Naples Federico II, Naples. https://doi.org/10.6092/1970-9870/125. Available at http://www.tema.unina.it

La Rocca RA (2014) The role of tourism in planning the smart city. TeMA. J Land Use Mobility Environ 7(3):269–283. University of Naples Federico II, Naples. https://doi.org/10.6092/1970-9870/2814. Available at http://www.tema.unina.it

Massa P, Campagna M (2014) Social media geographic information: recent findings and opportunities for smart spatial planning. University of Naples Federico II, Naples. TeMA. J Land Use Mobility Environ (SP):645–658. https://doi.org/10.6092/1970-9870/2500. Available at http://www.tema.unina.it

Mc Loughlin JB (1969) Urban and regional planning. A systems approach. Faber and Faber, London

Meyer D (2004) Routes and gateways: key issues for the development of tourism routes and gateways and their potential for Pro-Poor Tourism. Background paper, output of the ODI project: Pro-poor Tourism Pilots in Southern Africa. Available at http://www.odi.org.uk

Mitchell M, Hall D (2005) Rural tourism as sustainable business: key themes and issues. In: Hall D, Kirkpatrick I, Mitchell M (eds) Rural tourism and sustainable business. Channel View n. 2, Clevedon, pp 157–170

Orlando M (2013) Didattica e turismo 2.0 - Nuove tecnologie per la divulgazione del patrimonio culturale. Storia e Futuro, 32 June 2013 Available at http://storiaefuturo.eu

Osservatorio Nazionale sul Turismo (2013) Panoramica sul turismo culturale. Available at http://www.ontit.it. Accessed Jan 2017

Papa R, Fistola R (eds) (2016) Smart energy in the smart city. Urban planning for a sustainable future. Springer International Publishing, Switzerland

Papa R, La Rocca RA (2017) New forms of mobility for an alternative territorial fruition: the rediscovery of tourist footpaths. In: Dell'Acqua G, Wegman F (eds) *Transport Infrastructure and Systems*. Proceedings of the AIIT International Congress on Transport Infrastructure and Systems (Tis 2017), Rome, Italy, 10–12 April 2017. CRC Press, Balkema, P.O. Box 11320, 2301 EH Leiden, The Netherlands CRC Press, Balkema, pp 669–676. https://doi.org/10.1201/9781315281896-87

Richards G (2003) What is cultural tourism? In: van Maaren A (ed) Erfgoed voor Toerisme. Nationaal Contact Monumenten

Richards G (2011) Cultural tourism trends in Europe: a context for the development of Cultural Routes. In: Khovanova-Rubicondo K (ed) Impact of European Cultural Routes on SMEs' innovation and competitiveness. Council of Europe Publishing, Strasbourg, pp 21–39

Santagata W (2002) Cultural districts, property rights and sustainable economic growth. Int J Urban Reg Res 26:9–23. https://doi.org/10.1111/1468-2427.00360

Santagata W (ed) (2009) Libro Bianco sulla creatività. Per un modello italiano di sviluppo. Commissione sulla Creatività e Produzione di Cultura in Italia. Ministero per i Beni Culturali e le Attività Culturali. Università Bocconi Editore. Milano

Trimarchi M (2002) Dentro lo specchio: economia e politica della domanda di cultura. Economia della Cultura. Rivista trimestrale dell'Associazione per l'Economia della Cultura 2/2002, pp 157–170. https://doi.org/10.1446/7399

UNESCO (1994) Routes as part of our cultural heritage. Report on the meeting of experts. Madrid, 24–25 November 1994. Available at http://whc.unesco.org/archive/routes94.htm

UNWTO (2016) Annual report. World Tourism Organization (UNWTO), Madrid, Spain. Available at http://www.e-unwto.org/doi/pdf/10.18111/9789284418725. Accessed Sept 2017

UNWTO–UNEP–WMO (2008) Climate change and tourism: responding to global challenges. UNWTO, Madrid. Available at http://sdt.unwto.org/sites/all/files/docpdf/climate2008.pdf

von Bertalanffy L (1972) General system theory. Penguin Books, Harmondsworth

Estimation of Mobility Flows at Sub-regional Level: An Application to Piedmont Based on a Socioeconomic Scenario

Simone Landini and Sylvie Occelli

Abstract Notwithstanding considerable efforts are being made to gather transport and mobility data, by means of new technological devices, in Italy a major spatially comprehensive source of information is the population census. Data about journeys-to-work and journeys-to-school at municipality level have been made available by the Italian Central Bureau of Statistics (ISTAT) at 1991, 2001 and 2011 census years. Such data provide a sound information basis for drawing retrospective accounts of mobility. However, it is likely to be unsatisfactory for a number of planning activities, such as monitoring and forward looking investigations. They often require expensive individual surveys which may be inaccessible, either for public and private agencies. This study aims at alleviating the problem by developing a computational and analytic tool for estimating journeys-to-work at sub-regional level. The study proposes a strategy which links the reconstruction of yearly mobility flows, based on available spatially fine-grained socioeconomic information, with the generation of new flow matrices, depending on regional socioeconomic scenarios. In this application to Piedmont, we first reconstruct the mobility flows in the 2001–2013 period according to sub regional population and employment data as well as the census mobility tables. Econometric regression techniques have been used to estimate the origin and destination totals of mobility matrices; Wilson's entropy maximization approach to fully constrained spatial interaction models has been applied to compute the matrix cell values. The mobility deterrence parameter series associated with the flow matrices have been analyzed according to a set of socioeconomic variables, for which regional level demographic and economic trends from 2014 to 2020 are provided by ISTAT (population forecasts) and Prometeia (macro-economic studies). Parameters estimates have been obtained which, together with auto-regressively calculated origin and destinations totals of mobility tables, have been used to infer the regional flow matrices from 2014 to 2020. In discussing the main results of the approach attention focuses on its practical relevance for planning purpose and its portability.

S. Landini (✉) · S. Occelli
IRES Piemonte—Istituto di Ricerche Economico Sociali del Piemonte, Turin, Italy
e-mail: landini@ires.piemonte.it

R. Papa et al. (eds.), *Smart Planning: Sustainability and Mobility in the Age of Change*, Green Energy and Technology,
https://doi.org/10.1007/978-3-319-77682-8_19

Some methodological issues concerning the calculation of the deterrence parameter are also mentioned.

1 Introduction

Notwithstanding considerable efforts to gather transport and mobility data, by means of new technological devices (Hawelka et al. 2014; Gössling 2017; Kung et al. 2014; Neill et al. 2017).

In Italy a major source is still population census. Data about journeys-to-work and journeys-to-school at municipality level have been made available by the Italian Central Bureau of Statistics (ISTAT) at 1991, 2001 and 2011 census years.

In Italy, municipality is the lowest administrative level having government and planning responsibilities including the delivery of public transport. Information about commuters at this spatial level, therefore, is crucial for laying out mobility plans and managing transport services.

The ISTAT information platform provides a sound basis for drawing retrospective accounts of commuting patterns. However, they are unlikely to be satisfactory for planning activities entailing monitoring and forward looking investigations. These often require expensive individual surveys that may not be affordable for most local authorities as well as for and public and private agencies.

This study aims at alleviating the problem and develops an analytic tool prototype for estimating journeys-to-work mobility at sub-regional level in the long run. This chapter proposes a strategy which links the reconstruction of yearly mobility flows, based on available spatially fine-grained socioeconomic information, with the generation of new flow matrices, depending on regional socioeconomic scenarios.

In this application to the Piedmont region we refer to 2001 and 2011 ISTAT journeys-to-work data and reconstruct the yearly mobility flows in the 2001–2013 period. Econometric regression techniques are used to estimate the origin and destination totals of mobility tables according to sub regional population and occupation data. Wilson's entropy maximization approach for fully constrained spatial interaction models is operationalized to compute the matrix cell values of mobility tables. Finally, we consider two elementary socioeconomic scenarios at the 2020 time horizon and a time series of such tables is estimated for the 2014–2020 period.

In the following we briefly recall the premises that underpin the study approach and motivate the model application. Then we give an overview of the analytic strategy. A novel approach to the empirical estimation of the deterrence parameter in the spatial interaction model is also proposed.

A few main results of the model application are discussed, and their contribution to the investigations currently being undertaken for the new Regional Transport and Mobility Plan (Regione Piemonte 2016) is highlighted. Finally, we summarize the main undertakings of the study and advance some suggestions for future research.

2 Motivation of the Study and Main Goals

The study strategy stems from several stimuli which motivate the goals of the model undertaking and the design of the analytic strategy. They can be articulated under three main headings: (a) the need to refine the analysis of the shape of the regional mobility pattern, depending on some socioeconomic changes; (b) the application of a well established methodological background and (c) the opportunity to show the potential of modeling tools in policy activity.

In Piedmont, as in many other European regions, the mobility phenomenon is undergoing deep transformations. In particular, investigations carried out at IRES Piemonte for the 2001–2014 period (Occelli and Sciullo 2015a; Landini and Occelli 2016a) made it possible to highlight salient aspects of the modifications taking place in the region, such as: (a) a decrease of the total number of trips made by the resident population, (b) a slight increase of the propensity to use public transport and more active transport modes (e.g. biking and walking). (c) a greater permeability of the regional boarders to inter-regional mobility, and (d) a reshaping of commuting patterns at sub sub-regional level resulting in relatively denser and more clustered ones.

The opportunity to get deeper insights into this last aspect is a major motivation of this study. By applying a prototype model we want to answer the following basic question: *to what extent the spatial distribution processes of both resident population and activities currently taking place in Piedmont, is likely to result in longer or shorter travelled distance, thus contributing to less or more sustainable commuting patterns?*

To respond to the question the study adopts a meso-level observation lens, which makes it possible to look at the mobility phenomenon at an intermediate level between the urban and regional ones. It considers 186 zones, resulting from an aggregation of the 1206 Piedmont municipalities, according to a set of clustering criteria such a population size, connectivity and morphology. The 186 zones are the main units of analysis: they have a considerable extension, population density and connectivity: such areas are those the regional transport agency adopted to survey individual mobility data (Agenzia della Mobilità Piemontese 2013).

Therefore, in the light of the above research question, the purpose of the model application is to determine the number and distribution of the journeys-to-work among the 186 zones depending on: (a) the resident population who daily commute for working; (b) the spatial distribution of firms in, and (c) a measure of the distance between the residential and employment zones.

The proposed approach can be considered as hybrid as it jointly involves a variety of statistical techniques, such as regression methods, Montecarlo simulation, and time series auto-regressive analysis. This is carried out together with a spatial interaction entropy maximization modeling backcloth, pioneered by Wilson more than 50 years ago (Wilson 1967, 1968, 1971). This a consolidated modeling approach which has a longstanding tradition in activity location, population mobility and travel studies (Wilson 1987, 1998).

An additional aspect of applying a consolidated modeling approach is that it is likely to potentially have a greater possibility to rely on existing socioeconomic data. This is the case of this model application which furthermore could be repeated on a regular time interval as most of the official statistics used are regularly updated over time.

As many other travel models do, this application of the Wilson spatial interaction model is meant to provide an objective assessments of the impact on travel flows of possible alternative socioeconomic and demographic configurations for the regional local areas. By running the model with different sets of input assumptions, meaningful performance metrics are obtained which can provide more detailed insights into the commuting patterns.

Finally, when considering the study undertaking within the planning task broader perspective, it is worth reminding that the calculation of appropriate estimates of mobility flows has been a core topic in land use and transport studies (Acheampong and Silva 2015; Arentze and Timmermans 2000). In this respect, Landini and Occelli (2016b) note that in order to appreciate their achievements one should take into account three aspects such as methods, data and the ways they are applied in practice. They note that whereas methods and data usually are at the core in scientific undertakings the issues of their appropriation by urban and transport policy practices are rarely addressed (one exception is Givoni et al. 2012) Of course, appropriation of modeling tools by decision makers, and practitioners as well, depends on many factors such as users' technical competences, institutional mandates and, more broadly, the permeability of the cultural setting to changes and innovative thinking.

Strictly speaking, diffusion of scientific results in policy practices might be well considered as being outside the scope of scientific advancement and even be viewed as an extraneous matter. Eventually, because of the steady pace of technological and information progress, two cases may happen in the long run: (a) either the issue is no longer relevant, thus meaning that an appropriation process has occurred; or (b) the spreading of scientific progress is still late and likely to exacerbate the earlier difficulties encountered in transferring scientific achievements in policy practices.

A main argument of the present study is that in both cases the appropriation process by the transport stakeholders, and primarily by the staff working in the transport planning departments, deserves greater attention. This sets two main requirements for the operational tool: first, on the analytic ground it should produce reasonable metrics, to be included in performance and dashboard indicators to support planning tasks; second, on the technical ground, it should exploit as best as possible consolidated techniques and implement software components which could be extensible and easy to integrate with other types of model designs.

3 Analytic Strategy: Methodological Framework and Distance Parameter Estimation

3.1 Methodological Framework

As mentioned above, Wilson's spatial interaction maximum entropy approach is at the core of the model application.[1] In particular, the model is:

$$T_{i,j}(t) = [A_i(t) \cdot O_i(t)] \cdot [B_i(t) \cdot D_i(t)] \cdot f\left(d_{i,j}|\beta(t)\right) \tag{1}$$

where

1. $T_{i,j}(t)$ is the flow of commuters between residence zone i and employment zone j,
2. $O_i(t)$ is the total of flows from residence zone i,
3. $D_j(t)$ is the total of flows terminating at employment zone j,
4. $f\left(d_{i,j}|\beta(t)\right)$ is a function representing the deterrence of the distance between zones, $(d_{i,j})$. and $\beta(t > 0)$ is its parameter,
5. $A_i(t)$ and $B_j(t)$ are balancing factors ensuring that the total number of flows leaving i and the total one terminating in j is equals, respectively, to $O_i(t)$ and $D_j(t)$.

In this model application the distance is measured by kilometers on the road network between any two zones. Although this measure may appear unsatisfactory compared with other more refined formulations of travel costs, this metrics was the easiest to operationalize when the study was undertaken.

The methodological framework consists of a number of steps summarized in Fig. 1. The study period covers a relatively long time span, from 2001 to 2020 and focuses on intra-regional mobility.[2] Census data are the information source of the journeys-to-work among the 186 zones at 2001 and 2011, also useful for the validation of the model. Information for updating the flow totals have been gathered from ISTAT population and firm statistics,[3] both of which provide yearly data at municipality level for the whole 2001–2013 period. Therefore, 2013 represents the threshold in the implementation path which however can move forward as new data will be released.[4]

[1]The entropy procedure is based on the enumeration approach of combinatorial analysis, and it follows from statistical mechanics. The basic premise of these models is to enumerate all the possible zone-to-zone flow interchanges and pick the one with the highest uncertainty subject to constraints, see Wilson (1967, 1971).

[2]Although in some boarder zones of Piedmont the share of flows to and from other regions is not negligible, the issue was not considered relevant in the current version of the model application.

[3]Firms statistics involved are those of the ASIA database of ISTAT, the statistical archive of active firms.

[4]At the time the model has been developed the last updated information were at 2013. Currently, population data are available up to 2016 and firms' data are available up to 2014. Adding three

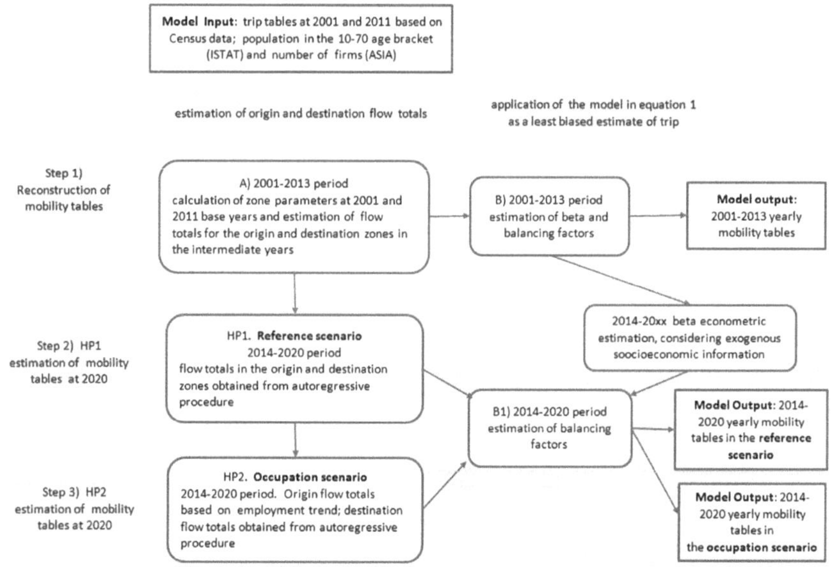

Fig. 1 Overview of the main steps in model application

Basically, the implementation procedure distinguishes three main stages (Fig. 1). The Step 1 reconstructs the mobility tables in the 2001–2013 period. These are obtained by applying Wilson's approach after having econometrically estimated the totals of each mobility table on the available population and firm's data. Step 1 mainly serves as an input for Step 2 and 3 in the second stage which aim to yield estimations of the mobility tables from 2013 onward.

More specifically, two hypotheses have been assumed to extrapolate the future trajectories of the origin and destination totals of the mobility tables. The hypothesis in Step 2 (Fig. 1), herein called as "reference scenario" (namely, Ref), assumes that the evolution of regional mobility is mainly triggered by its underlying trend. This hypothesis has been operationalized by applying a first order autoregressive procedure, according to which the totals of the mobility tables at time $t + 1$ only depend on those at time t.

The hypothesis in Step 3 (Fig. 1) relaxes the previous one by taking into account some additional information based on observed socioeconomic trends. More specifically, it assumes that the totals of outflows from residence zones reflect the trends in the labor market observed in the Piedmont provinces (indexed by p). From 2011 to 2016, say $\gamma_{t,p}$ from 2011 to 2016. An order 2 auto-regressive difference equation has then been econometrically estimated for each province to extrapo-

years more for population and one year more for firms would have led to an impairment of yields. Hence, to maintain the synchronization, 2014 would have been the last year at which both sources of information were available: adding one more year would have not improved the outcomes significantly.

late a future trend up to 2020, $\hat{\gamma}_{t+k,p}$ from 2016 onward: each zone within the a province has been given the province-specific growth rate as follows $\hat{D}_{j,p}(t+k) = \hat{D}_{j,p}(t+k-1) * \left(1 + \hat{\gamma}_{t+k,p}\right)$. In the following, this case will be referred to as "occupation scenario" (namely, Occ).

3.2 *An Indirect Method for Estimating the Distance Parameter*

A novel methodological feature in this study is the empirical estimation of the distance parameter in Eq. (1). In the current applications of Wilson's approach $\beta(t) > 0$ the deterrence function has been set as $f\left(d_{i,j}|\beta(t)\right) = \exp\left(-\beta(t)\, d_{i,j}\right)$ to account for the inverse or negative effect of the distance. For each time each t, the parameter value $\beta(t)$ has been estimated by means of an iterative procedure to fulfill the constraints

$$\sum_{i=1}^{186} T_{i,j}(t) = D_j(t), \quad \sum_{j=1}^{186} T_{i,j}(t) = O_i(t), \quad \sum_{i=1}^{186}\sum_{j=1}^{186} T_{i,j}(t) = N(t) \qquad (2)$$

associated with Eq. (1).

According to the maximum entropy principle and given the information associated with the system's constraints, the estimated flow values are those maximally consistent with the values for the observed or reconstructed totals of mobility tables. In this study the iterative procedure has been included in a Montecarlo simulation algorithm and used to estimate the values of $\beta(t)$ from 2001 to 2013 (Step 1 in Fig. 1). The main steps of the algorithm are the following: (i) K candidate values $\beta_k(t)$ are randomly drawn from a uniform distribution $U(0.5; 2.5)$; (ii) the k-th candidate is plugged in Eq. (1) and an *Iterative Proportional Fitting* procedure is used to adjust the matrix by rows and columns in order to fulfill constraints in Eq. (2) with an approximation error; (iii) the k-th approximation error is estimated and stored into an array; (iv) steps from (ii) are reiterated until all the candidate values are explored; (v) the value of $\beta_k^*(t)$ with the minimum error (i.e. best-performing) is selected and stored together with its error; (vi) all the steps from (i) are repeated M times. M replicates of the best-performing estimates $\beta_{k,m}^*$ are obtained to evaluate $\beta(t) = M^{-1} \sum_m^M \beta_{k,m}^*$ as the Monteacarlo estimate, together with a standard error and an empirical distribution for Bootstrap inference.

Notice that this procedure is applied at each t from 2001 to 2013. A time series $\{\beta(t) : t = 2001, \ldots, 2013\}$ is obtained and then used to calculate exogenously estimated values of $\beta(t)$ in the next stages of the implementation procedure.

Table 1 Impact of selected variables on the β time series: coefficient values and standard errors for Eq. (3)

Coefficient	Variable	Estimated value	Standard error
$\hat{a}$	Intercept	1.1938	1.8383
$\hat{b}$	β_{t-1}	−0.2699	0.2645
$\hat{c}$	Y_{t-1}	−0.0126	0.0053
$\hat{d}$	U_{t-1}	−0.0040	0.0011
$\hat{e}$	P_t	1.0039	0.5216

Source Estimates of IRES Piemonte on data from ISTAT and Prometeia

More specifically, first a correlation analysis has been carried out between the β values obtained in Step 1 and a set of socioeconomic variables, for which there exist credited scenario values from 2001 to 2020.[5] Estimation of β involves a regression equation, whose estimates are then used to extrapolate the β values in the 2014–2020 period as follows:

$$\beta_t = a + b\beta_{t-1} + cY_{t-1} + dU_{t-1} + eP_t + \varepsilon_t \tag{3}$$

where Y is the regional GDP, U is the share of population in search of first employment, P is the active population (in the 15–70 age bracket) and ε_t a residual term, with the usual statistical assumptions.

Results of Eq. (3) are presented in Table 1. Reminding that the parameter has a negative sign in the distance function ($-\beta\,(t)\,d_{i,j}$), the sign of the selected variables in Table 1 has to be reversed. For example, a growth of the share of population searching for a first employment, while it reduces the deterrence effect of the distance parameter, it positively affects the flows, $T_{i,j}\,(t)$.

In this application, the impact of distance parameter on the flow pattern is a combination of two opposite forces. The first, which has a sort of clustering effect, is produced by the active population: a growth of this variable, in fact, would produce a relatively denser flow pattern around the origin zones. All the other variables have an attenuating impact on the distance parameter and tend to favor a more dispersed pattern along with longer distance travels. It is worth noting that similar findings as for the sign of the socioeconomic variable have been observed in another empirical analysis of commuting in Greece (Polyzos et al. 2013).

To illustrate the results of this procedure Fig. 2 shows the β values obtained by applying the Wilson (classical) and econometric (exogenous) estimation approaches for the 2001–2025 period.

The variables' coefficients in Table 1 have then been used to extrapolate the values for $\hat{\beta}_{t+k}$ in the 2014–2025 period as follows:

[5]The regional population forecast is from ISTAT. The Piedmont socioeconomic scenario is from Prometeia.

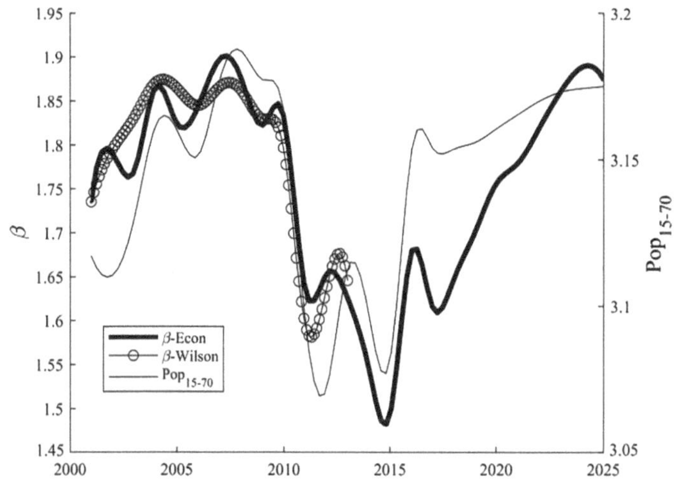

Fig. 2 Estimated values of the parameter, in the classical (Wilson) and econometric approaches, and population counts in the 2001–2025 period: discrete time series have been spline-interpolated

$$\hat{\beta}_{t+k} = \hat{a} + \hat{b}\hat{\beta}_{t+k-1} + \hat{c}\hat{Y}_{t+k-1} + \hat{d}\hat{U}_{t+k-1} + \hat{e}\hat{P}_{t+k} + \varepsilon_{t+k} \quad (4)$$

The trajectory for the active population has been added as a complementary information and reveals a coherent trend with that of the distance parameter. It can be observed that in the 2001–2013 period, when the estimation procedure can rely on existing socioeconomic data, the β values in the two approaches are very similar. This supports us in the decision to use the econometric estimated values for determining the mobility tables in stage 2 of the implementation procedure (see Fig. 1).

4 Results and Discussion

The model is very sensitive to the granularity of the spatial articulation. In this application the 186 zones have very different socioeconomic weights and spatial extensions: although they make it possible to draw meaningful accounts of commuters, the fact that they have very different population and firm sizes causes some disappointing results in the outputs produced by the models.

Figure 3 compares the observed (a) and estimated (b) flows in 2011. It can be seen that the model reproduces the overall regional pattern in a convenient way. However, it tends to over-estimate the zone outflows, e.g. there are 282 links in the observed mobility table compared with 402 in the estimated one. The phenomenon is even more evident in the zones around Turin, the main regional city, whose location, in Fig. 3, can be easily guessed by the denser cluster of flows.

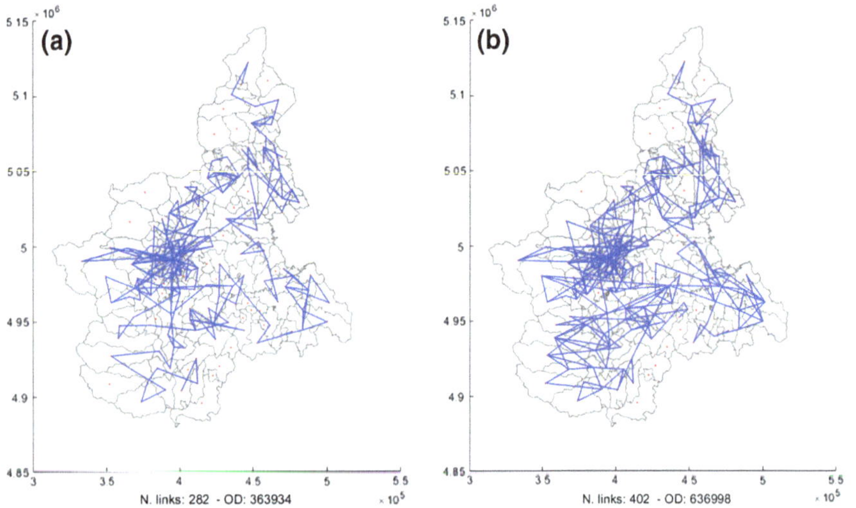

Fig. 3 Observed (a) and estimated (b) journeys-to work, 2011: connections with more than 500 commuters are represented

Notwithstanding this shortcoming will have to be conveniently addressed in future model applications, it does not seem to be a too severe limitation in the light of the goals of this study. In addition, as discussed in Landini and Occelli (2017), it is worth mentioning that in both the calibration years, 2001 and 2011, the fitting of the estimated totals of the mobility tables to the observed values is very satisfactory.

Figure 4 summarizes the regional observed and estimated trends of population, firms and mobility over the whole 2001–2020 period. From the time series three phases can be isolated. In the earlier, from 2001 to 2007, all the variables have a relatively smooth and increasing trend, which is relatively steeper for mobility and firms than for population.

The second period, from 2007 to 2013 is characterized by some oscillations in the trends of the variables. Whereas some of them may well depend on the different population metrics (census data and population registry counts) used in the model estimation for this time period, more substantial impacts cannot be excluded, as a result of the turmoil caused by the late economic crisis. This is particularly evident for the number of firms which in 2007 reversed its growth trend. This however did not seem to significantly affect the total number of journeys-to-works, which continued to growth although at decreasing rate.

The last period from 2014 to 2020, is the one covered by simulation. Here the population trend is based on the ISTAT values estimated in the demographic scenario. The mobility trends are those resulting from (Ref) and occupational (Occ) scenarios, described above, both of which apply econometric estimates (see Fig. 2) of the parameter β. In both scenarios we observe a slight increase in the number of journeys-to-work at 2020 compared with that observed in 2011: 0.9 and 1.7% in the reference

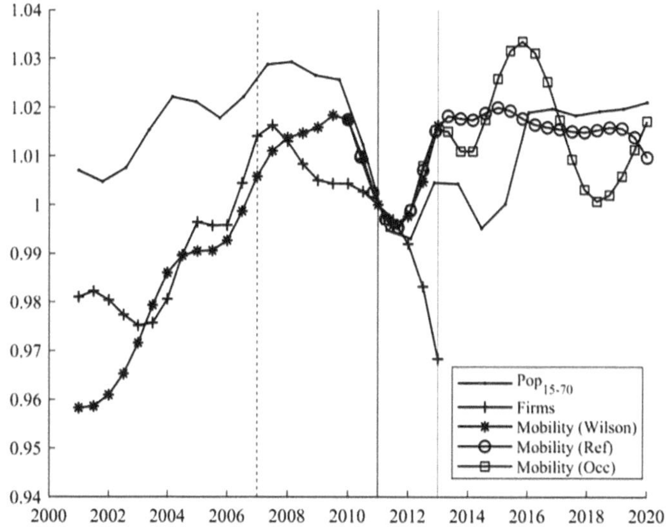

Fig. 4 Regional trends of the journeys-to-work in the reference and occupation scenarios, of population and firms, between 2001 and 2020. Index numbers (base = 2011, vertical solid line). Observed values before 2013 (dotted vertical line). Simulated values after 2013

and occupation scenario, respectively. The trend of the former however is relatively flat, while that of the latter oscillates: this reflects the impact of the input used in the simulation, which take into account the differences in the growth rate of the employment observed in the provinces from 2011 to 2016.

The regional mobility patterns at 2020 resulting from the Ref (a) and Occ (b) simulated scenarios are presented in the maps of Fig. 5 which show the connections having more than 500 commuters. The patterns in the two scenarios are very similar and much like the one in the reference year 2011 (Fig. 3b).

Given the meso-level observation lens adopted in this study, the result is not surprising. First, because, if compared with the changes that took place in earlier periods (Occelli 2006) those affecting mobility socioeconomic determinants in recent years are quantitatively more modest. Second, because much of the transformations occurring in Piedmont socioeconomic fabric have a more qualitative nature that cannot be properly captured in the current version of the model formulation which deals with aggregated variables.

Some changes are however apparent and can be appreciated by having a closer look at the 2011–2020 variations of the flows in the residence and employment zones and by analyzing the average distance travelled. Both these arguments offer valuable contributions in the preparatory studies for the new Regional Transport and Mobility Plan (Regione Piemonte 2016) and are an interesting topic for future investigation.

As for the spatial variations of flows between 2011 and 2020, the maps in Fig. 6 present the results for the reference and occupation scenarios.

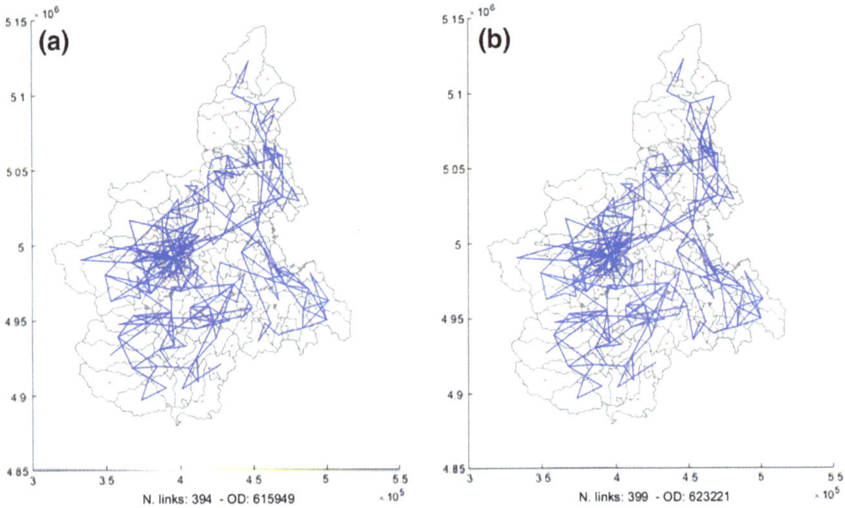

Fig. 5 Journeys-to work estimates at 2020 in the reference (a) and occupation (b) scenario: connections with more than 500 commuters are represented

First, it is worth emphasizing that the fluctuations we observed in the mobility trend associated with the occupation scenario over the 2013–2020 period, see Fig. 4, are also accompanied by a greater variability in the spatial changes of flows, especially in the residence zones (see Fig. 6a.2). More specifically, in the occupation scenario almost half of the (186) residence zones have a relatively higher increase (greater than 2%) in outflows compared with only one out of three in the reference scenario (see Fig. 6a.1). In the occupation scenario, one zone out of two remains almost unchanged, against 68% in the reference scenario. In both scenarios, few residence zones reduce their outflows between 2011 and 2020.

Second, when considering the employment zones, the differences between the scenarios are smaller. In both of them, more than half of the zones have a significant inflow increase, and slightly less than 40% do not vary appreciatively. The number of employment zones decreasing their inflow, however, is higher in the occupation scenario (13%) than in the reference one (7%).

Finally, it is worth noting, that the growth of flows in employment zones occurs in a relatively greater number of zones in the reference scenario. Given the assumption of this scenario, this is hardly surprising as a relatively higher growth of (in)flows in employment zones was also revealed when analyzing the spatial variations of journeys-to-work between 2001 and 2011 (Occelli and Sciullo 2015a).

Turning the attention to the distance issue, Fig. 7 provides an overview of the values of the average distance associated with the journeys-to-work in the 2001–2020 period. The distance trajectory is complemented by that of the values of the β parameter which has a symmetric path to that of distance: the shorter the average travelled

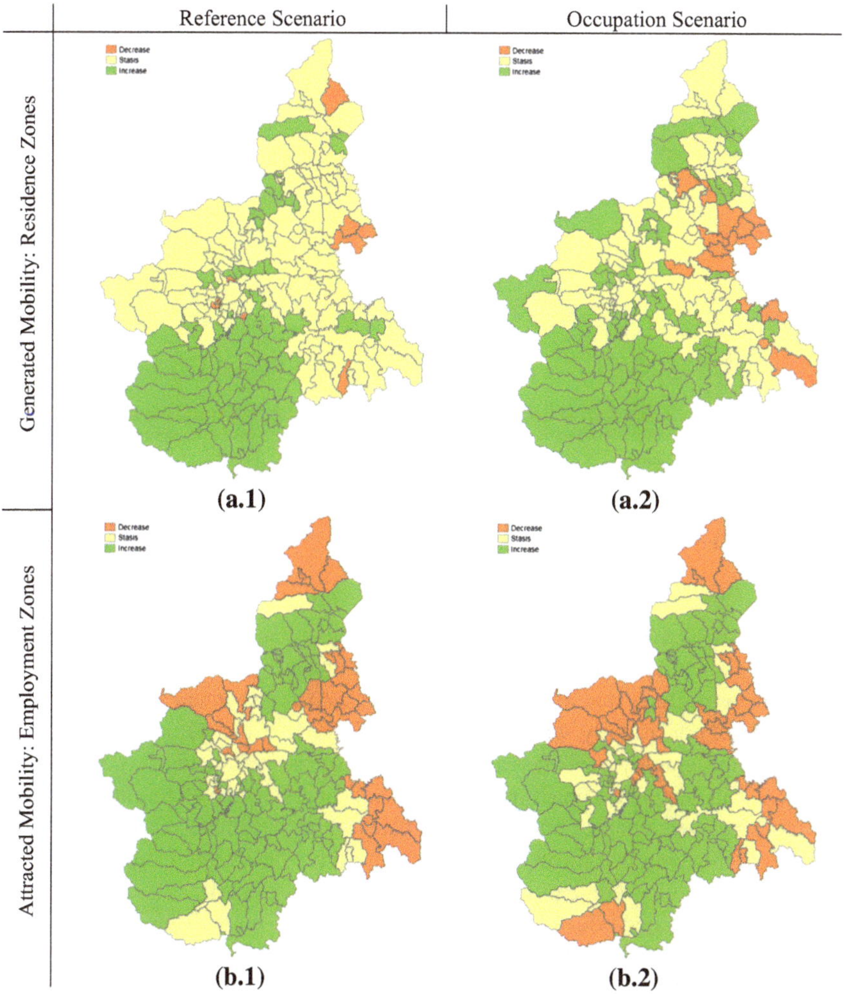

Fig. 6 2011–2020 variations in the total flows in the residence and employment zones in the reference scenarios (a.1 residence zone, b.1 employment zone) and occupation scenario (a.2 residence, b.2 employment zone). Stasis: variation between −2 and +2%. Decrease: variation lower than a −2%. Increase: variation greater than +2%

distance the higher the parameter β. This result is coherent with the assumptions of the underlying spatial interaction modeling.

Since $f\left(-\beta d_{i,j}\right)$ represents the deterrence function, and the distance $d_{i,j}$ is assumed as a proxy of travel-costs, it follows that increasing β values either reduce the negative impact of travel costs on commuting.

Figure 7 also shows the trajectory of the share of population in search of first employment U. Comparing the trajectories of the average distance (either in the

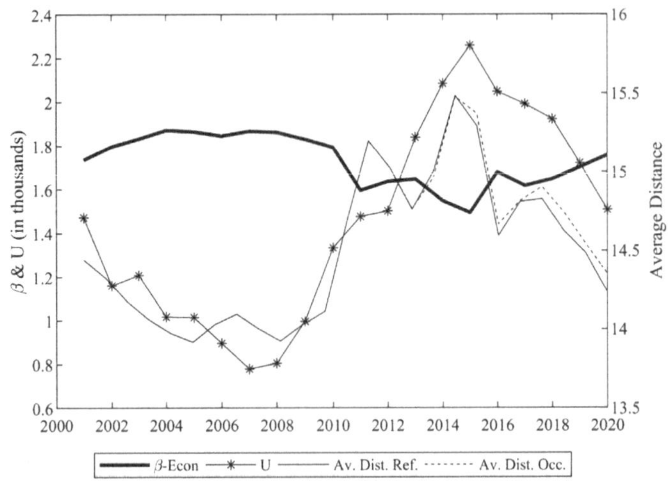

Fig. 7 Trajectories of the β parameter, the average distance (km) and the share of population in search of first employment (U, in thousands) in the 2001–2020 period

Ref and in the Occ scenarios) with that of the components of the (econometrically estimated and extrapolated) time series of the parameter β, reveals that the trend of the parameter follows that of people in search of first employment (U).

Therefore, consistently with the results in Table 1, if the number of people in search of first employment increases, the value of the parameter β reduces, thus attenuating its negative impact on travel-costs: as a result longer journeys-to-work might be expected. On the contrary, if the share diminishes in the short run, this will increase β, meaning that travel-costs will exert a more deterrent effect to commute, thus decreasing the average travel distance in journeys-to-work.

The distance trajectories plotted in Fig. 7 reveal that, notwithstanding the oscillations between 2011 and 2020, the values at the end of the period are likely to be lower than that in 2011. The observed reduction is slightly greater in the reference scenario than in the occupation scenario.

To better appreciate the results at sub regional level, the graph in Fig. 8 plots the 2001–2020 variations of the total number of journeys-to-work and of the total distance travelled (given by the sum of kilometers travelled in each journey-to-work) in a selection of regional sub-areas: the province capital cities and the rest of their province area, except for the metropolitan area where an aggregation of the more populated zones is distinguished.

The following remarks can be put forward. First, for Piedmont as a whole, the small increase in the total number of journeys-to-work (less than 2%) is accompanied by a slight decrease in the distance travelled. In the light of the main research question of this study, then, one might say that, for the region as a whole, a tendency may exist towards a more sustainable commuting pattern. Results at sub-regional level are more mixed.

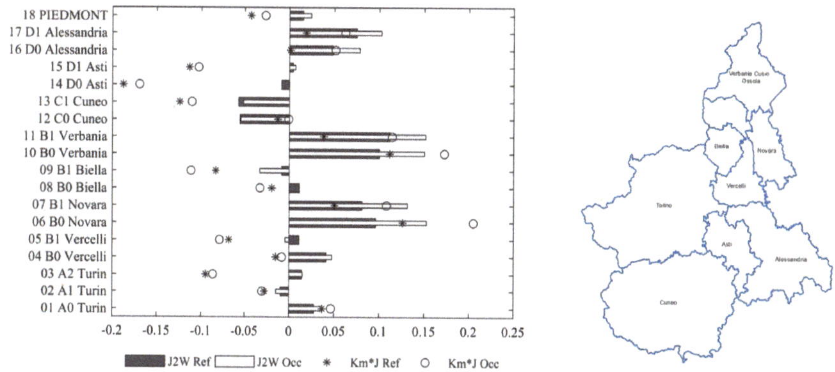

Fig. 8 2011–2020 Variations of the journeys-to-work (J2W) and of the distance travelled (Km*J) in the Reference (Ref) and Occupation (Occ) scenarios, in the province capital cities (those labeled with a 0) and in their outer area (those with 1). (*)The variations refer to the residence zones. A: Turin metropolitan province; A0 is the Turin city, the regional capital; A1 includes the zones closer to Turin and the main cities; A2 is the rest of the Turin province. B area: North-East it consists of the VCO, Biella, Vercelli and Novara provinces; C area: South-West: Cuneo province; D area. South-East: it consists of the Asti and Alessandria provinces

A reduction in both the journeys-to-work and distance travelled is observed in the South-Western area (Cuneo) of the region. The opposite occurs in most of the provinces in the eastern part of the region (areas B and D), where journeys-to-work and distance travelled increase. Some exceptions are observed in the Biella, Vercelli and Asti provinces, where the variations of the journeys-to-work and distance have opposite sign.

Not unexpectedly, a more differentiated combination is revealed in the Turin metropolitan area (A) where, in the last decade, a number of activity locations were created in the zones surrounding the city as a result of a relocation process from the core area and new settlements. In the city of Turin, as in most of the other province capital cities, there is an increase in both journeys-to-work and distance travelled. The opposite occurs in the area closer to the regional capital (A1).

In the outer area (A2) a positive variation in the journeys-to-work is accompanied by a reduction in the distance travelled.

5 Concluding Remarks

This study has developed a prototypical analytic tool for estimating journeys-to-work at sub-regional level, linking a backward and forward analytic strategy.

From an operational point of view, the current version of the model allows practitioners to take advantage of the available data at municipal and regional level to obtain reasonable metrics for the mobility determinants. The model can be used retrospectively to better understand the relationships between socioeconomic changes and impacts on journeys-to-work. It can also be applied in a prospective way, to explore the likely impacts of plausibly expected economic and demographic trends.

In particular, for the Piedmont region the investigation of elementary but plausible socioeconomic scenarios revealed that in spite of an increase in the number of journeys-to-work, the amount of travelled kilometers reduced. This an interesting finding for the sustainability debate in Piedmont. The evidence produced by the model application suggests in fact that, as for the residence-work functional relationships, the spatial organization of activities in the region is becoming more efficient and likely to accrue to a more sustainable mobility. The model outputs also show, however, that a variability exists across the sub- regional areas and that in some of them (those located in the Eastern part of the region) very small or no improvements at all are produced, at least, in terms of the total travelled kilometers.

Although limited in scope, these insights can be a main stimulus to engage into a more comprehensive assessment of the performance and sustainability of the daily commuting areas in the region.

Research efforts will have to be undertaken and address the challenges of long-term sustainable mobility (Bernardino et al. 2015; Dovey Fishmant 2012; Mitchell et al. 2010). They will require to fully address the wider impacts of the transport, such as reduction in greenhouse gasses, more livable and healthy settlements, and the ICT impact on activity patterns (Banister 2008; Debyseru 2014; Litman 2006; Litman and Burwell 2006; European Parliament 2014; Lord et al. 2015; Papa et al. 2015). Investigations will also have to acknowledge that nowadays the future development of mobility is uncertain both in scientific and non-scientific literature (Lyons and Davidson 2016; Gillis et al. 2016; Shuldiner and Shuldiner 2013). In this respect, as pointed out in the recent literature. modeling activity has an increasingly important role to play in supporting transport policy practices (De Bruijne et al. 2010; Givoni 2014; Timms et al. 2014).

Arguably, more refined formulations of this model version are expected to be among the likely candidates for this undertaking. As for the model's improvements, in particular, we can mention three directions.

The first relates to the reconstruction stage in the analytic strategy. It recognizes the possibility to update the mobility tables by exploiting the information about the place of residence and the place of work in the institutional registry which records people's job movements.

The second advance concerns a partial reshape of the model in order to improve the spatial adherence of the flow estimates. In particular, a more appropriate distance metrics is to be developed, which accounts not only for the cost of travel and congestion but also better reflects more intangible effects associated with the social and built environments (Feuillet et al. 2015; Graham and Patsy 1999). Furthermore, given the organization of the regional city system it might be advisable to have a two-tiered perspective in which model applications at the lower level distinguish the main commuting areas in the local sub regional areas and their relationships with the areas outside the region.

A final research development, which is broader in scope, relates to the formulation of socioeconomic scenarios to be used as a backcloth in the estimation of future mobility tables. This an important endeavor worth being undertaken for its own, and for the Piedmont region a few of them are already in use (see in particular the one provided by the Prometeia). The articulation of these scenarios at sub-regional level, though, is an open research challenge which is increasingly needed (Viguié et al. 2014), above all if collectivities bear a responsibility in engaging themselves to more sustainable mobility behaviors. In a policy context, smart enough to appreciate the full range of model potentials, the model applications can, themselves, be a spur to develop spatially articulated scenarios that may better reflect the socioeconomic evolution of local areas.

References

Acheampong RA, Silva AE (2015) Land use-transport interaction modeling: a review of the literature and future research directions. J Transp Land Use 8(3):11–38

Agenzia della Mobilità Piemontese (2013) Rapporti IMQ 2013. http://mtm.torino.it/it/dati-statistiche/indagine-imq-2013/rapporti-imq-2013

Arentze T, Timmermans H (2000) ALBATROSS: A learning based transportation oriented simulation system. Technische Universiteit Eindhoven, Eindhoven

Banister D (2008) The sustainable mobility paradigm. Transp Policy 15:73–80

Bernardino J, Aggelakakis A, Reichenbach M, Vieira J, Boile M, Schippl J, Christidis P, Papanikolaou A, Condeco A, Garcia H, Krail H (2015) Transport demand evolution in Europe—factors of change, scenarios and challenges. Eur J Futures Res 3(13):1–13

De Bruijne M, van de Riet O, de Haan A, Koppenjan J (2010) Dealing with dilemma's: how can experiments contribute to a more sustainable mobility system? Eur J Transp Infrastruct Res 3:274–289

Debyseru A (2014) Urban mobility: shifting towards sustainable transport systems. European Parliament Research Service. http://www.europarl.europa.eu/thinktank/en/document.html?reference=EPRS_IDA(2014)538224

Dovey Fishmant T (2012). Digital-age transportation: the future of urban mobility. Deloitte University Press. http://www.enterrasolutions.com/media/docs/2013/07/Digital-Age-TRANSPORTATION.pdf

European Parliament (2014) Urban mobility. http://www.europarl.europa.eu/RegData/etudes/IDAN/2014/538224/EPRS_IDA(2014)538224_REV1_EN.pdf

Feuillet T, Charreire H, Menai M, Salze P, Simon C, Dugas J et al (2015) Spatial heterogeneity of the relationships between environmental characteristics and active commuting: towards a locally varying social ecological mode. Int J Health Geogr. https://doi.org/10.1186/s12942-015-0002-z

Gillis D, Semanjski I, Lauwers D (2016) How to monitor sustainable mobility in cities? Literature review in the frame of creating a set of sustainable mobility indicators. Sustainability 8(12):29

Givoni M (2014) Addressing transport policy challenges through policy packaging. Transp Res 60:1–8

Givoni M, Shiftan Y, Beyazit E, Ishaq R, Tzu O (2012) The use of state-of-the-art models by policy makers to address global energy and environment challenges: the case of transport policy (Working Paper N° 1060). Retrieved March 2017, from http://www.tsu.ox.ac.uk/pubs/1060-givoni-beyazit.pdf

Gössling S (2017) ICT and transport behaviour: a conceptual review. Int J Sustain Transp. https://doi.org/10.1080/15568318.2017.1338318

Graham S, Patsy H (1999) Relational concepts of space and place: issues for planning theory and practice. Eur Plan Stud 7(5):623–646
Hawelka B, Sitko I, Beinat E, Sobolevsky S, Kazakopoulos P, Ratti C (2014) Geo-located Twitter as proxy for global mobility patterns. Cartogr Geogr Inf Sci 41(3):260–271
Kung KS, Greco K, Sobolevsky S, Ratti C (2014) Exploring universal patterns in human home-work commuting from mobile phone data. PLoS ONE 9(6):e96180. https://doi.org/10.1371/journal.pone.0096180
Landini S, Occelli S (2016a) La mobilità quotidiana e per cura negli ambiti di integrazione territoriale del Piemonte. http://www.ires.piemonte.it/reset/RESET2016_A4/RESET2016_cap3.1.pdf
Landini S, Occelli S (2016b) Mobility flow estimates at sub-regional level: an application to piedmont. In: Colombo G, Lombardi P, Mondini G (eds) Proceedings of the INPUT 2016 9th international conference on innovation in urban and regional planning, Turin, pp 533–537
Landini S, Occelli S (2017) Stima e simulazione delle tavole di mobilità casa-lavoro in Piemonte: esplorazione di scenari socioeconomici. Paper presented at the AISRE Conference, Cagliari 20–22 September. https://www.aisre.it/images/aisre/5965dac732e174.97480154/Landini_Occelli_Aisre2017.pdf
Litman T (2006) Changing travel demand: implications for transport planning. ITE J 76(9):27–33
Litman T, Burwell T (2006) Issues in sustainable transportation. Int J Glob Environ Issues 6(4):331–347
Lord S, Negron-Poblete P, Juan Torres J (eds) (2015) Mobilité et Exclusion, Quelle relations? Les Presses de l'Université Laval, Canada
Lyons G, Davidson C (2016) Guidance for transport planning and policymaking in the face of an uncertain future. Transp Res Part A: Policy Pract 88:104–116
Mitchell WJ, Borroni-Bird CE, Burns LD (2010) Reinventing the automobile, personal urban mobility for the 21st century. MIT, Cambridge
Neill G, Bright J, Hale SJ (2017) Estimating local commuting patterns from geolocated Twitter data. EPJ Data Sci 6:24. https://doi.org/10.1140/epjds/s13688-017-0120-x
Occelli S (2006) La mobilità in Piemonte nei primi anni del 2000, Quaderni di Ricerca, 110, Ires, Torino. www.digibess.it/fedora/.../openbess_TO082-01680.pdf
Occelli S, Sciullo A (2015a) Uno sguardo ai cambiamenti della mobilità (quotidiana) in Piemonte tra passato e futuro, http://www.ires.piemonte.it/reset/RESET2014_A4/RESET2014_cap3.1.pdf
Papa R, Galderisi A, Vigo Majello MC, Saretta E (2015) Smart and resilient cities. A systemic approach for developing cross-sectoral strategies in the face of climate change. TeMA 8(1):19–49. https://doi.org/10.6092/1970-9870/2883
Polyzos S, Tsiotas D, Minetos D (2013) Determining the driving factors of commuting: an empirical analysis from Greece. J Eng Sci Technol Rev 6(3):46–55. http://www.prd.uth.gr/uploads/publications/2013/7f20e656f4da760ae45938e0599f058a4b2aabd8.pdf
Regione Piemonte (2016) Piano Regionale della Mobilità e dei Trasporti. http://www.regione.piemonte.it/trasporti/piano_regionale.htm
Shuldiner A, Shuldiner PW (2013) The measure of all things: reflections on changing conceptions of the individual in travel demand modelling. Transportation 40:1117–1134
Timms P, Tight M, Watling D (2014) Imagineering mobility: constructing utopias for future urban transport. Environ Plann A 46(1):78–93
Viguié V, Hallegatte S, Rozenberg J (2014) Downscaling long term socio-economic scenarios at city scale: a case study on Paris. Technol Forecast Soc Chang 87:305–324
Wilson A (1967) A statistical theory of spatial distribution models. Transp Res 1:253–269
Wilson A (1968) The use of entropy maximising models in the theory of trip distribution-mode split and route split. Centre for Environmental Studies, London
Wilson AG (1971) A family of spatial interaction models, and associated developments. Environ Plann 3:1–32

Wilson AG (1987) Transport, location and spatial systems: planning with spatial interaction models. In: Bertuglia CS, Leonardi G, Occelli S, Rabino GA, Tadei R, Wislon AG (eds) Urban systems. Contemporary approaches to modelling. Routledge, pp 337–426

Wilson AG (1998) Land-use/transport interaction models: past and future. J Transp Econ Policy 32(1):3–26

Milan: The Configuration of a Metropolis

Valerio Cutini, Denise Farese and Giovanni Rabino

Abstract The configurational approach to the study of urban settlements has been largely debated in the last years, gaining a wide audience and a full recognition as a useful and reliable tool for the analysis and the comprehension of the inner geography of cities. Milan, whose diachronic genesis was never before analysed by means of a configurational approach, appears a peculiar and outstanding case study, as its complex structure, actually difficult to read, has undergone significant changes over time; a broad and varied sequence of projects, developments and planning choices have materialized and interiorized within the whole grid of the settlement, from its inner nucleus up to the external edges. Yet, the purpose of this research goes even beyond the mere understanding of the diachronic genesis of the inner geography of Milan, anything but easy, due to the dimension of the settlement and to its internal complexity. The configurational analysis is here also shown as an excellent planning tool, suitable for supporting the decision-making process of urban actions and projects, with a particular reference to those involved with to the theme of accessibility, so as to usefully orient the present and future planning of public spaces. The results show that the space syntax approach actually enhances the transparency of strategic decisions concerning the design and planning of urban spaces, making to emerge and highlighting their likely effects on a wide range of aspects and phenomena.

V. Cutini (✉)
DESTEC, University of Pisa, Pisa, Italy
e-mail: valerio.cutini@ing.unipi.it

D. Farese
Politecnico of Milan, Milan, Italy

G. Rabino
DASTU, Politecnico of Milan, Milan, Italy

R. Papa et al. (eds.), *Smart Planning: Sustainability and Mobility in the Age of Change*, Green Energy and Technology,
https://doi.org/10.1007/978-3-319-77682-8_20

1 Introduction

The configurational approach to the study of urban settlements has been largely debated in the last years, gaining a wide audience and a full recognition as a useful tool of urban space analysis. An amount of studies and researches have highlighted the innovative capacity of the method and demonstrated its reliability, as well as proving its potential in complementing more common and traditional methods of spatial analysis.

The inner geography of Milan, whose diachronic genesis was never before analysed by means of a configurational approach, appears a peculiar and outstanding case study, as its complex structure, actually difficult to read, has undergone significant changes over time; a broad and varied sequence of projects, developments and planning choices have materialized and interiorized within the whole grid of the settlement, from its inner nucleus up to the external edges.

But this is not merely an exercise of urban reading, aimed at understanding the diachronic analysis of the system, reconstructing the evolution and shifting of the levels of urban centrality over time, and interpreting the distribution of activities and centrality levels within the city of Milan; what moreover is anything but easy, nor simple, due to the dimension of the settlement and to its internal complexity. In this study, the configurational analysis is also shown as an excellent planning tool, suitable for supporting the decision-making process of urban actions and projects, with a particular reference to those involved with to the theme of accessibility, usefully orienting the present and future planning of public spaces. The results show that the space syntax approach actually enhances the capability of strategic decisions concerning the design and planning of urban spaces.

2 Methodology

The configurational approach to the analysis of urban settlements was introduced by Bill Hillier in the late 1970s to early 1980s under the name of space syntax, as a tool to help urban planners simulate the likely social effects of their designs; since then, the theory and the techniques have been advanced and developed by a large community of researchers all over the world.

The configurational analysis focuses on space, as it results defined by the blocks and building aligned along the streets that compose the grid of the urban settlement, which can be broken down into components, suitable to be analyzed as a network of discrete elements, to be represented by graphs that describe the relationships of each of those spaces with all the others. The general idea is that the grid of the urban paths is assumed as the primary element in determining the patterns of socio-economic behaviour of humans (Hillier 1996), which aims at taking benefit of the flows of through movement connecting its spaces; within such movement, a part, named natural movement, is assumed as merely depending on the grid configuration and not affected by the presence of the activities along its paths (Hillier et al. 1993). The analysis is expected to provide each element of the system with values that correspond to a set of numeric parameters, named configurational indices. Such indices are called configurational, in that they are calculated taking into account only the relationships between each element and all the others, thus excluding any interest in its geometric or morphologic features. Such aspect suggests to represent the spatial system by means of a graph, whose single node corresponds to a spatial element and each edge corresponds to a direct connection with another element. Such representation highlights that the computation of the connections and mutual depth between elements (or nodes) has to adopt a topologic approach, disregarding the metric features of the settlement.

Among the several configurational techniques so far introduced, distinguishing each other for the way of discretizing the urban space, the so-called axial analysis and visibility graph analysis are by far the most used. The research this paper is concerned with the use of axial analysis, whose system, named axial map, is constructed reducing the space of the settlement into the set of the longest and fewest accessible lines, called axial lines, that cover the whole grid and connect each of its convex spaces. The other technique distinguishes for reducing the grid into a system of vertices, called visibility graph, mutually related each other by visual interconnection.

The configurational indices extracted from the analysis of the grid have been demonstrated suitable for narrowly reproducing some significant urban phenomena or aspects (more or less directly) concerned with the presence of movement. A first parameter is the connectivity value, reproducing the number of elements (either lines or vertices) directly connected with the observed one, and corresponding to the parameter that in graph theory is usually defined as grade. Among the most significant of the other parameters, the integration value (corresponding to the closeness index in graph theory) is defined as the mean depth of the observed spatial element from all the others within the grid, that is:

$$D_M(x) = 1/n \sum_{i=1}^{n} D(x, i)$$

Also expressed (and calculated) by the mean number of turns along the shortest path connecting the element to all the others, the integration value was strongly demonstrated to correspond to the distribution of retail activities. The set of the elements provided with the highest integration values compose the so-called integration core, which can be used as an effective spatial indicator of urban centrality. The same analysis can also be declined in an angular version, called angular analysis, which takes into account, in addition to the number of intersections, also their respective mutual angle of incidence, as follows (Turner 2007):

$$D_{\theta M}(x) = 1/n \sum_{i=1}^{n} D_{\theta}(x, i)$$

Another significant configurational index is the choice value, corresponding to the parameter that in graph theory is commonly defined as betweenness, expressing the frequency of a spatial element (again, either line or vertex) along the shortest paths connecting all the couples of other elements, what in analytical terms can be written as follows (Turner 2007):

$$\mathrm{Ch_x} = \sum\nolimits_{i=1}^{n} \sum\nolimits_{j=1}^{n} \sigma\,(\mathrm{i, x, y}) \quad \text{being } \mathrm{i} \neq \mathrm{x} \neq \mathrm{y}$$

Both those parameters can be usefully expressed in relative terms, normalized with respect to the number of spatial elements, that is the dimension of the grid, and both them can also be calculated taking into account all the elements of the grid (obtaining global integration and global choice, respectively) as well as only the neighboring elements, so as to obtain the local integration and choice value. Observing the correspondence of global and local indices allows appraise the relationship between the whole urban system and the single parts that compose it. In particular, the R^2 determination coefficient of the correlation between global integration and local integration is commonly called synergy coefficient; a high value, approaching 1, is expected standing for a narrow anchorage of the local subsystems to the global one, in that the most integrated elements of the grid are also prime local integrators, that is central elements at a local scale.

An amount of researches actually demonstrates that choice value narrowly corresponds to the distribution of movement (either vehicular movement and pedestrian one) (among others: Hillier and Iida 2005).

Connectivity value, integration value and choice value can all be regarded as indices of centrality, assuming the notion of urban centrality under different meanings. Spatial elements provided with high connectivity values will be expected to be a prominent local pivot; high values of integration will stand for highly attractive areas with reference to the location of economic activities, to be regarded as an ideal destination of movement as meanly easily accessible from any part of the settlement; high values of choice will then characterize spatial elements interested by dense movement, and hence attractive locations for activities aimed at intercept through movement flows.

3 Diachronic Analysis of the City

In order to carry on the analysis of settlement genesis of Milan by means of a configurational approach, three different dates have been chosen, suitable for representing the development of spatial expansion of the system over time: 1704, 1884, 2014. The first map is a representation of the city of Milan at the time of the Austrian domination (Fig. 1); the second one shows engineer Beruto's town plan in 1884 (Fig. 2); the last map is representative of the city of Milan nowadays (Fig. 3).

Every map has been digitalized and analysed by space syntax; interesting outputs have come from the observation of integration core, in that it allows determine the evolution of urban centralities over time (Cutini 2001). Centrality is a key factor

Fig. 1 Integration core of Milan at 1704

Fig. 2 Integration core of Milan at 1884

in shaping both urban space and urban life; the integration core, that should reflect centrality, may present various features as Bill Hillier suggests: "The core takes a form typical of many types of town or urban area, which we call deformed wheel. A small semi-grid of lines in the heart of the settlement (the hub) is linked in several directions (the spokes) to lines on the periphery of the settlement (the rim), which also form part of the core" (Hillier 1996).

The city of Milan in 1704 is extremely restricted into a medieval form within a circle of about three kilometres; the integration core seems to be linear and developed within adjacent and central streets where historically commercial activities were concentrated. The urban expansion in the late 1800s led to a decentralization of the centrality due to the first urbanization expansion over the boundary walls; in the second map, the integration core appears composed of the main axes that connect the peripheral zones of the city to the centre. This spatial analysis reflects the aspiration and spatial strategy expressed by Cesare Beruto in his plan of 1884 of developing a radial structure inside the city of Milan.

Since the early 1900s, the city of Milan has undergone strong changes over time in terms of urban development, population growth and industrial progress, gaining an area of about 182 kilometres. The main characteristic highlighted by the spatial analysis of the city in Milan in 2014 is the multiple centrality structure due to the new urban approach based on a functional mix (Oliva 2002). Mixed-use development aims to reduce the need for travel, to increase walkability and generate street-life intensity; it blends residential, commercial, cultural and industrial uses, where functions are physically and functionally integrated. Differently from the previous maps, in the last one several centralities turn out as the image (Fig. 3) shows.

Integration Cores:

1. Duomo
2. Brera-Garibaldi
3. Centrale
4. Porta Romana
5. Washington-Pagano
6. Sarpi
7. B. Aires-XXII Marzo
8. Ticinese

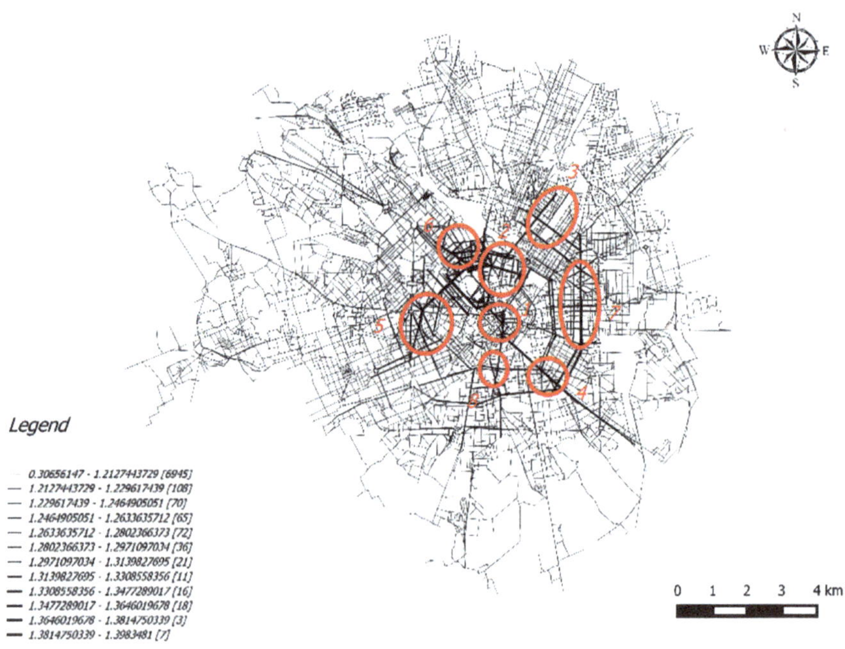

Fig. 3 Integration cores of Milan at 2014

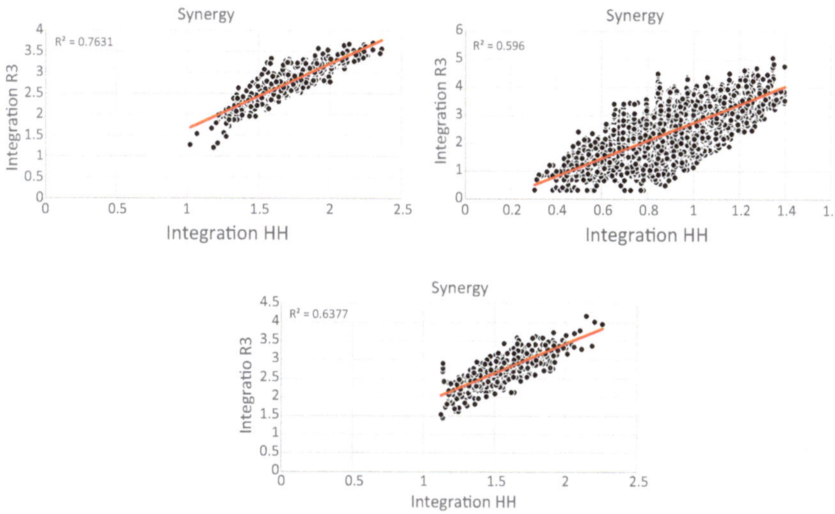

Fig. 4 Synergy coefficient of Milan at 1704, 1884, 2014 (from left to right)

The results above incentive to explore a second-level configurational index, the synergy value mentioned above, that is the coefficient R^2 of the correlation between local integration versus the global one. It measures the degree to which the internal structure of an area relates to the larger-scale system in which it is embedded. The figures represent the trend over time of the synergy coefficient, referred to the three maps, showing a progressive weakening. The synergy coefficient of Milan in 1704 is 0.763, what appears suggesting a quite resilient urban settlement, whose labyrinthian configuration provides various alternatives of movement from one site to another. As a consequence of the aspiration of Cesare Beruto to develop a radial structure, several axes gain higher relevance in movement distribution leading to a less flexible urban structure ($R^2 = 0.638$). In the last map, the synergy coefficient decreases as a result of the sprawl of the city of Milan and the growing of the metropolis (Fig. 4).

4 Correspondence to the Reality

In order to face urban issues by means of a configurational approach, it ought first to be verified if Space Syntax is well suited for the city of Milan. In general, since integration is expected to correspond to the actual distribution of retail activities, its value has been compared to the actual activities density, taking a sample of 30 streets, homogenously distributed and provided with a wide range of integration values. The correlation appears positive versus global integration as well as versus local integration (Fig. 5), computed taking account of the lines within a circle of radius 3. In order to filter these outcomes from local factors, the sample has been

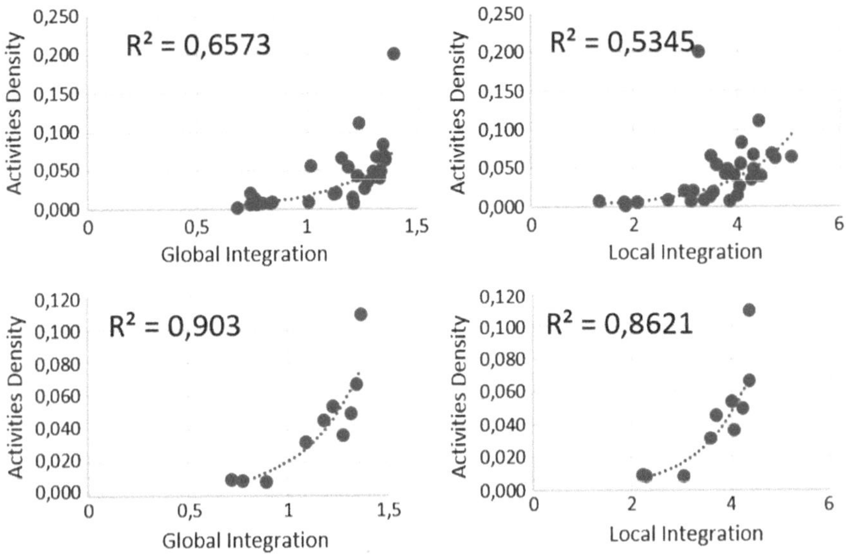

Fig. 5 Correlation between global integration and activities density, in the whole sample (1st row) and dividing the sample into 10 groups (2nd row)

divided into 10 groups of 3 elements. As a result, the correlation, as it is shown in Fig. 5, appears narrow and exponential.

5 Applications: Results and Discussion

The verification results so far briefly presented allows the provision of a positive overall assurance as to the suitability of this approach for the city of Milan; on such basis, the configurational analysis has been used to face several different issues of urban planning, especially referred to different kinds of accessibility. In fact, space syntax is capable to analyse pedestrian, cycle and vehicle movement networks, so as to narrowly reproduce their respective flows.

As the literature, this paper emphasizes the role of integration value as a key factor in the pedestrian movement across the city. The pedestrian movement is the mobility model that has the capacity of maintaining the most direct relationship and interaction with the city; consequently, there is a necessity of considering the pedestrian accessibility degree within urban settlement. The spatial configuration plays a primitive or principal role for the pedestrian mobility (Hillier et al. 1993) and affects to pedestrians when they have to take the decision about what route they select for their trips. Therefore, the spatial configuration could encourage or discourage the election of a route on which pedestrians can reach their destinations (Handy and Niemeier 1997). This effect on pedestrian mobility is based on the assumption of the

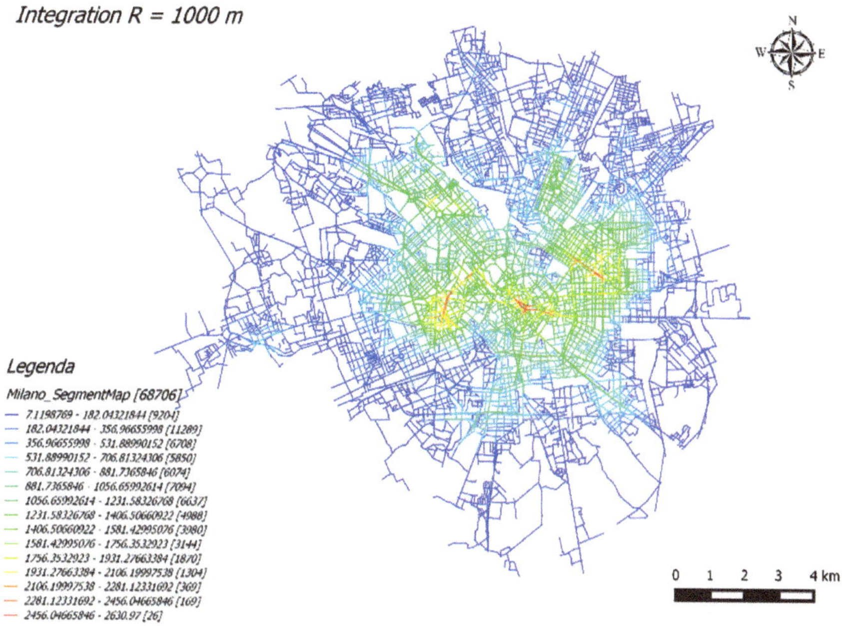

Fig. 6 Integration map of pedestrian accessibility

existence of natural movement, here mentioned above, that is the proportion of urban pedestrian movement determined by the grid configuration (Hillier et al. 1993).

Pedestrian map, shown below, is produced by angular segment analysis based on the radius set to 1000 meters, which corresponds to more or less 13 min, on foot. The pedestrian movement appears mainly concentrated within the central area, where integration cores are indicated to be by space syntax analysis, especially within 'Centrale', 'Duomo' and 'Washington-Pagano' (Fig. 6).

In Fig. 7, the map of cycling accessibility is represented: the accessibility is computed by angular segment analysis and the radius is set to 5000 meters, which corresponds to more or less 30 min by bike (Karlstrom and Mattsson 2009). From the integration map (Fig. 7), it is possible to observe the most accessible areas within a radius of 5000 meters, as well the most frequently visited that can be identified from the choice map (Fig. 8). Using these two maps, the most attractive areas for cyclists can be identified and this information might be useful in a possible decision-making process on localizations of public services aimed at encouraging sustainable mobility.

The third kind of accessibility that was analysed within the city of Milan is the vehicular accessibility. It is computed by axial analysis with radius n, in order to take into account the user's preference for straight street (axes). Considering space syntax capable to offer the possibility to test different strategic guidelines in urban planning, two diverse applications related to vehicle accessibility are described below.

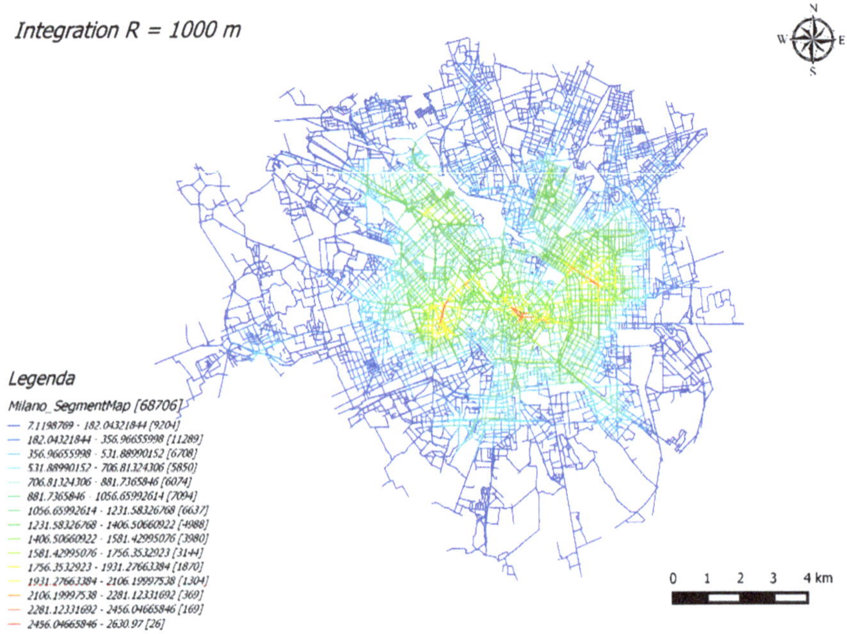

Fig. 7 Integration map of cycling accessibility

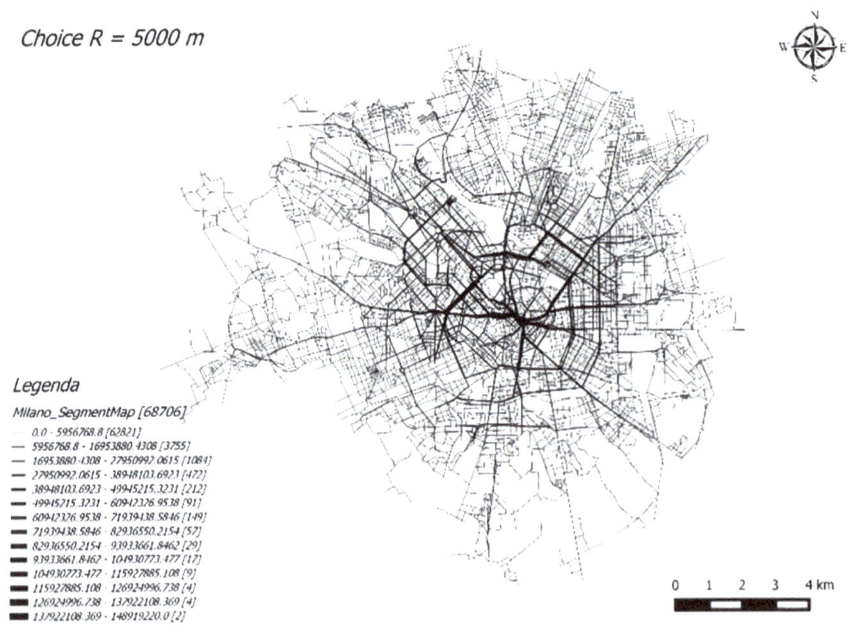

Fig. 8 Choice map of cycling accessibility

The first application consists on using the integration map based on n radius to identify potential car parking localizations: guided by a sustainable planning approach, they are localised along the boundary of the city providing interchange node in order to incentive intermodal transport behaviour. This analysis could result more efficient if integrated with the localizations of the existing park and ride and highlighting zones where they lack (Fig. 9).

The second application regards the study of traffic-restricted zone within the city of Milan, already defined as area C, and encompasses about 8.2 km^2 and 77,000 residents. The configurational analysis is able to study diverse scenarios of vehicular flow depending on traffic-restricted zone boundary, showing its influence in the distribution of local vehicular traffic flows and highlighting those roads that may assume a diverse role in traffic management. In this case study, the area C is completely close to the motorized vehicles: the main result is that ring road of the city of Milan gains more importance; the integration value along the streets that compose it increases making it more accessible from all the points in the study area.

The approach described has the potentiality to be a suitable planning tool for combining spatial topics and traffic concerns able to define the probable relationships between urban design, derived from the morphologic features of network, and traffic analysis, mainly connected to the distribution of movement flows within the network itself. Moreover, this kind of application can show the influence of traffic-closed zone on the resilience of the city through the study of the synergy coefficient. In this case study, it remains quite the same; that suggests a strong road structure over the medieval walls boundary. In fact, the ring road of the city of Milan assumes a relevance role in traffic distribution even without the closure of the traffic-limited zone as here explored (Fig. 9).

Other useful urban applications of space syntax consist in combining different configurational indicators in order to satisfy specific hypotheses and conditions. An example refers to a scenario for the location of areas for libraries, where two aspects are generally to be assumed as desirable features, if not obligatory requirements:

1. libraries have to be preferably located far from traffic flows, and hence in areas provided with low choice values;
2. libraries have to be easily accessible, and hence located in areas with high values of global integration.
3. These two requirements can be met in two ways:
4. selecting—by a specific query—elements showing both low choice values and high integration values, assuming arbitrary thresholds (Fig. 10);
5. creating a complex index, resulting from choice and integration indicators, suitably weighted (Fig. 10).

This kind of scenario can be implemented into a Geographic Information System (Jiang et al. 1999) so as to integrate the configurational results with the geographic information that indicate the location of the existing libraries in the city, as shown in Fig. 10.

As anticipated above, another relevant and innovative issue in urban planning is the resilience of the urban grid, here regarded in terms of network resilience, as the

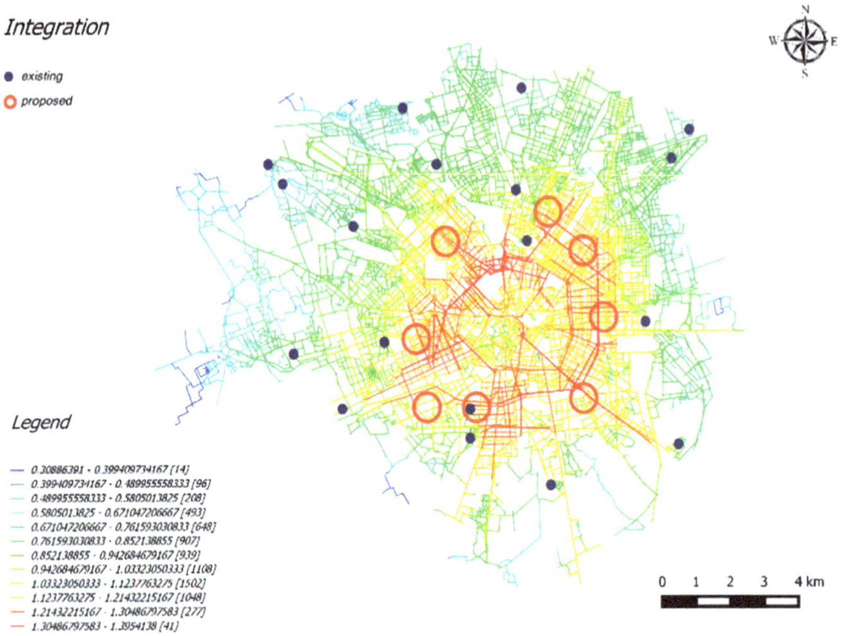

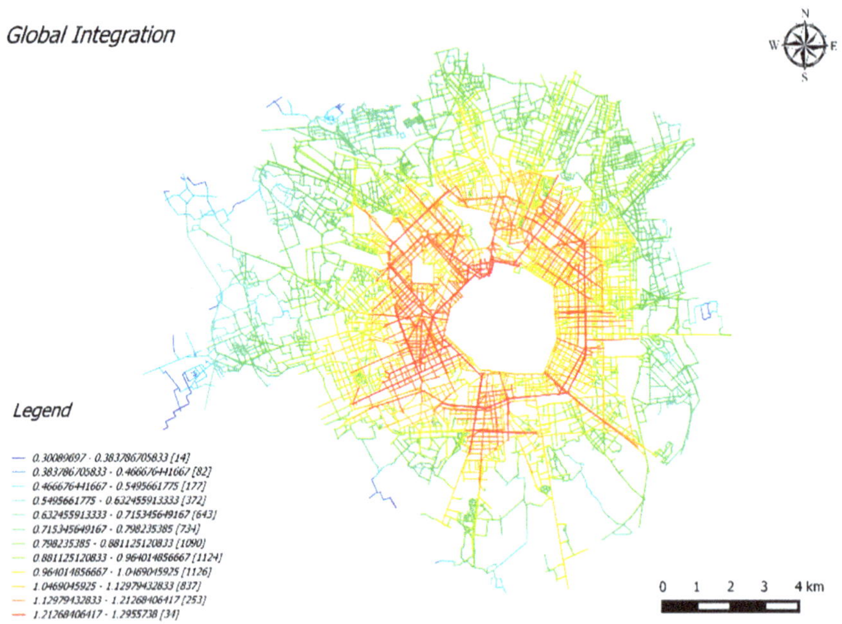

Fig. 9 Integration map of vehicular accessibility and potential location of park and rides (first image); integration map of vehicular accessibility based on prohibited traffic zone (second image)

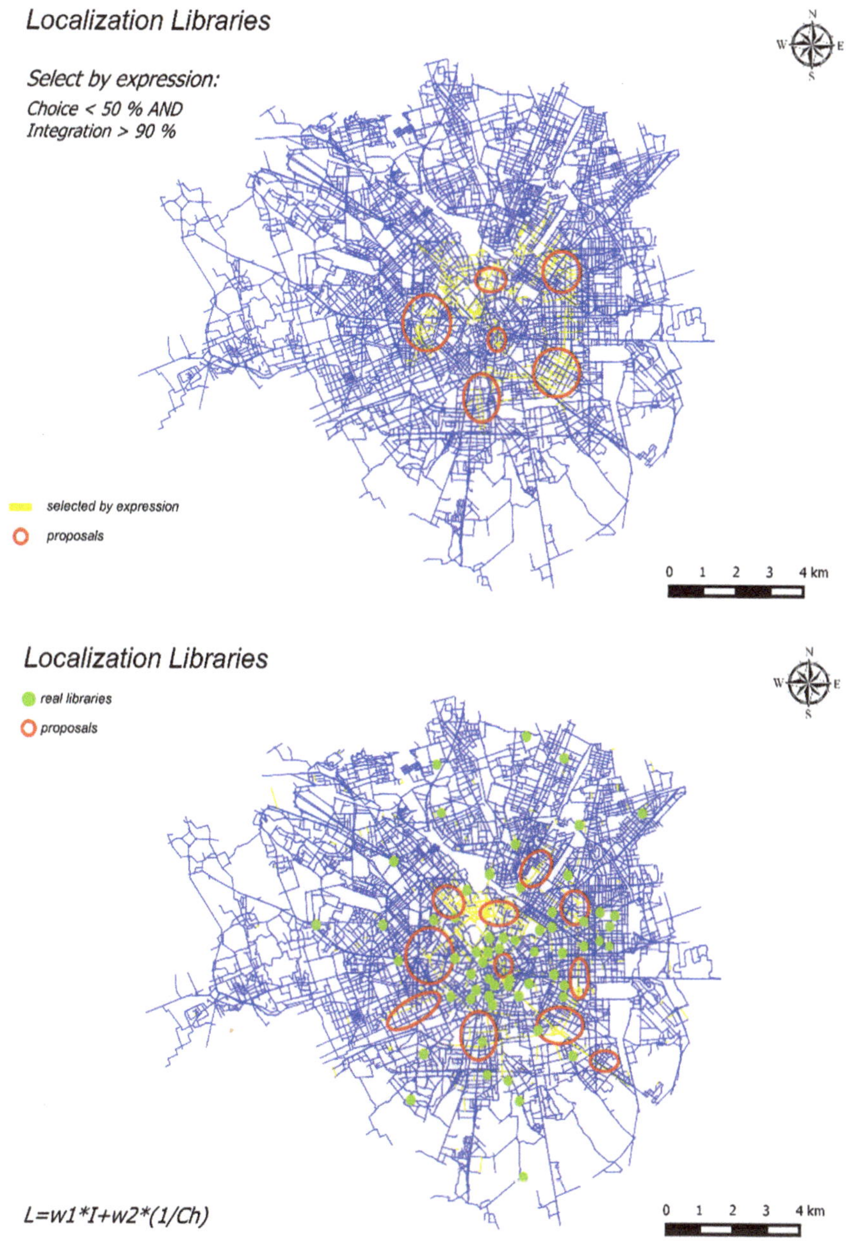

Fig. 10 Potential location of libraries within the city of Milan

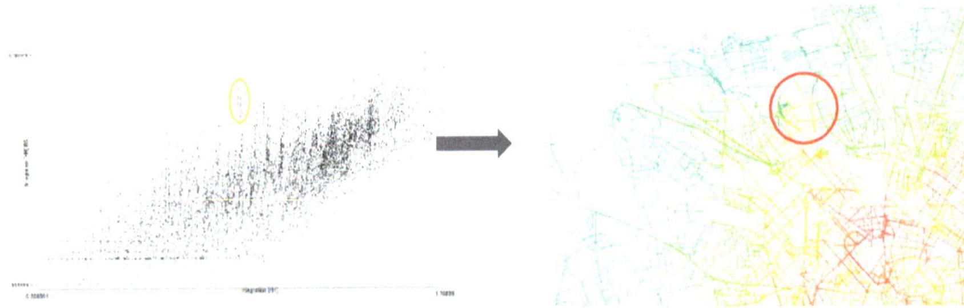

Fig. 11 From synergy scatterplot to integration map

capability of an urban system to absorb changes in their structure without suffering significant variations in its configurational state (Cutini 2013). In order to study the resilience in the case study of Milan, a punctual analysis has been carried out: in the scatterplot of synergy (correlation between global and local integration), a group of dots appears particularly separated from the cloud of the dots of the whole system, as characterized by high values of local integration and (relatively) low values of global integration.

Theoretically, it means that the corresponding streets appear rather segregated from the whole system, and still well integrated in their respective local neighbourhood. The presence of such areas has a negative effect on the global resilience of the city, as it makes the grid more disaggregated. The selected dots correspond to the Bovisa neighbourhood, which is defined as self-supporting quarter in Milan with a specific identity in its urban configuration (Erba et al. 2000). Therefore, this scenario reasserts the capability of space syntax to support the understanding of the inner geography of the city of Milan (Fig. 11).

6 Conclusions

In this study, the configurational analysis is shown as a reliable method, revealing an excellent tool for the diachronic analysis of the system and for the reconstruction of the evolution over time of centralities within the settlement. On the basis of the outcome of the research, space syntax techniques appear well suited for the interpretation of the distribution of activities and centralities within the city of Milan. Moreover, and even more interesting, the results point out the capability of the configurational techniques to support the decision-making process of planning and selecting urban actions and projects, with a particular reference to those involved with the theme of accessibility, usefully supporting the present and future location of public spaces. Furthermore, the results show that such kind of approach actually enhances the transparency of strategic decisions concerning the design and planning of urban spaces, making to emerge and highlighting their likely effects on a wide range of aspects and phenomena.

Possible future developments are plenty, such as adding some other historical dates that correspond to relevant urban transformations of the city; or, considering other socio-economic variables to be compared with the integration value, such as the distribution of social network (such as Facebook or Twitter) flows or the distribution of traffic flows over the whole map. Furthermore, it would be possible and interest to determine and discuss other—secondary—indices, resulting from the aggregation of two or more configurational parameters in order to point out other specific features of the system.

Acknowledgements This research is an outcome of a team cooperation between the all three authors. In particular, Farese has conducted the operative and computational part of this research; the analysis and the interpretation of the outcomes has been got by the collaboration between Cutini and Farese.

References

Cutini V (2001) Centrality and land use: three case studies on the configurational hypothesis. Cybergeo, Rev Eur Geogr 188

Cutini V (2013) The city when it trembles. Earthquake destructions, post-earthquake reconstruction and grid configuration. In: Kim YO, Park HT, Seo KW (eds.) Proceedings of the 9th international space syntax symposium, Sejong University, Seoul, pp 102.1–102.17

Erba V, Molon M, Morandi C (2000) Bovisa una riqualificazione possibile. Unicopli, Milano

Handy SL, Niemeier DA (1997) Measuring accessibility: an exploration of issues and alternatives. Environ Plann A 29(7):1175–1194

Hillier B (1996) Space is the machine. A configurational theory of architecture. Cambridge University Press, Cambridge

Hillier B, Iida S (2005) Network effects and psychological effects: a theory of urban movement. In: International conference on spatial information theory. COSIT, 2005: Spatial Information Theory, pp 475–490

Hillier B, Penn A, Hanson J, Grajewski A, Xu J (1993) Natural movement: or configuration and attraction in urban pedestrian movement. Environ Plann B: Plann Des 20:29–66

Jiang B, Claramunt B, Batty M (1999) Geometric accessibility and geographic information: extending desktop GIS to space syntax. Comput Environ Urban Syst 23:127–146

Karlstrom A, Mattsson LG (2009) Place, space syntax and attraction-accessibility. In: Koch D, Marcus L, Steen J (eds.) Proceedings of the 7th international space syntax symposium, Royal Institute of Technology, Stockholm

Oliva F (2002) L'urbanistica di Milano: quel che resta dei piani urbanistici nella crescita e nella trasformazione della città. Hoepli, Milano, pp 420–430

Turner A (2007) From axial to road-centre lines: a new representation for space syntax and a new model of route choice for transport network analysis. Environ Plann B: Plann Des 34:539–555

Shaping the Sustainable Urban Mobility. The Catania Case Study

Paolo La Greca and Francesco Martinico

Abstract The increasing complexity of cities will require a greater level of integration between land use policies and transport planning. In this chapter, we present the case study of Catania, a medium sized metropolitan area in Southern Italy where recent investments are changing the urban railway network considerably. In spite of the lack of integration between land use and transport planning, this example demonstrates that the construction of a new metro line has a substantial potential for revamping outdated planning practices, promoting a comprehensive approach towards a strategic planning of the metropolitan area. The level of traffic congestion and the related environmental problems of the examined area make the proposed case study particularly relevant within the framework of sustainable mobility policies.

1 Land Use and Transport Planning

Land use and transport planning are strictly interrelated since mobility is one of the key features of the city. Contemporary lifestyles and business practices depend on mobility. In order to thrive, places seek ever better connections to other places. (Bertolini 2012). This recalls the longstanding debates between *centrists* and *decentrists* that characterizes urban planning, since the H. Howard Garden City proposal (Breheny 1996). The relationship between transport and land use planning is being tackled, from different directions, by a wide range of disciplines, considering that mobility will also play a key role in CO_2 emission reduction (Hickman and Banister 2014). Urban planners are interested in which transport investments work best for the city and how urban planning might help encourage greater sustainability in travel behaviors (Hickman et al. 2015, p 4). Transport planners assume land use as an exogenous variable that generates mobility demand that has to be satisfied

P. La Greca (✉) · F. Martinico
Dicar, Università degli Studi di Catania, Catania, Italy
e-mail: paolo.lagreca@unict.it

R. Papa et al. (eds.), *Smart Planning: Sustainability and Mobility in the Age of Change*, Green Energy and Technology,
https://doi.org/10.1007/978-3-319-77682-8_21

with new transport infrastructures and services, neglecting the feedback effect of accessibility changes in land use dynamics (Ignaccolo et al. 2006).

In this chapter, only a few elements of this complex relationship will be examined in reference to the proposed case study: the construction of a metro line in medium sized conurbation in Southern Italy, where a relevant investments is taking place without taking into account the existing urban fabric.

Among all investments in transport infrastructure, urban railways play a fundamental role. Evaluating the impact of these investments requires a careful investigation at different levels, taking into account the relevant non transport benefits (Banister and Thurstain-Goodwin 2011). Impacts are considerably influenced by the initial economic conditions and the new infrastructure will affect a wide range of elements not only accessibility and property markets but also the overall quality of the city around the line and its liveability (Blanco et al. 2009).

A proper planning of urban railways requires a careful consideration of the relationship between residential locations and other activities. A key problem is the location of new developments in relationship with the commuters' attitudes. Headicar (2015), highlights that the development location of new residential settlement has often neglected the strategic dimension represented by commuting, with reference to Oxfordshire planning policy. This case confirms the existing trends, common to many city regions, towards a concentration of major employment in or around the core city and a centrifugal pattern of housing demand, extending over the surrounding hinterland. Survey results confirm that employment and housing location have relevant effects on commuting and car mileage (ibid p 62), showing that car commuting was motivated by a perceived deficiency in public transport that leaved commuters with no choice. In the study area, there is no obligation to integrate strategic location of new developments with the planning of transport networks, an approach that is also prevailing in Italy. The same concept can apply in considering infill or redevelopment aimed at optimizing an existing settlement structure.

Locating land uses that are major traffic generators (employment and services) at nodes within transport corridors, close to residential population, serves to support efficient public transport service. It favours modal shift, reduce traffic congestion and road transport investment (Curtis 2015, p 136).

The introduction of Transit Oriented Development (TOD) principles (Calthorpe 1993; Bernick and Cervero 1997) was an important turning point in order to integrate land use and mobility infrastructure. Although this approach cannot be overstated in its capacity of solving urban mobility problems,[1] it plays a major role in changing the attitude of inhabitants towards public transport, contributing to the increase of rail ridership and giving more mobility opportunities. Generally, accessibility must be ranked as the dominant criterion against which alternative settlement patterns must be judged, but this is rarely the case (Rubulotta et al. 2013). In current practice, it is uncommon to develop land use and transport plans in symbiosis (Curtis 2015, p 138).

[1]A detailed study conducted in the Greater London area shows that there is no clear cut evidence that increased density and mixed-use development will produce more sustainable travel reducing distances and car share of total travel (Echenique and Donald 2015).

In many Western countries, the areas around the stations of the new urban railways infrastructure are already developed, accordingly the new investment is likely to trigger an infill process. Infill development is intended here as the use of vacant sites in an already developed area or redevelopment of defunct or underused industrial, commercial or service sites. Infill around stations of new urban railway lines have several positive aspects if compared with greenfield development (Loo et al. 2017).

The large number of studies on TOD includes both the macro scale, at regional and metropolitan level, (Calthorpe and Fulton 2001) and the neighborhood scale. The latter brings to relevance the crucial role of public space in relationship with the role of public transport. The sustainable city is strengthened if a large part of the transport system can take place as green mobility. Good public space and a good public transport system are simply the two sides of the same coin (Gehl 1971, 2010). The healthy city is strengthened if walking and biking can be a natural part of the pattern of daily activities.

The quality of public space depends not only on the amount of investments in infrastructure like sidewalks, cycle tracks and pedestrian paths. Several factors have to be considered both at personal and community level, including motivation and pressure to conform to predominant behaviour (Handy 2015, p 199). The first step is the analysis of the quality of the environment for walking and bicycling, considering land use patterns, network configurations, road and sidewalk design, urban design qualities topography, weather and vegetation. All these factors influence feasibility, safety, comfort, enjoyment and desirability of walking and cycling (Handy 2015, p 198). In Mediterranean weather, topography and vegetation are key factors in influencing active travels due to high temperatures that last for longer periods. Outdoor comfort is very important to encourage walking and biking (Caprì et al. 2016; Evola et al. 2017; Inturri and Ignaccolo 2011).

In Italy, the current debate does not consider the relevance of integrating transport policies and land use planning. The OECD (2012) study on compact cities policy highlights that Italy does not have a national policy in this direction. This is coherent with the trend the country took from the 1970s, progressively abandoning the idea of a national land use planning guidance. The last attempt in this direction was made in those years with the *Progetto '80* proposal of a National Plan. Nowadays, Italy is a quasi-federal country and planning policies are depending from regional legislation.

National policy on transport is emphasizing the role of sustainable mobility but the idea of integration with land use planning is not clearly conveyed. The recent National Strategy for Transport Infrastructure and Logistics *Connettere l'Italia*[2] includes, among its objectives, the enhancement of the attractiveness of cities. The short analysis contained in this document highlights the relevant gap of urban railways endowment of main Italian cities compared with European ones. Cities are considered as a driving force for enhancing national competitiveness and one of the four objectives is "improving the quality of life and competitiveness of urban areas" by promoting the rapid mass transit and shared mobility, encouraging walking and

[2]Source: http://www.mit.gov.it. The strategy is the first step towards the new General Transport and Logistics Plan (*Piano Generale dei Trasporti e della Logistica*).

biking and promoting Intelligent Transport Systems. This objective includes a target of 20% increase in route length of rapid mass transit (tram and metro) per inhabitant in urban areas. There is only a generic reference to integration of transport and urban policies but no reference to land use policies or urban form, compactness or TOD. By the same token, the Italian National Energy strategy (SEN), issued in November 2017, does not take into account the land use transport relationship as far as energy savings are concerned. The recently approved[3] National Guidelines for Sustainable Urban Mobility Plans (SUMP) have included a better integration with land use policies among the objectives of SUMP. In addition, the Guidelines state that the contents of SUMP are legally binding for land use municipal plans, despite not providing details on the terms of this integration. For instance, there is no reference to the compactness of urban settlements, similarly to the National Strategy for Transport document.

The case study described in the following paragraphs confirms not only the relevance of the land use and transport integration but also the difficulty in considering this approach in the Italian planning practice.

2 The Catania Case Study

2.1 The Settlement System of the Metropolitan Area

The metropolitan area of Catania is located along the eastern coast of Sicily, the largest Italian region with a total population of 5.03 million inhabitants (August 2017). The conurbation includes several municipalities on the lower southern-eastern flanks of Mt Etna. The main city of the conurbation is the tenth largest in Italy and the fourth in *Mezzogiorno* (as Italian southern regions are known). From the first half of the 20th century, Catania has increased its role of main marketplace and supplier of services for a large region, namely South-Eastern and Central Sicily, an area that includes the provinces of Syracuse, Ragusa, and Enna. The favorable location of the city, along the coast, well connected to the motorway and railway system and the presence of a commercial port and of a busy airport gives to the city a strategic role within the island. After WWII, the city became extremely attractive especially to the inhabitants of the inner towns of eastern and central Sicily. As result of the ensuing migration its population reached 400,000 in 1971. The current settlement has developed around the historical center, extending mainly towards north and north-east, overcoming the city administrative Northern border. These developments covered progressively the rich agricultural areas (mainly vineyards, citrus fruit and olive orchards) north of the 19th century main city settlement, extending towards the hills that crown the city. After the 1960s the building activity slowed down in the main city and the immigrants found new dwellings in the surrounding villages that dotted the fertile

[3]Decree of Minister of Infrastructures and Transports (MIT) of 4th August 2017.

and hilly countryside around the main city (La Greca et al. 2011b). From 1960s to 1980s the population of the small municipalities around the main city towards north and east grew considerably. As a consequence, today the majority of the population is living outside the main city which hosts only 40% of the Metropolitan Area total population (Table 1).[4] North-East municipalities have been the preferred location of middle and upper class houses, since they enjoy the best climate conditions and scenic views. Even before urbanization, property prices were historically higher on this side, due to the fertility of volcanic soil. Low income settlements were mainly located in municipalities south and west of main city, in areas with less favorable climate conditions. Subdivisions for low income classes often took place on the sterile basalt soil of the 1669 lava flow. This land was traditionally sold at very cheap prices and used for extracting gravel and stone for buildings or considered as ideal location for land uses like dumps, warehouses o industrial sheds.

2.2 *Public Transports and the New Metro Line*

Today, the Catania Metropolitan Area is still the most important commercial, industrial, administrative, cultural and educational center of Eastern Sicily, extending its influence well beyond the administrative borders of its province. However, the poor quality of land use planning at the municipal level, the absence of both a metropolitan strategic plan and of integrated transport planning is causing major environmental problems. Traffic congestion is uncontrolled, since the majority of inhabitants use their private cars for any trip. Prevailing commuters' flows are directed towards Catania where the most important functions (public administration, commerce, healthcare, secondary and higher education, leisure) at a metropolitan level are still located (La Greca et al. 2011a). The majority of trips from outside the main city (87%) are made by car.[5] Public transport (13% of total mobility) is constituted mainly by low frequency, unreliable bus services (50% urban buses and 50% long distance buses), rail share is negligible.

The railway network of Sicily is significantly lacking both in the infrastructure and the service provided. The deficiencies of the infrastructure network are present along the main Tyrrhenian and Ionian lines and also in the inner lines. The role of the railway system in the Catania Metropolitan Area is marginal. The Catania Central

[4]In May 2017 total population of the 27 municipalities included in the 1995 designation was 787,478 the main city was 312,255 (Source; Istat- Italian National Institute of Statistics http://www.demo.istat.it).

[5]Source: City of Catania, General Plan of Urban Traffic (*Piano Generale del Traffico Urbano—PGTU*), available at https://www.comune.catania.it (in Italian).

Table 1 Recent variations in metropolitan area population in the 27 municipalities of the 1995 designation

Municipality	Population 2001	Population 2017	Variation %
Aci Bonaccorsi	2.549	3.527	38.4
Aci Castello	18.272	18.638	2.0
Aci Catena	27.058	29.647	9.6
Aci Sant'Antonio	15.389	18.001	17.0
Acireale	50.190	52.374	4.4
Belpasso	20.358	28.128	38.2
Camporotondo Etneo	3.007	5.106	69.8
Catania	313.110	312.255	−0.3
Gravina di Catania	27.343	25.506	−6.7
Mascalucia	24.483	32.158	31.3
Misterbianco	43.995	49.833	13.3
Motta sant'Anastasia	10.244	12.176	18.9
Nicolosi	6.197	7.536	21.6
Paternò	45.725	47.992	5.0
Pedara	10.062	14.404	43.2
Ragalna	3.103	3.938	26.9
S. Giovanni la Punta	20.850	23.369	12.1
S. Gregorio di Catania	10.366	11.872	14.5
S. Pietro Clarenza	5.863	7.915	35.0
Santa Maria di Licodia	6.760	7.641	13.0
Santa Venerina	7.901	8.536	8.0
Sant'Agata li Battiati	10.378	9.564	−7.8
Trecastagni	8.212	10.992	33.9
Tremestieri Etneo	20.442	20.312	−0.6
Valverde	7.246	7.845	8.3
Viagrande	6.591	8.645	31.2
Zafferana etnea	8.139	9.568	17.6
Total metropolitan area	733.833	787.478	7.3

Source Istat

Station has an average of 4000 daily passengers with only 30 regional trains per day (Caprì et al. 2015). *Ecosistema Urbano* 2017[6] the annual survey promoted in Italy by *Legambiente*, a prominent environmental association, shows that public transport use

[6]Legambiente 2017.

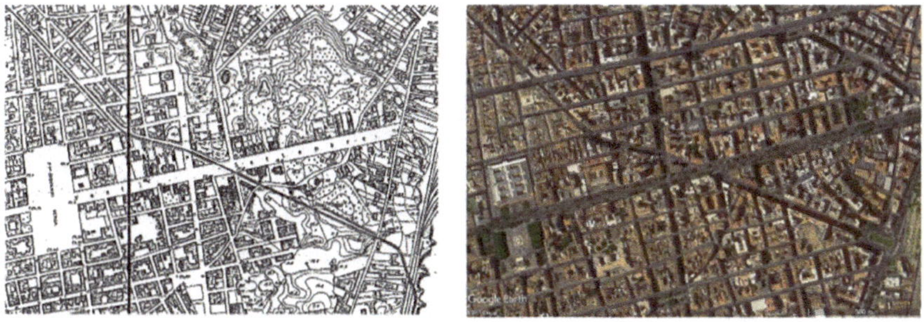

Fig. 1 Urban section of the Circumetnea in 1928 map (above) and in 2017 satellite image (below) (*Source* Google earth)

in Catania is still marginal as well as the provision of infrastructure for sustainable mode of transport.[7]

The main relevant novelty in the city public transport is the development of a metro line: the *Metropolitana di Catania*. This metro line was initially built as a replacement and upgrade of the urban section of the old *Circumetnea*,[8] a narrow gauge railway line, built from 1889 to 1895, which runs for 113 km around the volcano, terminating back to the coast at the town of Riposto. This historic railway was initially operated with 10 steam locomotives, 40 passenger coaches and 170 freight railcars, with a maximum speed of 27 km/h and a commercial speed of 15 km/h (Caprì et al. 2015). The urban section crossed at grade the northern borders of the compact city, in the area where the 1960 city expansion took place (Fig. 1).

The new metro line opened in 1999, with a very short section (3.8 km) that ran between the stations *Borgo* and *Porto,* with only six stations (Fig. 2). The service was not successful since it was too short to connect relevant city districts and the frequency was scarce.[9] Only in December 2016, an underground branch was completed to serve the very heart of the historic city center (station *Stesicoro*). In March 2017, a extension towards west was opened (from *Milo* to *Nesima*), making the line about 7 km long, with 11 stations. Trains now run every ten minutes and after the opening of these two extensions passenger rapidly increased by 323%, reaching 1,320,000, in the first

[7]The city ranked 100th among the 104 provincial capitals. Results of the survey include the following: only 44 trips/inhabitants/year in public transport (Milan: 469), 29 km vehicle/inhabitants/year as supply of public transport (Milan: 96); 69 cars/100 inhabitants (Milan: 51), 2.06 meters/100 inhabitants of bike lanes (Milan; 4, 5); 0.19 sqm/inhabitants of pedestrian areas (Milan: 0.46) (cfr. https://www.legambiente.it/sites/default/files/docs/ecosistema_urbano_2017_dossier.pdf, downloaded 4 November 2012.

[8]A word that means "around mount Etna".

[9]Trains initially ran every 20 min.

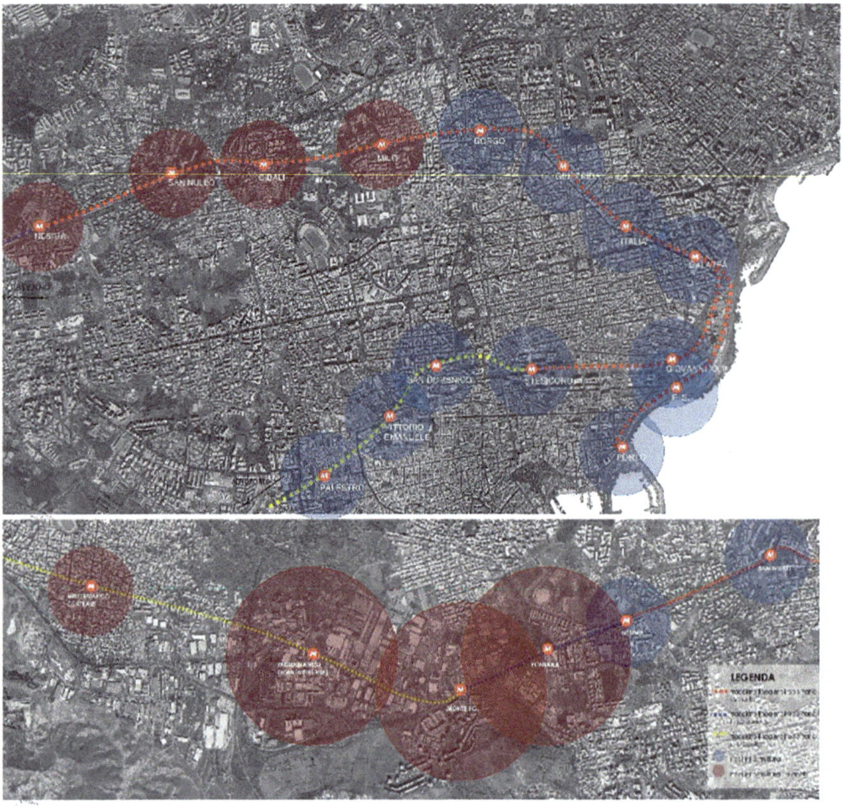

Fig. 2 Metropolitana di Catania alignment

semester of 2017, and about 3 million by the end of the year.[10] Two new sections (1.8 km towards west, and 2.2 towards south) are currently under construction and two more have been scheduled for funding, one of which would reach the airport.

The history of *Circumetnea railway* its origins and purpose has influenced considerably the planning and the design of the current infrastructure. It was initially built to serve the agro towns located to the west of the main city. The new metro line alignment outside the city center follows very closely the one of the original line. As a consequence, it does not accommodate for the growth of the residential settlements that took place during the 1960–1970 which was mainly oriented towards north-east.[11] The alignment was also constrained by the use of traditional rail traction technology, given that the topography features the steepest slopes in the areas

[10]Source: www.circumetnea.it/news.

[11]The three municipalities directly served by the metro line have a total population of about 126,000, 16% of the official Metropolitan Area.

where the population is more densely concentrated. Observing more in detail the areas around the metro stations, it becomes more evident that an engineering based approach has influenced considerably the alignment choice. For instance, *Milo* station, which opened in March 2017 in the new west section, is located about 1 km away from the compound that hosts the University Campus and the main University Teaching Hospitals. These facilities are located in a hilly site at an average of 100 meters in height difference from the station entrance. The chosen alignment avoided the need of building a deep station, equipped with elevators, but reduced substantially the accessibility to the metro line by a large number of students and patients.

The adopted temporary solution is a shuttle bus, whereas a more permanent one would be an automated people mover. The shuttle bus appears to have been quite successful,[12] confirming the significant demand of public transport in this area, due to the unsustainable level of traffic congestion. Another bus shuttle service has been implemented to compensate the lack of proper planning in determining the new metro alignment.[13]

2.3 *Population and Land Use Around Metro Line Stations*

An in depth analysis of the current land use and demographic situation around the stations of the new metropolitan line has been conducted.[14] Population data were obtained from the 2011 National Census, with census tracts being the smallest unit available. Land uses were mapped with GIS by using recent vector cartography, orthophotos and satellite images.[15]

Table 2 shows that, in the west section, the new stations, as well as the planned ones (from *Milo* to *Misterbianco Industriale*, Fig. 2) are located in areas with low or medium population density. The highest density values are found around *Misterbianco Centro* station, located in the historical center of Misterbianco, a 50.000 inhabitant town East of the main city, and around *Milo*. The latter is located in the middle of one of the residential districts built during the expansion of the city that took place during the 1960s. Only in four stations residential land use, within the 10 min isochrones (Fig. 3), is prevailing (more than 50%, Table 3), whereas it is below 25% around three other stations. The mixed use (residential with retail, convenience

[12]2000 passengers/day, Cfr. http://catania.mobilita.org/2017/10/20/metro-shuttle…

[13]http://catania.mobilita.org/2017/06/08/…

[14]This paragraph shows a brief synthesis of a larger study. Data analysis was conducted by Agnese Strano as part of her Master Thesis, in July 2017.

[15]The land use maps are one of the results of the research activity conducted, from 2001, by the Laboratory for Territorial and Environmental Planning at University of Catania (http://www.lapta.dicar.unict.it).

Table 2 Population density within 10 min isochrone around metro line stations

Station name	Population size (2011)	Area (sqm)	Population density (inh/ha)
Milo	6557	652,810	100
Cibali	3806	701,469	54
San Nullo	3531	678,939	52
Nesima	4404	768,254	57
Fontana	1966	378,516	52
Monte Po	4452	836,790	53
Misterbianco industriale	135	855,253	1.6
Misterbianco centro	9316	657,008	142

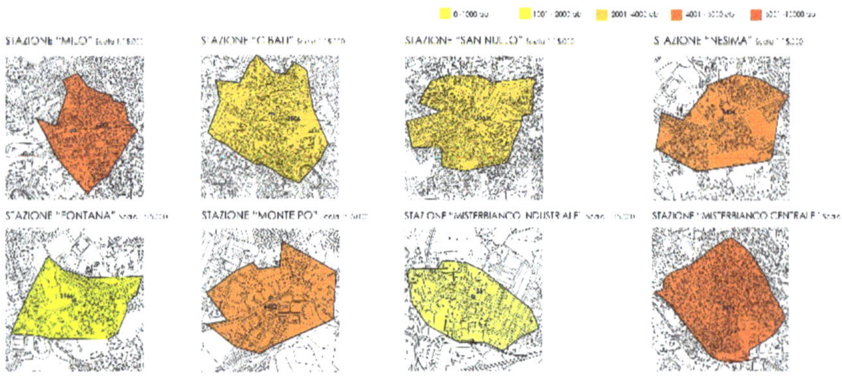

Fig. 3 10-minutes isochrones and population density

stores, service to individuals and other employment activities) is very low around all stations, except *Misterbianco Centro*, since it is located in the middle of the historical center. Even the areas around stations within the main city are characterized by low values of mixed use. This can be considered the effect of the Catania Master Plan, based on modernist principles. In *Misterbianco Industriale* residential land use is almost zero and large retail stores are prevailing as well as small and medium sized manufacturing facilities, partially abandoned. These data confirm that the location choice of new stations was based only on engineering considerations. However, the large availability of undeveloped or underused land around many stations can be considered an opportunity for redevelopment or infill.

Table 3 Land uses around new stations (10 min isochrone)

Station name	Residential %	Mixed %	Commercial %	Services %	Undeveloped %	Industrial %	Other %
Milo	56.1	9.4	0.0	24.8	3.4	0.0	6.3
Cibali	70.8	2.5	1.3	10.9	13.7	0.0	0.9
San Nullo	74.1	6.7	1.1	6.7	14.1	0.0	1.4
Nesima	46.2	0.4	0.5	4.7	42.9	0.0	5.3
Fontana	20.3	5.0	7.6	11.9	41.8	0.0	13.4
Monte Po	15.3	0.0	22.5	8.6	29.6	17.5	6.5
Misterbianco industriale	0.5	0.0	53.2	0.0	8.7	37.2	0.5
Misterbianco centro	58.9	36.4	0.0	4.7	0.0	0.0	0.0

2.4 Current Planning Situation

Land use planning of areas around metro stations does not take into account the new infrastructure. New stations are located mainly within the administrative borders of the main city, but they will reach the adjacent municipality of Misterbianco, where the west extension will become operational after 2018.

The Catania Master Plan was approved in 1969 but it is still legally binding for planning consents. This plan is mainly car oriented and therefore rail-based public transport options were limited to the National Railways infrastructure.[16] According to this Master Plan the *Circumetnea* alignment would have been completely dismissed, ending in a new terminus station outside the city north west municipal border. The eastern section of the *Circumetnea* was planned to be changed into an urban motorway, called *Asse Attrezzato*, that was considered as the key component of the new comprehensive road system.[17] Accordingly, the current master plan does not take into account the new metro line. The consequences of this lack of consideration are quite negative, especially for the areas around the stations in the western part of the line. These stations are located in the urban fabric that results from this 1960s modernistic plan[18] that has produced a settlement pattern where "life between buildings" (Ghel 1971, 2010) is lacking. For instance, areas around the stations *San Nullo* and *Cibali* are currently zoned for housing and the provision of services and public open space is limited or absent (Fig. 4).

The Misterbianco Master Plan includes old fashioned zoning choices for the area around the new station *Misterbianco Industriale,* despite being recently adopted (in 2017). The station is located in the middle of the industrial estate, originally zoned for Small and Medium Sized Enterprises, that was a thriving commercial area up to 10 years ago. Nowadays, the estate is decaying both as an industrial and as a commercial location, featuring a considerable amount of brownfield plots. The current master plan is basically confirming the previous zoning provisions that allow only industrial or commercial uses. The only provision that takes into account the new Metropolitan Line is a parking area adjacent to the new station. The TOD concepts are not considered at all in this new plan, confirming that integration between transport and land use planning is an approach that is still foreign to local planners.

2.5 Potentialities to Be Developed

The Metropolitana di Catania case confirms that a careful urban planning is necessary to reap the full benefits of relevant investments in Rail Rapid Transit. Sustainable

[16]At that time the long range regional and national transport was mainly by rail.

[17]The implementation of this disruptive road has been limited to the south west section, due to the high costs of demolishing large parts of the existing dense urban fabric.

[18]The 1969 Master Plan was designed by Luigi Piccinato (1899–1983) one of most acclaimed Italian planners who designed several master plans in Italy and abroad.

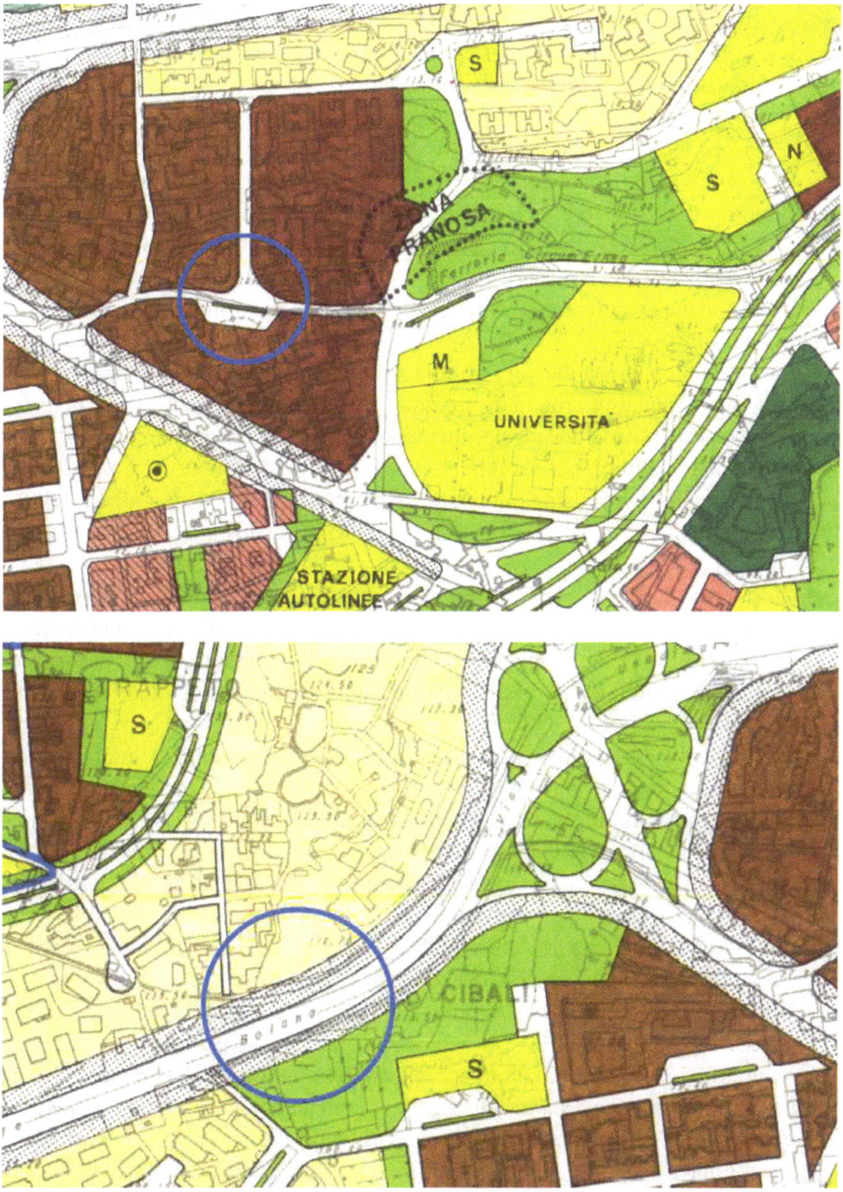

Fig. 4 Current Catania land use plan. Metro station areas are marked with a blue circle. In the upper figure a section of the *Asse Attrezzato* is visible (down, right)

cities need a wise use of urban land, even when the need of new development is reduced. In the examined case study, redevelopment of brownfields for mixed use could bring several advantages, the main one being the improvement of the overall quality of the settlement. Another benefit would be the influence it could have on commuting attitudes, an objective which requires not only an efficient public transport system but also an overall planning approach that "fosters predictable built results and a high-quality public realm, by using physical form rather than separation of uses".[19] Mixed land uses and high quality public space that encourage walking and cycling, also favored by public participated transport decisions (Le Pira et al. 2016), are key components of this unavoidable change of paradigm that is required for urban planners. It is likely that in a near future the street design will have to take into account new advances in transport technology, including the presence of self-driving vehicles, but in the examined case study the issue is still providing basic accessibility to metro stations.

In the study area, given the very limited amount of high quality public transport, the upgrading of land uses around the new metro line stations is even more crucial for several reasons. These include the optimization of investments, considering also their impacts on sustainability, but also the reinstatement of the trust of the community towards the possibility of using public transport. Passenger data after six months of service show that there are great expectations in the improvement of public transports in this area.

A short term objective is changing the design of streets around metro stations. A considerable effort is needed to design streets that prioritize pedestrians and dedicate more space to bicycle infrastructure and car sharing.[20] Integrating different forms of public transport, walking and biking will also enhance road safety and livability of public realm.

A long term objective is enhancing the land uses mix by changing the zoning codes. In this sense, the most promising areas are the ones around the stations *San Nullo, Nesima, Fontana, Monte Po and Misterbianco Industriale.* In particular, the area around *Misterbianco Industriale* requires the approval of a regeneration scheme of the decaying industrial estate. This area shows the greatest potentialities for developing functions like innovative manufacturing and new services that should be mixed with residential areas, including research and higher education facilities (Fig. 5). The conversion of a portion of the abandoned sheds allows the settlement of more than 4000 new inhabitants within the 10 min isochrone, compared with the current 135.

The presented case study confirms that transport infrastructures can contribute to the upgrading of metropolitan areas, especially where the infrastructural gap is relevant and the settlement pattern is highly fragmented. Concentrating residential settlements and other functions around urban railway stations can play a key role in the overall improvement of the quality of urban settlements.

[19]https://formbasedcodes.org/definition/.

[20]A private car sharing service started in Catania in June 2016 with 200 vehicles.

Fig. 5 A redevelopment and infill proposal around the new station of the metro line *Misterbianco Industriale* (based on Agnese Strano master thesis)

References

Banister D, Thurstain-Goodwin M (2011) Quantification of the non-transport benefits resulting from rail investment. J Transp Geogr 19:212–223
Bernick M, Cervero R (1997) Transit villages in the 21st century. McGraw-Hill, New York
Bertolini L (2012) Integrating mobility and Urban development agendas: a manifesto. disP—Plann Rev 48(1):16–26
Blanco H, Alberti M, Forsyth A, Krizek KJ, Rodríguez DA, Talen E, Ellis C (2009) Hot, congested, crowded and diverse: emerging research agendas in planning. Prog Plann 71(4):153–205
Breheny M (1996) Centrists, decentrists and compromisers: views on the future of Urban form. In: Burton E, Jenks M, Williams K (eds) The compact city: a sustainable Urban form?. E&FN Spon, London, pp 13–35
Calthorpe P (1993) The next American metropolis. Princeton Architectural Press, New York
Calthorpe P, Fulton W (2001) The regional city: planning for the end of sprawl. Island Press, Washington
Caprì S, Ignaccolo M, Inturri G (2015) The feasibility assessment of the new 'Circumetnea' railway through a multimodal network model with elastic demand. Ingegneria Ferroviaria 70(2):133–174
Caprì S, Ignaccolo M, Inturri G, Le Pira M (2016) Green walking networks for climate change adaptation. Transp Res Part D: Transp Environ 45:84–95
Curtis C (2015) Public transport-orientated development and network effects. In: Hickman R, Givoni M, Bonilla D, Banister D (eds) Handbook on transport and development. Edward Elgar Publishing, Cheltenham, pp 136–148
Echenique M, Donald A (2015) New Urbanism and Travel. In: Hickman R, Givoni M, Bonilla D, Banister D (eds) Handbook on transport and development. Edward Elgar Publishing, Cheltenham, pp 104–117

Evola G, Gagliano A, Fichera A, Marletta L, Martinico F, Nocera F, Pagano A (2017) UHI effects and strategies to improve outdoor thermal comfort in dense and old neighbourhoods. Energ Procedia 134:692–701
Gehl J (1971) Life between buildings. Danish Architectural Press, Copenhagen
Gehl J (2010) Cities for people. Island Press, Washington, Covelo, London
Handy S (2015) Community design and active travel. In: Hickman R, Givoni M, Bonilla D, Banister D (eds) Handbook on transport and development. Edward Elgar Publishing, Cheltenham, pp 198–207
Headicar P (2015) Home jobs and commuting. Development location and travel outcomes. In: Hickman R, Givoni M, Bonilla D, Banister D (eds) Handbook on transport and development. Edward Elgar Publishing, Cheltenham, pp 59–72
Hickman R, Banister D (2014) Transport, climate change and the city. Routledge, London, New York
Hickman R, Givoni M, Bonilla D, Banister D (2015) The transport and development relationship. In: Hickman R, Givoni M, Bonilla D, Banister D (eds) Handbook on transport and development. Edward Elgar Publishing, Cheltenham, pp 3–18
Ignaccolo M, Caprì S, Giunta U, Inturri G (2006) Discrete choice model for defining a parking-fee policy on Island of Ortigia, Siracusa. J Urban Plann Dev 132(3):147. https://doi.org/10.1061/(asce)0733-9488
Inturri G, Ignaccolo M (2011) The role of transport in mitigation and adaptation to climate change impacts in Urban areas. In: Zimmermann O (ed) Resilient cities: cities and adaptation to climate change—proceedings of the global forum 2010, Book series: local sustainability, vol 1, pp 465–478
La Greca P, Barbarossa L, Ignaccolo M, Inturri G, Martinico F (2011a) The density dilemma. a proposal for introducing smart growth principles in a sprawling settlement within Catania metropolitan area. Cities 28:527–535
La Greca P, La Rosa D, Martinico F, Privitera R (2011b) Agricultural and green infrastructures: the role of non-urbanised areas for eco-sustainable planning in a metropolitan region. Environ Pollut 15:2193–2202
Le Pira M, Ignaccolo M, Inturri G, Pluchino A (2016) Modelling stakeholder participation in transport planning. Case Stud Transp Policy 4:230–238
Legambiente (2017) Ecosistema Urbano,https://www.legambiente.it/sites/default/files/docs/ecosistema_urbano_2017_dossier.pdf. Accessed 3 Nov 2017
Loo BPY, Cheng AHT, Nichols SL (2017) Transit-oriented development on greenfield versus infill sites: some lessons from Hong Kong. Landscape Urban Plann 167:37–48
OECD (2012) Compact city policies: a comparative assessment, OECD green growth studies. OECD Publishing, https://doi.org/10.1787/9789264167865-en
Rubulotta E, Ignaccolo M, Inturri G, Rofè Y (2013) Accessibility and centrality for sustainable mobility: regional planning case study. J Urban Plann Dev 139(2):115–132